THE SPIRIT OF MEDIÆVAL PHILOSOPHY

THE SPIRIT OF
MEDIÆVAL PHILOSOPHY

(Gifford Lectures 1931-1932)

By Henry

ÉTIENNE GILSON, 1884—

Translated by

A. H. C. DOWNES

NEW YORK

CHARLES SCRIBNER'S SONS

1940

TO

JOHN LAIRD

PROFESSOR OF PHILOSOPHY AT
THE UNIVERSITY OF ABERDEEN

Παρὰ Θεοῦ περὶ Θεοῦ μαθεῖν
Athenagoras, Legatio pro Christianis, VII.

PREFACE

THE twenty lectures contained in this volume were delivered as Gifford Lectures in the University of Aberdeen in 1931 and 1932. I had been asked to undertake a sufficiently difficult task, namely, to define the spirit of mediæval philosophy : but I accepted it nevertheless in view of the widespread notion that however great the achievement of the Middle Ages in the fields of literature and art, they altogether lacked a philosophy that could be called their own. To attempt to bring out the spirit of this philosophy would be to commit oneself to the production of some proof of its existence—or to the admission that it never existed. In the effort to define its essence I found myself led to characterize it as the Christian philosophy *par excellence.* Here, however, I found myself face to face with the same kind of difficulty although on another plane ; for if the existence of a mediæval philosophy has been denied, the very idea of a Christian philosophy has been held to be impossible. It will be found, then, that all these lectures converge to this conclusion : that the Middle Ages produced, besides a Christian literature and a Christian art as everyone admits, this very Christian philosophy which is a matter of dispute. No one, of course, maintains that this mediæval philosophy was created out of nothing, nor yet that all mediæval philosophy was Christian—just as no one maintains that mediæval literature and art were created out of nothing or were wholly Christian. The true questions are, first, whether we can form the concept of a Christian philosophy. and secondly, whether mediæval philosophy, in its best representatives at any rate, is not precisely its most adequate historical expression. (As understood here, then, the spirit of mediæval philosophy is the spirit of Christianity penetrating the Greek tradition, working within it, drawing out of it a certain view of the world, a *Weltanschauung*, specifically Christian.) There had to be Greek temples and Roman basilicas before there could be cathedrals ; but no matter how much the mediæval architects owed to their predecessors, their work is nevertheless distinctive, and the new spirit that was creative in them was doubtless the same spirit that inspired the philosophers of the time. To see how much truth there may be in this hypothesis we shall have to examine mediæval thought in its nascent state, at that precise point, namely, where the Judeo-Christian graft was inserted into the Hellenic tradition. Thus the demonstration attempted is purely historical : if, very occasionally, a more theoretical attitude is provisionally adopted, it is merely because an historian who deals with ideas is bound at least to make them intelligible to his readers, to suggest how doctrines which satisfied the thought of our predecessors for so many centuries may still be found conceivable to-day.

CONTENTS

CHAPTER   PAGE

PREFACE   vii

I. THE PROBLEM OF CHRISTIAN PHILOSOPHY .   I

II. THE CONCEPT OF CHRISTIAN PHILOSOPHY .   20

III. BEING AND ITS NECESSITY 42

IV. BEINGS AND THEIR CONTINGENCE . . . 64

V. ANALOGY, CAUSALITY AND FINALITY . . 84

VI. CHRISTIAN OPTIMISM 108

VII. THE GLORY OF GOD 128

VIII. CHRISTIAN PROVIDENCE 148

IX. CHRISTIAN ANTHROPOLOGY 168

X. CHRISTIAN PERSONALISM 189

XI. SELF-KNOWLEDGE AND CHRISTIAN SOCRATISM . 209

XII. KNOWLEDGE OF THINGS 229

XIII. THE INTELLECT AND ITS OBJECT . . . 248

XIV. LOVE AND ITS OBJECT 269

XV. FREE-WILL AND CHRISTIAN LIBERTY . . 304

XVI. CHRISTIAN LAW AND MORALITY . . . 324

XVII. INTENTION, CONSCIENCE AND OBLIGATION . 343

XVIII. THE MIDDLE AGES AND NATURE . . . 364

XIX. THE MIDDLE AGES AND HISTORY . . . 383

XX. THE MIDDLE AGES AND PHILOSOPHY . . 403

NOTES 427

INDEX TO PROPER NAMES 487

ix

CHAPTER I

THE PROBLEM OF CHRISTIAN PHILOSOPHY

It would be hard to imagine any expression more naturally apt to occur to the mind of the historian of mediæval thought than *Christian Philosophy ;* none, it would seem, raises fewer difficulties, and we need not therefore be in the least surprised to find it in general use. And yet, when we come to reflect about it, few turn out to be so obscure and so hard to define. For the question is not simply this : whether the historian of mediæval thought is justified in considering separately, first those philosophies elaborated in the course of the Middle Ages by Christians, and then those constructed by Jews or Mussulmans. If we put it in that form the problem is purely historical and can be settled very easily. We have no right to isolate in our history things that in fact were united in reality. Christian thought, Jewish thought, and Mussulman thought acted and reacted on each other as we know, and it would not be at all satisfactory to study them as so many closed and isolated systems. Of course, as a matter of fact and practice, historical research proceeds by way of abstraction, and we all map out for ourselves a certain limited domain which extends as far as our competence will warrant. The chief thing here is not to take the limitations of our method for the limits of reality.

The real problem is quite a different one ; it belongs to the philosophical order and is much more serious. Reduced to its simplest terms, it consists in asking whether the very concept of *Christian Philosophy* has any real meaning, and then, as a subsidiary matter, whether there was ever any corresponding reality. Of course we do not ask whether there were any philosophic Christians, that is to say Christians who happened to be philosophers ; the point is, were there ever any Christian philosophers, or, for that matter, any Mussulman or Jewish philosophers? We are all well aware of the extraordinary importance attached to religion in the mediæval civilization, and we know too that Judaism, Islam and Christianity each produced a certain body of doctrine in which philosophy went more or less happily hand in hand with religious dogma, a body of doctrine rather vaguely known as scholasticism. Now the precise question is whether these various scholasticisms, Jewish, Mussulman, or, more especially, Christian, really deserved the name of philosophies. And as soon as the question is put in that way, the existence, even the possibility of a Christian philosophy, far from being obvious, becomes at once problematical ; so much so in fact that to-day, even in the most opposite philosophical quarters, we find something like a general agreement to refuse the expression all positive significance.

It encounters in the first place the criticism of historians who, without discussing *a priori* whether a Christian philosophy is possible or not, are clear at any rate that even in the Middle Ages there never was one in fact.[1] Shreds of Greek thought more or less clumsily patching up theology— that, we are told, is about all the Christian thinkers have left us. Sometimes they borrow from Plato, sometimes from Aristotle, that is to say when they are not engaged on something considerably worse, an impossible synthesis of Plato and Aristotle, an effort to reconcile the dead who never ceased to differ when alive—as John of Salisbury already remarked in the twelfth century. Never do we meet with a genuine impulse of thought which at one and the same time is thoroughly Christian and really creative ; and it follows that Christianity has contributed nothing to the philosophic heritage of humanity.

If the historians state the fact the philosophers go on to add the reason—if no Christian philosophy has appeared on the stage of history that is because the very concept of a Christian philosophy is contradictory and impossible. In the front rank of those who share this opinion we must put those whom we may call the pure rationalists ; whose position would hardly need description were it not that its influence is a good deal more widespread than we usually think. They maintain that religion and philosophy are so essentially different that no collaboration between the two is possible at all. They are very far from agreement as to what constitutes the essence of religion, but all agree in affirming that it does not belong to the order of reason, and that reason, on its side, is quite independent of religion. Now the order of reason is precisely that of philosophy ; and the latter therefore is essentially independent of all that is not itself and particularly of this irrational thing called Revelation. Nobody to-day would dream of talking of Christian mathematics, Christian biology, or Christian medicine. But why ? Because mathematics, biology and medicine are so many sciences, and because science, in its conclusions no less than in its principles, is altogether independent of religion. To speak of Christian philosophy is equally absurd and the expression should be simply discarded.

Of course we should find no single neo-scholastic philosopher to-day who would admit an absence of all relation between philosophy and religion ; but we should deceive ourselves if we supposed that all neo-scholastics are in absolute opposition to the rationalist position as just described. On the contrary, while they would expressly maintain certain necessary relations on another plane, there are those among them who would admit the premises of the rationalist argument, and some who would even have the hardihood to accept the conclusion. What these neo-scholastics would certainly deny is that no Christian has ever successfully constructed a philosophy, for they maintain that St. Thomas Aquinas actually founded one ; but it does not require much pressure to extract the admission that his philosophy is the only example [2] and that if it is the only example it is precisely because it stands alone in being constructed on a purely rational basis. Thus it is rather on facts than on principles that they disagree with the rationalists, or, if there is any difference of principle, it is not concerned with the concept of philosophy, but with the place it occupies in the hierarchy of the sciences. While the pure rationalist puts philosophy in the highest place, and identifies it with wisdom, the neo-scholastic subordinates it to theology which alone, as he holds, fully deserves that name ; but why then do certain neo-scholastics imagine that even when thus subordinated to theology, their philo-sophy remains precisely of the same nature as any other that recognizes no Wisdom higher than itself ? How is this attitude to be explained ?

If we could ask the mediæval thinkers by what right they called themselves philosophers we should obtain some very different answers. Some, without doubt, would reply that they felt no interest in the title at all—they would be quite content with that of Christian—what better one could they possibly have ? Here we might cite such resolute opponents of dialectic as St. Bernard and St. Peter Damian, but even if we put aside such extreme cases, we should find hardly any, save the Averroists, who would admit the legitimacy of an exercise of reason that would be purely philosophical and systematically withdrawn from the influence of faith. The normal view, as expressed in the twelfth and thirteenth centuries, would be well represented by St. Anselm and St. Bonaventure, who, for the rest, would justly claim descent from St. Augustine. They would certainly regard an exercise of pure reason as a possibility—after Plato and Aristotle who could doubt it?—but they would view the matter not so much from the standpoint of the mere definition of reason as from that of the actual conditions of fact under which it has to work. Now it is a fact that between ourselves and the Greeks the Christian revelation has intervened, and has profoundly modified the conditions under which reason has to work. Once you are in possession of that revelation how can you possibly philosophize as though you had never heard of it? The errors of Plato and Aristotle are precisely the errors into which pure reason falls, and every philosophy which sets out to be self-sufficing will fall into them again, or perhaps into others still worse; so that henceforth the only safe plan is to take revelation for our guide and make an effort to understand its contents—and this understanding of the contents of revelation will be philosophy itself. *Fides quaerens intellectum*: that is the basic principle of all mediæval speculation, but is it not also a mere confusion of philosophy with theology? And, if so, do we not run a risk of ruining the former altogether?

It is precisely to avoid this danger that certain neo-scholastics have felt bound to adopt, at any rate partially, the position of their opponents. They concede the principle, and then set out to show that the Middle Ages could boast of no philosophy really worthy of the name save that of St. Thomas. St. Anselm and St. Bonaventure take up their stand on faith, and therefore they shut themselves up in theology. The Averroists, on the other hand, shut themselves up in pure reason; but since they will not bind themselves to maintain the truth even of the most necessary rational conclusions, they exclude themselves from philosophy. In Thomism alone we have a system in which philosophic conclusions are deduced from purely rational premisses. (Theology remains in its proper place, that is to say at the head of the hierarchy of the sciences ; based on divine revelation, from which it receives its principles, it constitutes a distinct science starting from faith and turning to reason only to draw out the content of faith or to protect it from error. Philosophy, doubtless, is subordinate to theology, but, as philosophy, it depends on nothing but its own proper method ; based on human reason, owing all its truth to the self-evidence of its principles and the accuracy of its deductions, it reaches an accord with faith spontaneously and without having to deviate in any way from its own proper path. If it does so it is simply because it is true, and because one truth cannot contradict another.)

Doubtless a fundamental difference remains between such a neo-scholastic and a pure rationalist. For the former the faith is always there and any conflict between his faith and his philosophy is a sure sign of philosophic error. When such a conflict declares itself he must re-examine his principles and check his conclusions until he discovers the mistake that vitiates them. If, however, even then he fails to come to an understanding with the rationalist it is not for lack of speaking the same language. Never will he commit the unpardonable error of a St. Augustine or a St. Anselm, and, when asked for proofs of the existence of God, invite us first of all to believe in God. If his philosophy is true, it is solely in virtue of its own rational evidence ; if he fails to convince his opponent it would be lack of candour on his part to appeal to faith for his justification, and this not merely because his opponent does not share his faith but because the truth of his philosophy in no wise depends on that of his faith.

As soon as we look at the philosophy of St. Thomas in this light some rather surprising but no less inevitable consequences begin to appear. We are reminded in the first place of all those vehement protests made by the Augustinians of all ages against the paganization of Christianity by Thomism. If certain modern Thomists deny that Augustinianism is a philosophy, the mediæval Augustinians were beforehand with them in denying that Thomism is faithful to the Christian tradition. Whenever they had to contest some Thomist thesis which they considered questionable they backed up their purely dialectical attack with objections of a much more general nature impugning the very spirit of the doctrine itself. If Thomism was mistaken on the problem of illumination, on the *rationes seminales*, or on the eternity of the world, was it not because it was first and more fundamentally mistaken on the relation between reason and faith ? For when we refuse to follow St. Augustine who, for his part, professed to be guided by faith, and prefer to follow instead the principles of some pagan philosopher or his Arabian commentators, then reason is no longer able to distinguish truth from error. Forced to rely on its own light, it is only too easily blinded by doctrines that in fact are false, and it is this very blindness that hides the falsity.

But something still more curious follows. Just as certain Augustinians regard Thomism as a false, because not a Christian philosophy, so certain Thomists reply that it is true but not in the least because it is Christian. They are forced, in fact, into this position ; because, once reason, as regards its exercise, has been divorced from faith, all intrinsic relation between Christianity and philosophy becomes a contradiction. If a philosophy is true it is simply because it is rational ; but if it is rational, it is not at all because it is Christian. We must therefore choose. Never will a Thomist admit that there is anything in the doctrine of St. Thomas contrary either to the letter or the spirit of the faith, for he expressly maintains that the accord between revelation and reason is an accord of truth with itself ; but there is no great cause for surprise if some of them accept without flinching the classic reproach of the Augustinians : then your philosophy lacks all intrinsic Christian character ! And how could it possibly have such a character without ceasing to be ? The philosophical principles of St. Thomas are those of Aristotle, that is to say of a man who knew nothing of any revelation, whether Christian or Jewish ; if Thomism took up the doctrine of Aristotle and purified, completed it, gave it precision, it did not accomplish this by means of any appeal to faith, but simply by a more correct and complete deduction of the consequences implied in his own principles than Aristotle was able to achieve for himself. Thomism, in short, regarded from the standpoint of philosophic speculation, is nothing but Aristotelianism rationally corrected and judiciously completed ; and there was no more need for St. Thomas to baptize Aristotelianism in order to make it true, than there would be to baptize Aristotle in order to discuss philosophy with him. Philosophical discussions pass between man and man, not between man and Christian.

The logical upshot of this attitude is a pure and simple negation of the whole concept of Christian philosophy, and, strange as it may seem, there are some who are quite ready to face the fact. Not only are there historians who deny that Christianity has seriously influenced the course of philosophic speculation,[3] but there are neo-scholastic philosophers who maintain that the concept of Christian philosophy is quite obviously void of meaning.[4] We may

make use of philosophy to facilitate the acceptance of religious dogma, and then we assimilate philosophy to apologetics ; or we may allow rational conclusions to be judged
by their accord with dogma, and then we are at once in
theology ; or perhaps, in order to get over these difficulties,
we make up our minds to say that " Christian philosophy "
simply means " true philosophy," and then there is no more
reason why this philosophy should be discovered and professed by Christians than by unbelievers or anti-Christians ;
or we may finally call a philosophy " Christian " merely
because it is compatible with Christianity, but then, if this
compatibility be a mere fact, and due to nothing but a
purely rational working out of first principles, the relation
of this philosophy to Christianity remains just as extrinsic
as in the preceding case, and if, on the other hand, the
compatibility results from some special effort to achieve
it, we are back at once in theology or apologetics. And so
we go round. It looks as if we were trying to define, in
distinct terms, a contradictory concept ; that, namely, of
a philosophy, that is to say a rational discipline, which at
the same time would be religious, and thus depend, either
in essence or in exercise, on non-rational conditions. Why
not abandon a notion that satisfies nobody ? Augustinianism
would accept a Christian philosophy were it content to
renounce philosophy and be simply Christian ; neo-
Thomism would accept a Christian philosophy were it
content to abandon the claim to Christianity and be merely
a philosophy. Would it not be simpler to disassociate the
two notions altogether, to hand philosophy over to reason
and restore Christianity to religion ?

When observation of facts and analysis of ideas are so far
in agreement it would seem useless to inquire further did we
not opportunely call to mind the complexity of the threads
that bind them together. It is very true that history, as a

simple collection of facts, decides no question of right, it is for ideas to arbitrate in that sphere ; but it is equally true that ideas are inferred from facts and that these in their turn are judged in the light of the ideas inferred. Now it is a fact that around the definition of Christian philosophy there has been a good deal more of deductive than of inductive reasoning—and especially we may add, in Christian quarters. How in fact have philosophic thought and Christian faith conceived their interrelations ? What has each been conscious of giving to and receiving from the other ? These are immense questions, to which decisive replies have not been lacking ; but any methodical investigation of them would exhaust the powers of thought, and that not merely on account of the multiplicity of particular problems involved, but because our solution of each of these will partly depend on the view we take of the facts themselves. But it is important at least to get these problems stated if there is to be any discussion of the concept of Christian philosophy resting on serious bases, and if, supposing that the corresponding historical reality exists, we are to have any hope of defining it.

But does this historical reality exist ? Is it even conceivable that it ever existed ? Good historians have denied it, relying on what they conceived to be the exclusively practical character of primitive Christianity, a stranger, as they considered, to all speculation. Harnack has done much to diffuse this idea, and authors of a very different stamp have followed his lead. What I think of it will sufficiently appear in the course of these lectures ; but it raises a preliminary question that threatens to put a kind of ban on the whole discussion, and, for the moment, I would merely try to dispose of this.

What is meant by the assertion that, at the outset at any rate, Christianity was altogether unspeculative ? If it means that Christianity is not a philosophy nothing could be more obvious ; but if it is proposed to maintain that even in the properly religious field Christianity carried with it no " speculative " elements, that it was no more than an effort of mutual aid, at once material and spiritual, in communities,[5] then that is going somewhat farther than history will warrant. Where shall we find this eminently practical and unspeculative Christianity ? Well, we shall have to go back beyond St. Justin ; to tear a good many pages of the Apostolic Fathers out of primitive Christian literature, to suppress the First Epistle of St. John, and all the speculative mysticism of the Middle Ages that sprang out of it ; to reject the Pauline preaching of grace, which was soon to give birth to Augustinianism ; to abridge the Gospel of St. John, not forgetting the doctrine of the Word set forth in the Prologue ; we shall have to go behind the Synoptics and deny that Jesus Himself taught the doctrine of the Heavenly Father, preached faith in Divine Providence, announced eternal life in an everlasting Kingdom ; above all we shall have to forget that in proportion as it was the more primitive, Christianity was the more closely allied with Judaism, and that the Bible is full of ideas about God and His divine government which, although not properly philosophical in character, only needed to fall into the right soil to become fruitful of philosophic consequences. The fact that there is no philosophy in Scripture does not warrant the conclusion that Scripture could have exerted no influence on the evolution of philosophy ; if Christian life, from its beginnings, contained speculative as well as practical elements—even if they were only speculative in a properly religious sense—the possibility of such influence becomes at once conceivable.

If then history presents no obstacle to a study of this kind, we may add that there is nothing that renders it absurd *a priori* from the philosophical standpoint. Nothing at all, not even the secular quarrel that is always setting certain Augustinians and certain Thomists at odds. If they fail to see eye to eye it is simply because, under the same name, they are labouring at two different problems. The Thomists will accept the Augustinian solution of the question as soon as the Augustinians recognize that even for a Christian, reason is essentially distinct from faith, and philosophy from religion ; and, since St. Augustine himself recognized it, the distinction seems quite sufficiently Augustinian. The Augustinians, on the other hand, will accept the Thomist solution when the Thomists recognize that, for a Christian, reason is not to be divorced from faith in the sphere of its exercise ; now St. Thomas recognized it, and there seems to be nothing to prevent a Thomist doing likewise. If the case stands thus, then, although we do not yet know what goes to make up a Christian philosophy, there seems to be nothing theoretically contradictory about the idea ; there is at least a standing-ground on which it would not be impossible, that, namely, of the conditions of fact under which the reason of Christians is to be exercised. There is no such thing as a Christian reason, but there may very well be a Christian exercise of reason. Why should we refuse to admit *a priori* that Christianity might have been able to change the course of the history of philosophy by opening up to human reason, by the mediation of faith, perspectives as yet undreamt of? The thing might well have failed to happen, but that does not warrant the assumption that it could not possibly happen. And we may go a step further and say that a cursory glance at the history of modern philosophy would strongly suggest that it did happen.

We should hardly go wrong if we connected the rise of the classical philosophy of the seventeenth century with the development of the positive sciences generally, and especially of mathematical physics; it is this factor that makes Cartesianism so different from mediæval metaphysics. But why do we so seldom consider the contrast between Cartesianism and Greek metaphysics? There is no question of representing Descartes as a " Christian philosopher," but can it be seriously maintained that modern philosophy from Descartes to Kant would have been just what in fact it was, had there been no " Christian philosophers " between the end of the Hellenistic epoch and the beginning of modern times? In other words, the Middle Ages could hardly have been so barren of philosophy as they have been represented, and modern philosophy owes perhaps more than one of its directive principles to the preponderating influence of Christianity during this period. A summary examination of the philosophic output of the seventeenth, eighteenth, and even nineteenth centuries will at once reveal characteristics very difficult to explain unless we take into account the great work of rational reflection that went on between the end of Hellenistic times and the beginning of the Renaissance.

Open for example the works of René Descartes, the reformer of philosophy *par excellence*, of whom Hamelin went so far as to say that " he is in succession with the ancients, almost as if—with the exception of a few naturalists —there had been nothing but a blank between." What are we to make of this " almost " ? Consider, to start with, the title of the *Méditations sur la métaphysique*, " in which are demonstrated the existence of God and the immortality of the soul." Consider, again, the close kinship of Descartes' proofs of God with those of St. Augustine or even those of St. Thomas. It would not be at all difficult to show that his doctrine of liberty owes a great deal to the mediæval speculations on the relations between grace and free-will [6]— a Christian problem, if ever there was one. But it will suffice perhaps to note that the whole Cartesian system is based on the idea of an omnipotent God who, in a way, creates himself and therefore, *a fortiori*, creates the eternal truths, including those of mathematics, creates also the universe *ex nihilo*, and conserves it by an act of continuous creation, without which all things would lapse back into that nothingness whence His will had drawn them. We shall have to ask shortly whether the Greeks knew anything at all about this doctrine of creation : but even the fact that this is open to question irresistibly suggests that Descartes here directly depends on Biblical and Christian tradition, and that his cosmogony, taken in its essence, does little more than develop the teaching of his masters concerning the origin of the universe. Who, after all, is this Cartesian God, this Infinite, Perfect, Omnipotent Being, Creator of heaven and earth, Who makes man in His own image and likeness, and conserves all things by an act of continuous creation—who is He if not the God of Christianity quite easily recognizable with all His traditional nature and attributes ? Descartes claims that his philosophy in no wise depends either on theology or on revelation, that he sets out simply from those clear and distinct ideas that our natural reason discovers in itself when it attentively analyses its own content ; but how then does it happen that these ideas of purely rational origin turn out to be the same in all essentials as those which Christianity, during sixteen centuries, had taught in the name of faith and revelation ? The coincidence is suggestive in itself, and becomes still more so when, with the case of Descartes, we bring together all the similar cases that surround him.

Compared with that of his master, the personality of Malebranche falls into the second rank ; nevertheless, he is not to be neglected if the history of modern metaphysics is to remain intelligible. His doctrine of the indemonstrability of the existence of the external world, combined with that of vision in God, led directly to the idealism of Berkeley ; his occasionalism, which presupposes the impossibility of proving any transitive action of one substance on another, led directly to the criticism directed by Hume against the principle of causality ; and we have merely to read Hume to see how conscious he is of following Malebranche. This was the important moment, perhaps the decisive moment, in the history of modern philosophy. Now to whom does Malebranche appeal ? To St. Augustine quite as much as to Descartes. He protests, to be sure, against scholasticism and even more vigorously than Descartes himself, but this modern philosopher does not reproach it, as we might expect, for confusing philosophy and religion, but rather for being insufficiently Christian. The offence of St. Thomas Aquinas consisted in following Aristotle and Averroes, a pagan and his " wretched commentator," instead of the perfect representative of Christian tradition, St. Augustine. And this is no accidental or external criticism of the system, but a blow aimed at its heart. Had scholasticism been more Augustinian it would have been more religious, and, consequently, truer.[7]

What in fact ought a Christian philosophy to be in order to be really worthy of the name ? First and foremost an exaltation of the power and glory of God. He is Being and Efficiency, in this sense, that all that is exists by Him alone and all that is made is made by Him alone. But what, on the other hand, are Aristotelianism and Thomism ? Philosophies of nature, systems, that is to say, which suppose the existence of substantial forms or natures, entities endowed with efficiency, and themselves productive of all the effects commonly attributed to the action of bodies. It can be readily understood, of course, that a pagan system like Aristotle's should attribute this subsistence, independence,

and efficacy to finite bodies ; and if it goes on to attribute our knowledge of bodies to their existence and action on the soul, we need not feel any surprise. But a Christian, surely, should be more happily inspired ! Knowing that to cause is to create, and that creative action is proper to God, it was St. Thomas' business to deny the existence of natures or substantial forms, to ascribe all efficiency to God alone, and therefore to situate in God not only the origin of our actions but the origin of our knowledge as well. In short it is as essential elements in any philosophy truly Christian and based on the idea of omnipotence that Malebranche maintains the truth of occasionalism and vision in God.

Examples could easily be multiplied to show how the imagination of the classical metaphysicians was absolutely possessed by the idea of the Biblical Creator-God. It would not be altogether fair to cite Pascal since, to a large extent, we should be merely citing St. Augustine ; but consider the case of Leibniz. What would be left of his system if the properly Christian elements were suppressed ? Not even the statement of his own basic problem, that, namely, of the radical origin of things and the creation of the universe by a free and perfect God. The *Discours de métaphysique* opens with this idea of perfect being, and the whole treatise, incontestably of capital importance in the work of Leibniz, concludes with a justification of divine providence, and even with an appeal to the Gospel. " The ancient philosophers knew but little of these truths : Jesus Christ alone expressed them most divinely, and in so clear and familiar a manner that the simplest mind can follow ; and His gospel wholly changed the face of the earth." These are not the words of a man who considered himself a successor of the Greeks with nothing but a blank between. We might say as much of Kant if we were not so much in the habit of identifying him with the *Critique of Pure Reason* and of forgetting the

existence of the *Critique of Practical Reason* altogether. And indeed, it might very well be said even of some of our contemporaries.

It is a curious fact, and well worth noting, that if our contemporaries no longer appeal to the *City of God* and the Gospel as Leibniz did not hesitate to do, it is not in the least because they have escaped their influence. Many of them live by what they choose to forget. I mention only one example, but highly characteristic, that of W. P. Montague whose *Belief Unbound* has only recently been published.[8] After noting how crude hypotheses seem to well up spontaneously in the consciousness of primitive times, he points to the strange phenomenon that is seen to follow " perhaps the strangest and most retrogressive in all human culture, which consists in the translation of the crude hypotheses of our ignorant ancestors into dogmas proclaimed by divine omnipotence." The Christian Bible, for instance, which claims the nature and authority of a divine revelation, is really nothing but a body of popular belief, a kind of sacrosanct folklore. And so belief turns to slavery : a slavery from which Mr. Montague would like to see us free and would help us to free ourselves. But for that he would need, to use his own expression, a new Promethean god, and this Promethean conception of god " means that the holy spirit of God, could one but feel it, would not only be courage to hearten us in weakness, and solace to comfort us in sorrow . . . but power and light and glory beyond what we had however much we had."

No preaching could be better, and for good reason. If this is what the new faith, liberated at last from folklore and the Christian Bible, has to offer us, the University of Yale might well substitute a few public readings from St. Paul and St. John for the *D. H. Terry Lectures :* and if, moreover, as Mr. Montague believes, the new god differs from the old inasmuch as he affirms life instead of denying it, we cannot help asking what sort of notion of Christianity survives in the contemporary mind. The truth is that Mr. Montague's new religion is a very pretty example of Biblical folklore complicated with Greek folklore : its author has certain vague recollections of the Gospel absorbed in his childhood, and these he takes for new philosophical ideas ; something deep down in him refuses to forget.

There are therefore good historical reasons for doubting the radical divorce of philosophy and religion in the centuries that followed the Middle Ages ; at least we may reasonably ask whether the classical metaphysic was not nourished on the substance of Christian revelation to a far greater extent than we usually imagine. To put the question in this form is simply to re-state the problem of Christian philosophy in another field. If pure philosophy took any of its ideas from Christian revelation, if anything of the Bible and the Gospel has passed into metaphysics, if, in short, it is inconceivable that the systems of Descartes, Malebranche and Leibniz would be what in fact they are had they been altogether withdrawn from Christian influence, then it becomes highly probable that since the influence of Christianity on philosophy was a reality, the concept of Christian philosophy is not without a real meaning.

For anyone convinced of this, two different attitudes remain open. We may either admit with Comte that metaphysics are destined to sink into oblivion along with the theologies of which they are nothing but the shadow, or, since theology seems to survive its own funeral oration, we may suppose that it will long continue to inspire metaphysics. The latter forecast seems the more probable ; it accords better with the persistent vitality of Christianity, and there appears to be no reason why it should perturb those who look forward to the future of metaphysics. But whatever the future may have in store, there, certainly, is the lesson of the past. As Lessing profoundly remarked—" The great religious truths were not rational when they were revealed, but they were revealed so that they might become so " [9]—not quite all of them perhaps, but some. In that formula we have the whole meaning of the question to which the lectures that now follow will attempt an answer. Our first task will be to interrogate the Christian philosophers themselves as to their own idea of Christian philosophy ; and this we shall do by putting the following question : what intellectual advantages were to be gained by turning to the Bible and the Gospel as sources of philosophic inspiration ?

CHAPTER II

THE CONCEPT OF CHRISTIAN PHILOSOPHY

·WHEN the problem is put as we have just put it, the simplest method of dealing with it will be to see how it was that so many cultivated men, versed in the systems of antiquity, could suddenly make up their minds to become Christians. This of course has happened in every age, and it would be no less interesting to know why, even in our own days, so many philosophers seem to turn to Christianity for a more satisfying solution of philosophic problems than they find in philosophy itself. But if we wish to look at the matter objectively, in a form which does not touch our personal interests, or not in so immediate a manner, it will be best to go back to the origins. If Christianity ever helped philosophers to more rational truths than they found in philosophy itself the reality of the help could never have been so clear as in the moment when it was given. Let us therefore approach the first philosophers to turn Christian, and ask them what they gained, as philosophers, in so doing.

For a really satisfactory discussion of the matter we must go back beyond even the first Christian philosophers. The oldest available witness was not a philosopher, but none the less his thought dominated the whole subsequent evolution of Christian ideas. We refer, of course, to St. Paul. He may be said to have laid down the principle on which the whole matter rests, and later Christian thinkers will do little more than draw out its consequences. According to the Apostle, Christianity is in no wise a philosophy but a religion. He knows nothing, he preaches nothing,

save Jesus Christ crucified and the redemption of sinners by
His grace. It would be altogether absurd therefore to speak
of a philosophy of St. Paul, and if we find certain fragments
of Greek philosophy embedded in his writings these are
either wholly adventitious, or, more often, integrated with a
religious synthesis which transforms their meaning altogether.
The Christianity of St. Paul is not a philosophy added to
other philosophies, nor a philosophy which would replace
other philosophies, it is a religion which supersedes all that
is ordinarily called philosophy and absolves us from the
trouble of seeking one. For Christianity is a way of salvation,
and therefore something other than, and more than, a
scheme of knowledge ; and no one, we may add, was ever
more conscious of this than St. Paul.

As he puts it in the First Epistle to the Corinthians, the
new revelation was set up as a rock of offence between
Judaism and Hellenism. The Jews sought salvation by way
of a literal observance of a Law and by obedience to the
commands of a God Who made His power manifest in
miracles of glory ; the Greeks sought a salvation to be
achieved by way of the rectitude of the will and the certitude afforded by the natural light of reason. What had
Christianity to offer either ? Salvation by faith in Christ
crucified, that is to say a scandal to the Jews who asked for
a sign of power and were offered the infamy of an humiliated
God : a folly to the Greeks who sought after the intelligible,
and were offered the absurdity of a God-man, dead on a
cross and risen again from the dead to save us. Christianity
had nothing to oppose to the wisdom of the world but the
scandalous and impenetrable mystery of Jesus : " For it is
written, I will destroy the wisdom of the wise ; and the
prudence of the prudent I will reject. Where is the wise ?
Where is the scribe ? Where the disputer of this
world ? Has not God made foolish the wisdom of the

world ? For seeing that in the wisdom of God the
world by wisdom knew not God, it pleased God, by the
foolishness of preaching, to save them that believe.
For both the Jews require signs, and the Greeks seek after
wisdom : but we preach Christ crucified, unto the Jews a
stumbling-block and unto the Gentiles foolishness ; but
unto them that are called, both Jews and Greeks, Christ
the power of God and the wisdom of God. For the foolishness of God is wiser than men, and the weakness of God is
stronger than men."

Nothing, at first sight, could be more explicit or more
decisive than these utterances ; they seem purely and simply
to dismiss Greek philosophy in favour of the new faith.
That, moreover, is why we should not be wrong in summing
up the thought of St. Paul on this central point by saying
that, according to him, the Gospel is not a wisdom but a
salvation. However, we must add that in another sense
this interpretation would be hardly exact ; for in the very
act of proclaiming the bankruptcy of Greek wisdom, St.
Paul proposes to substitute another wisdom, namely the
Person of Jesus Christ. His real intention, therefore, is to
set aside the apparent wisdom of the Greeks which is
really folly, so as to make way for the apparent folly of
Christianity which is really wisdom. Instead, then, of
saying that according to St. Paul the Gospel is salvation
and not wisdom, we should rather say that in his eyes the
salvation he preaches is the true wisdom, and that precisely
because it is salvation.

If we admit this interpretation, and it seems to arise
naturally from the text itself, it becomes clear that the
problem of Christian philosophy, resolved in principle,
remains entirely open as regards the consequences that are
to follow. What St. Paul has laid down, what no Christian
can henceforth dispute, is this : that to have faith in Jesus

Christ is *a fortiori* to achieve wisdom, in this sense at least, that as far as concerns the interests of salvation, faith really and totally absolves us from all need of philosophy. We might say once more that St. Paul defines a position which is the exact antithesis of that expressed in Goethe's 136th Proverb :

> *Wer Wissenschaft und Kunst besitzt*
> *Hat auch Religion ;*
> *Wer jene beide nicht besitzt*
> *Der habe Religion.*

—On the contrary, if we have religion we have also the essential truth of science, art and philosophy, all of them admirable things no doubt, but rather poor consolations when religion is wanting. But then, if this be true, if to possess religion is to possess all the rest, the thing must nevertheless be demonstrated. An Apostle like St. Paul may be content to preach it, a philosopher will wish to assure himself of the fact. It is not enough to assert that a believer can dispense with philosophy because the whole content of philosophy, and much more beside, is given implicitly in his belief ; some kind of proof is required. Now although to prove it would undoubtedly be one way of suppressing philosophy, the proof might very well, in another sense, turn out to be the best way of philosophizing. What philosophic advantages then did our earliest witnesses conceive themselves to gain from their own conversion to Christianity ?

The oldest, and, at the same time, the most typical, witness is that of St. Justin, who, in his *Dialogue with Trypho*, gives us a very living and picturesque account of his own conversion. The aim of philosophy, as he had always conceived it, is to bring us into union with God. Justin first of all made trial of Stoicism, but appears to have come across a Stoic who was more interested in practical moral life than in theory ; he admitted in fact that he did not regard the knowledge of God as altogether indispensable. The Peripatetic who followed him began by insisting on an agreement about the fee for his lessons, an attitude that Justin did not regard as particularly worthy of a philosopher. The third professor was a Pythagorean, and he, in his turn, soon bowed him out on the ground that Justin had not studied music, astronomy and geometry, all of them necessary preliminaries to any study of philosophy. A Platonist, who came next, did better. " I spent as much time with him as I could," writes Justin, " and thus I made progress, every day I advanced further. The understanding of the incorporeal world entirely captivated me ; the contemplation of the ideas lent wings to my mind so that after a little time I seemed to myself to have become wise. I was even foolish enough to hope that I was about to look on God, such being the aim of the philosophy of Plato." [1] Everything was thus going on well when Justin fell in with a venerable old man, who, questioning him about God and the soul, showed that he was involved in strange contradictions, and when Justin inquired how he came by so much knowledge of these matters, he answered thus : " In the most remote times, long before the day of any of these pretended philosophers, there lived certain men, happy, just and beloved by God, who spoke by the Holy Spirit and foretold many things that have since come to pass. We call them prophets. . . . Their writings still remain and those who read them with faith draw much and various profit, concerning both the beginning and the end, and all a philosopher ought to know. They did not deal in demonstrations ; for far above all demonstration they were worthy witnesses to the truth." [2] At these words the heart of Justin suddenly burned within him, and, says he, " revolving all these things in my mind it seemed to me

that here was the only sure and profitable philosophy.)
That is how and why I became a philosopher." [3]

Οὕτως δὴ χαι διὰ ταῦτα φιλόσοφος ἐγώ. It is hardly possible to exaggerate the importance of these words ; and if
I have recounted the personal experience of Justin in some
detail it is because, there already in the second century, so
strong a light is cast on all those factors without which there
will be no solution of the problem of Christian philosophy.
A man seeks the truth by the unaided effort of reason and is
disappointed ; it is offered him by faith and he accepts ;
and, having accepted, he finds that it satisfies his reason.
But Justin's experience is no less instructive in another
aspect, for it raises a problem to which Justin himself could
not refuse attention. What he finds in Christianity, along
with many other things, is the attainment of philosophic
truths by non-philosophic methods. Disordered reason is
reduced to order by revelation ; but precisely because they
had experimented in every direction without being afraid
of contradictions, the philosophers had managed to say,
along with much that was false, a good deal also that was
true. Doubtless it was all in a very fragmentary form—but
how was it, after all, that they came to know so much ?

A first solution of this problem was proposed by Philo
the Jew, and it immediately struck the Christian imagination and for a long time held it captive. It was facile, and
that made its fortune. Why not turn to account the fact
that the Bible is chronologically prior to the pagan systems ?
With some hesitation at first, more decisively after Tatian,
it was suggested that the Greek philosophers had more or
less directly profited by the contents of the inspired books,
owing such truths as they taught to these, not, of course,
without mingling a good deal of error of their own. However, the absence of any direct proof of such borrowing told
against the success of this rather over-simple solution, and

although it had a long life and is probably not quite dead even yet, it was bound gradually to give place to another, much deeper, and moreover, almost as old, since it is found already in Justin.

It may be found in fact even in St. Paul, in germ at least, and, so to speak, preformed. In spite of his disdainful condemnation of the false wisdom of the Greeks, the Apostle does not condemn reason ; on the contrary, he is concerned to recognize a certain natural knowledge of God even in the Gentiles. When, in the Epistle to the Romans (i. 19–20), he affirms that the eternal power and divinity of God may be known from created things, he affirms by implication the possibility of a purely rational knowledge of God in the Greeks, and at the same time lays the foundation of all the natural theologies which will later arise in the bosom of Christianity. No philosopher, from St. Augustine to Descartes, but will make use of this text. Moreover, in declaring in the same Epistle (ii. 14–15) that the Gentiles, deprived as they are of the Jewish law, are a law unto themselves, since conscience will either accuse or excuse them in the day of Judgment, St. Paul implicitly admits the existence of a natural moral law, or rather of a natural knowledge of the moral law. Now although the Apostle himself does not put the purely speculative question it becomes henceforth impossible not to put it : what is the relation between our rational God-given knowledge of the true and the good, and the revealed knowledge that the Gospel adds ? That is precisely the question that Justin put and answered.

He asks himself : since Jesus Christ was born some hundred and fifty years before my time, ought I to consider those who lived before Christ, and therefore without the help of His revelation, as all guilty or all innocent ? The Prologue to the Gospel of St. John suggests the answer. Jesus Christ is the Word, and the Word is God ; now the

Gospel says that the Word enlightens every man who comes into the world ; consequently, and on the testimony of God Himself, we must admit a natural revelation of the Word, universal and antedating the revelation given when He took flesh and dwelt amongst us. Moreover, since the Word is Christ, in participating in the light of the Word all men participate in the light of Christ. Those who lived according to the Word, whether pagans or Jews, were therefore, properly speaking, Christians ; while on the other hand, those who lived in error and vice, neglecting what was taught them by the light of the Word, were really enemies of Christ before He came. If then the case stands thus, the Pauline position, while remaining materially the same, undergoes a spiritual transformation ; for where the Apostle invoked against the pagans a natural revelation which condemns them, St. Justin gives them the benefit of a natural revelation which saves them. Socrates became so faithful a Christian that it is not surprising that the devil made him a martyr for the truth ; and Justin, in fact, is not far from exclaiming like Erasmus : St. Socrates, pray for us !

From this decisive moment, therefore, Christianity accepts the responsibility for the whole preceding history of humanity ; but then it also claims the benefit. All the evil that is done is done against the Word, but since, conversely, all the good that is done is done by help of the Word, who is the Christ, all truth is as by defini- tion, Christian. Whatever has been well said is ours : ὅσα οὖν παρὰ πᾶσι χαλῶς εἴρηται, ἤμων χριστιανῶν ἔστιν.[4] There, already formulated in the second century in definitive terms, we have the perpetual charter of Christian humanism. Hera- clitus is one of us ; Socrates belongs to us, for he knew Christ in part thanks to the effort of a reason which had its source in the Word ; ours also are the Stoics, and, along with the Stoics, all those other genuine philosophers in whom already

lay the seeds of that truth which revelation discovers to us
to-day in its fulness.

If we decide to adopt this outlook on history we can
still say with St. Paul that faith in Christ dispenses with
philosophy and that revelation supersedes it ; but supersedes it, nevertheless, only because it fulfils it. Hence a
reversal of the problem as curious as inevitable. If all that
was true in philosophy was but a presentiment and foreshadowing of Christianity, then the Christian, just because
he is a Christian, is in possession of all that ever was or ever
will be true in philosophy. In other words, and however
strange it may seem, the most favourable rational position
is no longer that of the rationalist, but that of the believer ;
and the most favourable philosophical position is not that
of the philosopher, but that of the Christian. If we would be
assured of it we have only to consider the advantages it
offers.

We note in the first place that the great superiority of
Christianity lies in the fact that it is no mere abstract knowledge of the truth, but an efficacious way of salvation.
To-day, perhaps, we should not consider a way of salvation
as having much direct connection with philosophy, because
we regard this latter as belonging more or less to the sphere
of science ; but for the Plato of the *Phaedo* or the Aristotle
of the *Nichomachean Ethics*, philosophy, although essentially
science, was not merely a science but also a life ; so much
so indeed did this become the case with the Stoics and their
successors that they adopted a distinctive dress just as the
priesthood does to-day. Now it is a fact that to the Christians
of the second century the Greek systems appeared in the
light of interesting speculations, occasionally even true,
but wholly without influence on the conduct of life. Christianity, on the contrary, with its prolongation of the natural
order by the supernatural, and its appeal to grace as an

inexhaustible source of power for the apprehension of truth and the realization of the good, offered itself at one and the same time as a doctrine and as a practice, or, more precisely, a doctrine that at the same time carried with it the means by which it was to be put into practice.

It would be only too easy to accumulate historical examples in support of this view of Christianity, but it will doubtless suffice to recall that it forms the whole essence of St. Paul's doctrine on sin, redemption, and grace. What man would do he does not ; what he would not do, he does. It is one thing to *will* to act rightly, quite another thing to have the *power* to perform the good act ; one thing is the law of God that reigns in the interior man, another the law of sin that reigns in the members. And who then will make the law of God to reign also over the outward man, save only God Himself by the grace of Jesus Christ ? [6] The doctrine is sufficiently familiar, and yet on the other hand we often forget that it lies at the heart of St. Augustine's work, and thereby of all Christian thought. A great deal of discussion has raged over the testimony of the *Confessions*, some considering that Augustine was converted to neo-platonism rather than to Christianity, others that his conversion was genuinely Christian. Personally, I have no doubt that this latter view is the correct one ; but if some have thought themselves able to maintain the former on the strength of a long array of texts and arguments, it is precisely for lack of a realization that Christianity is essentially a way of salvation, and that consequently, to be converted to Christianity, is essentially to enter on this way. Now if one thing emerges from the *Confessions* more clearly than another, it is the fact that in the eyes of St. Augustine the radical vice of neo-platonism consists in its ignorance of the twofold doctrine of sin and the deliverance from sin effected by grace. One might show that the purely *intellectual*

evolution of St. Augustine was completed by his adhesion
to neo-platonism, and nevertheless, with St. Augustine
himself, we should have to add a good many restrictions ;
but his whole doctrine makes it impossible to confuse that
adhesion with his conversion. That Plotinus should advise
us to rise above sense, to rule our passions, and to adhere to
God, that is all well and good ! But will Plotinus give us
the strength to follow this excellent advice ? And what does
it avail to know the good without power to put it into practice ? What kind of physician is this who recommends
health, and knows neither the nature of the illness nor the
name of the remedy ? What really completed the conversion of St. Augustine was the perusal of St. Paul and the
revelation of grace : "For the law of the Spirit of life in
Jesus Christ freed me," he says, " from the law of sin and
death." It was not an intellect that agonized in the night
in the garden of Cassiciacum : it was a man.

Let us return, however, to the plane of purely speculative
philosophy and abstract knowledge ; here again it has to
be recognized that in the eyes of the first Christian thinkers
many advantages accrued from the side of religion. One of
their favourite arguments in support of their faith was
drawn from the contradictions of the philosophers. The
fact is familiar ; but whether the significance of the argument is always rightly grasped is rather more doubtful.
What seems to have struck Justin and his successors was not
merely the incoherence of the speculative philosophies
but more especially the coherence of the answers given to
philosophic problems by a doctrine, which, instead of
giving itself out as one philosophy among others, claimed
to be nothing less than the one true religion.

When this view was turned to polemical account it gave
rise to the classical argument ' by the contradictions of the
philosophers.' In the first centuries of Christian thought

we find it everywhere ; in Justin, in whom I have indicated its starting-point ; in Tatian's *Discourse to the Greeks*,[7] where it receives its full development ; in the anonymous *Derision of the Philosophers* [8] ; in Arnobius, whose scepticism and fideism it is made to justify ; but perhaps we should say above all in Lactantius, because this very sensible man measured its exact range and marked it in definitive terms. In spite of the abuse which he did not spare them on occasion, Lactantius frequented the philosophers. Persuaded that much good was to be found in Socrates, in Plato, in Seneca, this Christian came to recognize that each had seized some fragment of the truth and that if these fragments were reunited the truth might be reconstituted in its totality : *particulatim veritas ab iis tota comprehensa est.* Suppose then that somebody could gather up these fragments dispersed throughout the writings of the philosophers and bring them together in a body of doctrine, and you would have something like the whole truth ; but—and here is the essential point—no one could effect this separation of the true and the false in the philosophic systems without knowing the full truth in advance, and no one could know it in advance unless God first gave it him by revelation, that is to say unless he accepted it by faith.

Lactantius, then, conceived the possibility of a true philosophy, but conceived it as an eclecticism on the basis of faith. On the one side we have the philosopher pure and simple, who, with nothing but his own reason to rely on, would discover the truth by his own unaided powers ; all his labour avails but to seize a tiny fragment of the total truth submerged beneath a mass of contradictory errors from which he is quite unable to extricate it. On the other side we have the Christian philosopher ; his faith provides him with a criterion, a norm of judgment, a principle of discernment and selection, allowing him to restore rational

truth to itself by purging away the errors that encumber it.
Solus potest scire qui fecit, says Lactantius : God, Who made
all, therefore knows all. If then He condescend to teach us,
let us listen. Having to choose between the uncertainties
of a reason wandering about without a guide and the certitude of a reason directed by revelation, he will not hesitate
for an instant ; and just as little will St. Augustine after
him.

For indeed the same experience is always observed to
recur, until at last we find it abstractly formulated by the
mediæval thinkers, and even rediscovered by more than one
modern. When the young Augustine adheres to the sect of
Mani he does so precisely because the Manichæans are bold
enough to attempt to explain everything without appealing
to faith. In spite of all their wild and puerile cosmogony
these men are rationalists who would lead the mind to faith
by way of understanding. If, weary of a church in which
the promised understanding was never in the end forthcoming, Augustine finally broke with the sect, it was to give
himself up to the amiable scepticism of Cicero ; and when,
thanks to Plotinus, he emerges from this, it is soon to discover that everything that was true in neo-platonism was
already contained in the Gospel of St. John and the Book
of Wisdom ; and many another truth besides that remained
hidden from Plotinus. Thus the philosophy that he vainly
sought by reason was offered him by faith. Those all too
uncertain truths which Greek speculation reserved for an
intellectual *élite*, had already been brought together,
purified, justified, completed by a revelation which put
them within the reach of all the world.[10] In this sense, we
might not incorrectly sum up the whole experience of St.
Augustine in the title he gave to one of his own works :
De Utilitate credendi—On the Advantage of Believing—even
for the very purpose of assuring the rationality of reason.

If he so incessantly repeats the words of Isaiah as he finds them in his Latin version : *nisi credideritis non intelligetis*, it is because they so exactly express his personal experience ; and when St. Anselm, in his turn, would define the beneficent effect of faith on the reason of the philosopher he will have nothing more to add.

The attitude of St. Anselm in this matter has been described as a Christian rationalism. The expression is somewhat equivocal, but has at least the merit of indicating that when St. Anselm appeals to reason he will have nothing to do with anything except reason. Not only Anselm, but his hearers themselves, demand that nothing shall be allowed to mediate between the rational principles from which he sets out and the rational conclusions he proposes to deduce. We need only recall, for instance, the famous preface to the *Monologium*, where, yielding to the insistence of his pupils, he undertakes to vindicate nothing in Scripture by the authority of Scripture, but to establish by rational evidence and the sole natural light of truth, all that can be shown to be true without the help of revelation. And yet it was certainly Anselm who supplied the definitive formula for the primacy of faith over reason, for if reason would be fully reasonable, if it would satisfy itself as reason, there is only one sure procedure open to it, and that is to examine the reasonableness of faith. Faith, as faith, is self-sufficing, but it aspires to become an understanding of its own content ; it does not depend on rational evidence, but, on the other hand, it engenders it. We learn from Anselm himself that the original title of his *Monologium* was *Meditations on the Reasonableness of Faith* and that the title of his *Proslogion* was none other than the famous formula : *A Faith seeking to Understand.* Nothing could express his thought more justly, for he does not seek to understand in order to believe, but to believe in order to understand ; and he pushes it to such lengths that this very primacy of faith over reason is itself something which he believes before he understands, and believes in order to understand. Is it not proposed by the authority of Scripture : *nisi credideritis, non intelligetis ?*

St. Justin, Lactantius, St. Augustine and St. Anselm, four witnesses only—but then, what witnesses ! Their high authority, and the perfect accord of their several experiences will permit me to dispense, I trust, with the numerous others that might be added. Before leaving the point, however, I would ask the reader to listen to one more voice that echoes theirs across the centuries, attesting how perennial is the question, how inevitable the answer. To bring forward the final conclusions of Maine de Biran is to throw the experience of a whole lifetime into the scale. He, too, like Augustine and so many others, tried to resolve the riddles of philosophy by reason alone, and the last words he wrote in his *Journal intime* are the *Vae soli* of Scripture : " It is impossible to deny to the true believer, who experiences in himself what he calls the effects of grace, who finds his repose and all the peace of his soul in the intervention of certain ideas or intellectual acts of faith, hope and charity, *and who thus succeeds in satisfying his mind on problems which all the systems have left unsolved*, it is impossible, I say, to contest his experience, and to fail to recognize how well founded either in himself or in his religious beliefs, are those states of soul which make his consolation and his happiness." [11] It is a fact, then, that for the Christian, reason alone does not satisfy reason and it was not merely in the second century that philosophers became Christian in the interests of their philosophy. To the *fides quaerens intellectum* of St. Anselm and St. Augustine, corresponds the *intellectus quaerens intellectum per fidem* of Maine de Biran. *Optavi et datus est mihi sensus, invocavi et venit in me spiritus sapientiae ;* this effort of truth believed to transform itself into truth known, is truly the life of Christian wisdom, and the body of rational truths resulting from the effort is Christian philosophy itself. Thus the content of Christian philosophy is that body of rational truths discovered, explored or simply safeguarded, thanks to the help that reason receives from revelation. Whether this philosophy ever really existed or whether it is nothing but a myth, is a question of fact on which we shall have to turn to history for a decision ; but before entering on this I would dissipate a misunderstanding which, by obscuring the meaning of the *fides quaerens intellectum* would make the very concept of Christian philosophy unintelligible.

Unless the expression is to be emptied of all positive content it must be frankly admitted that nothing less than an intrinsic relation between revelation and reason will suffice to give it meaning. And it is important that this meaning should be exactly defined. There is no question of maintaining—no one has ever maintained—that faith is a kind of cognition superior to rational cognition. It is quite clear, on the contrary, that belief is a succedaneum of knowledge, and that to substitute science for belief, wherever possible, is always a positive gain for the understanding. For Christian thinkers the traditional hierarchy of the modes of cognition is always faith, understanding, and vision of God face to face : *Inter fidem et speciem*, wrote St. Anselm, *intellectum quem in hac vita capimus esse medium intelligo.*[12]

Nor is there any question of maintaining the absurdity that you can accept the major premiss of a syllogism by faith and know the conclusion as science. If you start from belief and deduce its content you can never get anything more than belief. When those who define the method of Christian philosophy by the *fides quaerens intellectum* are accused of confusing philosophy with theology the accuser merely shows how little he understands their position, and gives us reason to suspect, besides, that he has no very clear notion of the nature of theology. For although theology is a science, it does not propose as its end the transformation of the belief by which it adheres to its principles into understanding ; to do that would be to destroy its proper object. Nor will the Christian philosopher on the other hand, any more than the theologian, attempt to transform faith into science, as if by some queer chemistry you could combine contradictory essences. What he asks himself is simply this : whether, among those propositions which by faith he believes to be true, there are not a certain number which reason may know to be true. In so far as the believer bases his affirmations on the intimate conviction gained from faith he remains purely and simply a believer, he has not yet entered the gates of philosophy ; but when amongst his beliefs he finds some that are capable of becoming objects of science then he becomes a philosopher, and if it is to the Christian faith that he owes this new philosophical insight, he becomes a Christian philosopher.

The present discord between philosophers as to the meaning of this expression thus becomes easier to explain. Some are considering philosophy in itself, in its formal essence as philosophy, abstraction being made from the conditions which rule either its constitution or its intelligibility. In this sense it is clear that a philosophy cannot be Christian, nor, for that matter, Jewish or Mussulman, and that the idea of Christian philosophy has no more meaning than " Christian physics " or " Christian mathematics." [13]

Others, taking account of the evident fact that, for a Christian, faith plays the part of an extrinsic regulative principle, admit the possibility of a Christian philosophy ; but, anxious to safeguard the formal purity of its essence as philosophy, they consider as Christian any philosophy that is true, any philosophy which presents " a conception

of nature and reason open to the supernatural."[14] And
that, undoubtedly, is one of the essential characters of
Christian philosophy, but by no means the only one nor,
perhaps, the deepest. A philosophy open to the supernatural would certainly be compatible with Christianity,
but it would not necessarily be a Christian philosophy. If
it is to deserve that name the supernatural must descend as
a constitutive element not, of course, into its texture which
would be a contradiction, but into the work of its construction. Thus I call Christian, *every philosophy which, although
keeping the two orders formally distinct, nevertheless considers the
Christian revelation as an indispensable auxiliary to reason.* For
whoever understands it thus, the concept does not correspond to any simple essence susceptible of abstract definition ; but corresponds much rather to a concrete historical
reality as something calling for description. It is but one of
the species of the genus philosophy and includes in its
extension all those philosophical systems which were in fact
what they were only because a Christian religion existed
and because they were ready to submit to its influence. As
concrete historical realities these systems are distinguished
from each other by individual differences ; as forming a
species they present common characteristics, and thus may
be grouped together under the same denomination.

In the first place, and it is perhaps his most obvious trait,
the Christian philosopher is one who effects a choice between
philosophic problems. Like any other philosopher, he has a
perfect right to interest himself in the whole circle of these
problems ; but in fact he is interested uniquely or above
all in those which affect the conduct of his religious life.
The rest, indifferent in themselves, become the objects of
what St. Augustine, St. Bernard and St. Bonaventure
stigmatize as curiosity : *vana curiositas, turpis curiositas.* Even
Christian philosophers like St. Thomas, whose interest

extended to the whole of philosophy, did their creative work only in a relatively restricted sphere. And nothing could be more natural. Since the Christian revelation teaches us only truths which are necessary to salvation, its influence could extend only to those parts of philosophy that concern the existence and nature of God, and the origin, nature, and destiny of the soul. In the very title and in the first lines of his treatise *De la connaissance de Dieu et de soi-même*, Bossuet held to the teaching of sixteen centuries of tradition : " Wisdom consists in knowing God and in knowing oneself. From the knowledge of self we rise to the knowledge of God." Everyone will recognize in these formulæ the *noverim me, noverim te* of St. Augustine, and although St. Thomas did not expressly make them his own he put them into practice. There is no question of minimizing his merits as a commentator and interpreter of Aristotle ; it is not in that field, however, that he is greatest, but rather in those genial views in which he prolonged and surpassed the philosophic effort of Aristotle. And these views are almost always to be found when he is speaking of God and of the soul and of the relations between God and the soul. The deepest of them have often to be disentangled from the theological contexts in which they are embedded, for it is there, in the bosom of theology, that they effectively come to birth. In a word, faith has a simplifying influence on all Christian philosophers worthy of the name, and their originality shines forth especially in the sphere directly influenced by faith, that is to say in the doctrine concerning God and man, and man's relations with God.

From the very fact that faith eliminates vain curiosity, the influence of revelation on philosophy facilitates the work of its constitution. From any Christian point of view the merely curious man is engaged on an interminable enterprise. He takes all knowledge for his province, every reality falls within it, and of none is he entitled to say that, if he knew it, it would not transform his knowledge of all the rest. Reality is inexhaustible and the attempt to synthetize it under principles is consequently impracticable. It may even be, as Comte was later on to suggest, that natural reality is not synthetic, and that it can be unified only by considering it from the point of view of a subject. Choosing man in relation to God as his central theme, the Christian philosopher acquires a fixed centre of reference which helps him to bring order and unity into his thought. That is why the tendency to systematization is always so strong in a Christian philosophy : it has less to systematize than any other and it has the necessary centre for the system as well.

It has also the necessary material for its completion, and this, in the first place, even in the field of natural philosophy. It seems to be sometimes supposed that only the Augustinians were convinced of this. In fact, in the *Summa Contra Gentiles* (Book I, Chapter IV) St. Thomas has left us a luminous *résumé* of the whole teaching of the Fathers of the Church on this fundamental question. He asks : is it fitting that God should reveal philosophical truths which are in fact accessible to reason ?—and he answers : yes !— provided that the knowledge of these truths is necessary to salvation. Were it otherwise, both these truths, and the salvation that depends on them, would be the exclusive prerogative of a very few ; the bulk of mankind would have to go without them, either for lack of intellectual light, or of leisure for research, or of a taste for study. He adds that even those capable of attaining them would do so only with a great deal of labour, after long thought, and would be running, moreover, all the risks of ignorance for the greater part of their lives. What, he asks, would be the state of mankind, if all our knowledge of God depended

only on human reason ? *In maximis ignorantiae tenebris*. And
this he confirms with a third consideration of no less weight
than the other two. The weakness of the human intellect,
in its present state, is such, that without the aid of faith
what to many would seem to be clearly demonstrated would
to others be exceedingly dubious ; and the spectacle of the
philosophic conflicts thus arising would contribute not a
little to breed scepticism in the generality of mankind, who,
for the most part, would view the discussion from outside.[15]
To overcome this *debilitas rationis* man has therefore need of
divine aid ; and this is what faith offers him. Like St.
Augustine and St. Anselm before him, St. Thomas situates
the reason of the Christian philosopher in an intermediate
position between the faith which guides his first footsteps,
and the full knowledge which the beatific vision will bring
hereafter ; like Athenagoras he thinks that man can aspire
to no perfect knowledge of God without first putting himself to
school with God, *qui est sui perfectus cognitor*. Faith, taking him
so to speak by the hand, puts him on the right road [16] and
goes along with him as long as he needs protection from error.

This, as will be seen, is no very bright picture of the results
attainable by human reason alone in the field of natural
theology—yet it is the picture drawn by the most thoroughly
intellectualist of all Christian philosophers. Why then should
we hesitate to follow where so many concordant indications
point the way, especially if it is possible to do so without
losing sight of the necessary distinctions, the fruit of so many
years of laborious reflection, and which reason, moreover,
clearly demands ? A true philosophy, taken absolutely and
in itself, owes all its truth to its rationality and to nothing
other than its rationality : that is indisputable, and St.
Anselm and even St. Augustine would be the first to admit
it. But the constitution of this true philosophy could not in
fact be achieved without the aid of revelation, acting as an

indispensable moral support to reason ; that is equally certain from the standpoint of the Christian philosophers, and, as we have just seen, St. Thomas himself asserts it. Now if he was right, or if we merely admit that he may have been right, the problem of Christian philosophy acquires a positive meaning. Doubtless in the abstract philosophy professes no religion, but we may very well ask whether it is altogether a matter of indifference that a philosopher should profess one. We may ask especially whether it is indifferent to the history of philosophy as such that there have been philosophers who were Christians, and whether, in spite of the purely rational texture of their systems, we cannot still to-day discern the mark of the influence of their faith on the conduct of their thought.

This hypothesis, since for the present that is all it is, is not in itself impossible or contradictory. Let us then grant that St. Augustine, St. Anselm and St. Thomas had a true appreciation of what they were doing. Let us admit provisionally that when they spoke of the debt of reason to revelation they had in mind the moving memory of those moments when, as in the meeting-place of two convergent rays, the opacity of faith suddenly gave way within them to the transparency of understanding. And let us go further. Let us ask whether they were not often more original than they knew, innovating with unconscious boldness even there, where they thought themselves to be no more than faithful followers of Plato and Aristotle. Thus to reveal in history the presence of an influence exerted on the development of metaphysics by the Christian revelation would be to demonstrate, so to speak experimentally, the reality of Christian philosophy. The task is immense and full of pitfalls—who can doubt it ? But it is possible to make the venture even without any very high hope of achievement ; and after all we aim at nothing but a sketch.

CHAPTER III

BEING AND ITS NECESSITY

WERE we asked to name the severest of all critics of the Middle Ages and their culture, our thoughts might very naturally turn to Condorcet. However, even this irreconcilable enemy of the priests was very willing to admit that their efforts in the sphere of philosophy were not wholly without merit. In the picture he draws of the Seventh Age of the progress of the human mind, we find him writing in terms which are somewhat remarkable coming from one who had such a lively hatred for all established religion. " To these scholastics we owe much of the precision of our ideas on the Supreme Being and his attributes ; on the distinction between the First Cause and the universe he is supposed to govern ; on that between mind and matter ; on the various meanings of the word ' liberty ' ; on the notion of *creation ;* on the distinctions between the various operations of the mind and the classification of its ideas concerning real things and their properties." [1] In spite of all his ill-humour Condorcet recognized that the scholastics gave a new precision to all the essential ideas of metaphysics and epistemology—a generous testimony that might easily be developed into an apologetic. Let us be content for the moment to consider what Christian thought has made of the idea of God, the very keystone of all metaphysics.

In adopting the expression " Supreme Being," not, to be sure, a very exact one, Condorcet merely spoke the language of his time ; but that time had learnt to condense into two words the results of a secular effort of reflection on Christian teaching. To speak of a Supreme Being, and to give the words all their weight, is first of all to admit that there is but one being really worthy of the name of God, and then, in the second place, that the proper name of this God is Being, a name that is applicable to this unique being in a sense in which it is applicable to no other. Now, can it really be said that the Christian thinkers derived this monotheism from the Hellenic tradition ?

It is not at all easy to say precisely how far the Greeks had travelled in this direction, and historians who try to decide it are by no means always in agreement. It may be observed, however, in the first place, that wherever monotheism has met with full acceptance, for instance in the Christian world, it has at once taken a central place as the principle of all principles. This results from the very nature of the case, for if there is one God and only one God, everything else must be ultimately referred to Him. Now we know of no system of Greek philosophy which reserved the name of God for a unique being, and made the whole system of the universe revolve round this single idea ; and it is therefore not very probable *a priori* that Hellenic speculation ever really succeeded in grasping a principle which, being but one principle, *the* principle, nevertheless failed to play this *rôle* of principle in their thought. Let us see whether this supposition is borne out by the facts.

It is clear from the evidence that first comes to hand that if the Greek poets and thinkers waged a successful warfare against anthropomorphism in natural theology, they never succeeded in eliminating, and hardly even dreamt of eliminating, polytheism. Xenophon teaches the existence of a great god, but that merely means a supreme god among gods and men. Neither Empedocles nor Philolaus went any further, and as for Plutarch, it is well known that the plurality of gods was one of his dogmas. Never, it seems, did Greek thought rise higher than this ; for it failed to do so even in the natural theologies of Plato and Aristotle.

If we keep our attention fixed on the precise question before us, without confusing it with others more or less closely akin, the answer is not in the least doubtful. Plato's doctrine unquestionably provided Christian speculation with many important elements—notably the Idea of the Good as developed in the *Republic*—which helped later on to elucidate the philosophical notion of the Christian God. But we are concerned with a different question, namely, what did Plato think of God, and whether or not he admitted a plurality of gods. Now with him the concept of God is very far from corresponding to the higher and perfect type of existence, and that is why divinity belongs to a class of multiple beings, perhaps even to all beings whatsoever in the precise measure in which they are. The *Timaeus* (28 C) represents a considerable effort to rise to the idea of a god who shall be the cause and father of the universe ; but no matter how great this god may be supposed to be he has rivals in the intelligible order of Ideas, and is moreover comparable with all the members of the whole vast family of Platonic gods. He does not exclude the sidereal gods, whose author he is (*Timaeus*, 41 A–C), nor even the divine character of the world he fashions ; first among the gods he is but one among them nevertheless, and if, in virtue of that primacy, the Demiurge of the *Timaeus* has been represented as " almost analogous to the Christian God " [2] we must say at once that a nuance of this kind is not allowable here. Either there is one God or there are many, and a god who is " almost analogous " to the Christian God is not the Christian God at all.

The case is the same when we turn to Aristotle ; and if the assertion seems surprising it is only because Christianity invaded the history of philosophy at the same time as it invaded philosophy itself. However, certain incidents in the life of Aristotle should focus our attention on this aspect of his doctrine. The man who gave testamentary disposition that the image of his mother should be consecrated to Demeter and that there should be erected at Stagira, on account of a vow he had made to the gods, two marble statues, one to Zeus Soter, and the other to Athene Soteira [3] —this man certainly had not shaken off the traditional polytheism. Here again, be it noted, we do not dream of disputing Aristotle's undoubted contribution to the philosophic idea of the Christian God. What is really surprising on the contrary is, that having gone so far along the right road he should have failed to follow it to the end. There however is the fact : he stopped on the way.

When we speak of Aristotle's god for the purpose of comparison with the Christian God, we refer of course to the unmoved mover, separate, pure act, thought of thought, set forth in a celebrated text of the *Physics* (VIII, 6). How this text is to be taken we shall have to consider later on ; for the moment I would simply observe that the first unmoved mover is very far from occupying in Aristotle's world the unique place reserved for the God of the Bible in the Judeo-Christian world. Returning to the problem of the cause of movements in *Metaphysics* (XII, 7–8), Aristotle begins by glancing back to the conclusions previously established in the *Physics :* " It is clear, then, from what has been said, that there is a substance which is eternal and immovable, and separate from sensible things. It has been shown also that this substance cannot have any magnitude, but is without parts and indivisible. . . . But it is also clear that it is impassive and unalterable ; for all the other kinds of change are posterior to change of place. It is clear, then, why this first mover has these attributes." Well, and what more could we ask ? An immaterial, separate, eternal and

immutable substance—is this not precisely the God of
Christianity ? Perhaps—but- read on, read the next sentence : " We must not ignore the question whether we have
to suppose one such substance or more than one, and, if
the latter, how many ? " Then at once he plunges into
astronomical calculations in order to determine whether,
under the first mover, we ought not to admit forty-nine, or
perhaps even fifty-five other movers, all separate, eternal
and unmoved. Thus although the first unmoved mover
stands alone in being first, he is not alone in being an
unmoved mover, that is to say a divinity. And were there
but only two that would be enough to prove that " in spite
of the supremacy of the first Thought, the mind of the
Philosopher is still profoundly impregnated with polytheism." [4] In short, Greek thought, even in its most eminent
representatives, did not attain to that essential truth which
is struck out at one blow, and without a shadow of proof,
by the great words of the Bible : *Audi Israel, Dominus Deus
noster Dominus unus est* (Deut. vi. 4).

It may very well be that to the minds of those to whom
they were addressed these words carried no such full and
clear meaning as they convey to-day to a Christian philosopher. Perhaps the people of Israel only gradually became
conscious of the profound truth of monotheism ; [5] but there
is no doubt at all that if Jewish thought moved only slowly
on this point the movement had long been completed when
Christianity inherited the Bible. Asked to indicate which is
the greatest commandment of the Law Jesus answers at
once with the fundamental assertion of Biblical monotheism,
as if everything else hung on that : " The first commandment is this, Hear O Israel, the Lord thy God is One God "
(Mark xii. 29). Now this *Credo in unum Deum* of the Christians,
this first article of their faith, appeared at the same time as
an irrefragable rational truth. That if there is one God there

is only one God is something which after the seventeenth
century will be taken for granted as self-evident, nobody
will any longer trouble himself to prove it. The Greeks
did not realize it however. What the Fathers had never
ceased to affirm as a fundamental belief because God
Himself had said it, is also one of those rational truths,
and the most important of all, which did not enter philosophy by way of reason. The nature of this phenomenon,
which had a decisive influence on the development of
philosophic speculation, will be understood better if we
link up the problem of the unicity of God with the problem
of His nature.

The two questions, of course, are closely connected. If
the Greeks were never quite sure how many gods there
were, that was precisely because they lacked that clear
idea of God which makes it impossible to admit more than
one. The best of them, with an admirable effort, shook off
all that was materialistic in Greek polytheism ; we see them
even ranging the gods in hierarchical order, subordinating
the gods of fable to the gods of metaphysic ; ordering these
in turn under a supreme god ; but having done so much
why did they fail to reserve a proper and exclusive divinity
to this supreme god ? The answer is to be found in the
conception they formed of his essence.

The interpretation of Plato's natural theology certainly
raises difficult questions. Excellent Hellenists, philosophers
into the bargain, have maintained with considerable force
that Platonism achieved an idea of God practically indistinguishable from that of Christianity. According to the
most resolute defender of this thesis, the true thought of
Plato is that " the degree of divinity is proportionate to the
degree of being ; that, therefore, is most of all divine which
most of all is being ; now that which most of all is being is
the universal Being or the All of being." How, after that,

can we fail to see that τὸ παντελῶς ὄν, in Plato, is universal being, that is to say God, this same God of Whom Fénélon will say in his *Traité de l'existence de Dieu* (II, 52) that He gathers up in Himself " the fullness and totality of being," and of Whom Malebranche will say in his *Recherche de la vérité* (IV, 11), that His idea is " the idea of being in general, of unrestricted being, of infinite being " ?

There is a literal resemblance in the texts thus brought together, but it is quite misleading. The παντελῶς ὄν of the *Sophist* (248 E) is, no doubt, the totality of being in so far as this is intelligible and, consequently, real ; but what this really signifies is a refusal to follow Parmenides of Elea in his attempt to deny the reality of movement, of becoming, of life. In this sense it is true to say that Plato restored to being all that which, possessing a certain degree of intelligibility, possesses also a certain degree of reality. But, in the first place, Plato does not say that his " universal being " is God, and even supposing that it is identified with God, in spite of Plato's silence on the point, all that one can draw from this formula is that the Platonic god gathers up in himself the totality of the divine, just as he gathers up in himself the totality of being. Set the two thoughts side by side and see how profound is the divergence of meaning underlying the common formulæ. According to Plato " the degree of divinity is proportionate to the degree of being " ; but for a Christian there are no degrees of divinity ; God alone is divine. Again for Plato, " what is most of all being is most of all divine," but for a Christian it is only by way of analogy or metaphor that beings can be more or less divine ; properly speaking there is but one God Who is Being, and beings, which are not God. The radical difference of the two traditions lies in the fact that, for Plato, there is no sense of the word " being " reserved exclusively for God. That is why his god possesses divinity only in a supreme degree, not at all as an unique prerogative : wherever there is being there is divinity, because there is no single being that claims the whole fullness and privilege of divinity.

There, moreover, we have the hidden cause of the difficulties encountered by all interpreters of Plato who try to bring his conception of the divine into line with the Christian God. Much ingenuity has been wasted on the attempt. Sometimes the Demiurge of the *Timaeus* is identified with the Idea of the Good in the *Republic*, and the only result of this is that the Demiurge becomes, not Being, but the Good [6] —a thing, by the way, that Plato never made of him.[7] Sometimes the whole sum-total of divinity' is attributed to one being—who, for the rest, does not exist in Plato—and then we no longer know what to make of this diffused divinity existing everywhere in beings, particularly in the Ideas, as if, according to this doctrine, the gods were other than what is most divine. But a difficulty of the same kind awaits the interpreters of Aristotle, and this too has to be considered. Did he effect the difficult operation of finding a place in the ranks of Greek polytheism for this unique Being, the Christian God ?

Certainly there is no lack of texts that might be alleged in support of the affirmative. Does not Aristotle speak of an essence sovereignly real, transcending the order of physical things, situated, consequently, beyond nature, and thus none other than God ? It would seem that we really have to do here with a natural theology of which the proper object would be, as Aristotle says himself, " being as being " (*Metaph.*, Γ, I, 1003a 31), being *par excellence* (A² 2994b 18), a necessary substance always in act (Λ, 1071b 19 and 1072b 10), in short, this God that St. Thomas will so easily recognize in the Aristotelian formulæ without having to modify them at all. And indeed if these formulæ contained no hint of the Christian God St. Thomas would never have found Him there. We might almost say, in a sense, that it would have been difficult for Aristotle to go further without reaching the true conception, but then that is no ground for saying that he reached it. What is true is, that Aristotle clearly understood that God is, of all beings, the one that deserves the name of being *par excellence ;* but that his polytheism prevented him from conceiving divinity as anything but an attribute of a class of beings. We cannot say of him, as we can of Plato, that he regards all that exists as divine, for he reserves the name of divinity to the order of necessity and pure actuality ; but although his First Unmoved Mover is, of all beings, in the highest degree divine and in the highest degree being, he remains nevertheless but one of these " beings as being." It cannot possibly be denied that his natural theology has for proper object, a plurality of divine beings, and that is enough to distinguish it radically from Christian natural theology. For him, the necessary is always a collective ; for Christians it is always a singular.[8] But we need not stop there. Even were it granted, in the face of all the texts, that Aristotle's being as being is a unique being, it would still be true that this being is none other than the pure act of thought thinking itself. It would be all that, no doubt, but it would be only that, and it is incidentally for this reason that the attributes of Aristotle's god are strictly limited to those of thought. In good Aristotelian doctrine the first name of God is thought, and pure being is reduced to pure thought ; in good Christian doctrine the first name of God is being, and that is why we can refuse to Being neither thought, nor will, nor power, and why the attributes of the Christian God overflow the attributes of Aristotle's in every direction. But no one rises to the Christian conception of Being who sets up statues to Zeus and Demeter.

Compared with all these laborious gropings how straightforward is the method of the Biblical revelation, and how startling its results !

In order to know what God is, Moses turns to God. He asks His name, and straightway comes the answer : *Ego sum qui sum, Ait : sic dices filiis Israel ; qui est misit me ad vos* (Exod. iii. 14). No hint of metaphysics, but God speaks, *causa finita est*, and Exodus lays down the principle from which henceforth the whole of Christian philosophy will be suspended. From this moment it is understood once and for all that the proper name of God is Being and that, according to the word of St. Ephrem, taken up again later by St. Bonaventure, this name denotes His very essence.[9] Now to say that the word *being* designates the essence of God, and the essence of no other being but God, is to say that in God essence and existence are identical, and that in God alone essence and existence are identical. That is why St. Thomas Aquinas, referring expressly to this text of Exodus, will declare that among all divine names there is one that is eminently proper to God, namely *Qui est*, precisely because this *Qui est* signifies nothing other than being itself : *non significat formam aliquam sed ipsum esse.*[10] In this principle lies an inexhaustible metaphysical fecundity ; all the studies that here follow will be merely studies of its results. There is but one God and this God is Being, that is the corner-stone of all Christian philosophy, and it was not Plato, it was not even Aristotle, it was Moses who put it in position.

To realize the importance of this perhaps the shortest way would be to read the first lines of the *De primc rerum omnium principio* of Duns Scotus : " O Lord our God, when Moses asked of Thee as a most true Doctor, by what name he should name Thee to the people of Israel ; knowing well what mortal understanding could conceive of Thee and unveiling to him Thy ever blessed name, Thou didst reply : *Ego sum qui sum ;* wherefore art Thou true Being, total Being. This I believe, but if it be in any wise possible this I would also know. Help me, O Lord, to seek out such knowledge of the true being that Thou art as may lie within the power of my natural reason, starting from that being which Thou Thyself hast attributed to Thyself." Nothing can surpass the weighty fullness of this text, since it lays down at once the true method of Christian philosophy, and the first truth whence all the others derive. Applying the principle of St. Augustine and St. Anselm, the *Credo ut intelligam,* Duns Scotus, at the very outset of his metaphysical speculation, makes an act of faith in the truth of the divine word ; like Athenagoras, it is in the school of God that he would learn of God. No philosopher is invoked as inter-mediary between reason and the supreme Master ; but forthwith, after the act of faith, philosophy begins. Whoever believes by faith that God is being sees at once by reason that He can be nothing but total being, true being. But now we have to see for ourselves how these consequences are implied in the principle.

When God says that He is being, and if what He says is to have any intelligible meaning for our minds, it can only mean this : that He is the pure act of existing. Now this pure act of existing excludes all non-being *a priori.* Just as non-being is absolutely void of all being and of all the con-ditions of being, so also Being is wholly unaffected by non-being, both actually and virtually, both in Itself and from our point of view. Although therefore in our language it bears the same name as the most general and abstract of all our concepts, the idea of Being signifies something radically different. Perhaps—and this is a point to which we shall soon have to return—our very power of conceiving abstract being is not unconnected with the ontological relation in which we stand to God, but it is not as a concept

that God would have us think of Him nor even as a being
whose content would be that of a concept. Beyond all
sensible images, and all conceptual determinations, God
affirms Himself as the absolute act of being in its pure
actuality. Our concept of God, a mere feeble analogue of
a reality which overflows it in every direction, can be made
explicit only in the judgment : Being is Being, an absolute
positing of that which, lying beyond every object, contains in
itself the sufficient reason of objects. And that is why we
can rightly say that the very excess of positivity which hides
the divine being from our eyes, is nevertheless the light
which lights up all the rest : *ipsa caligo summa est mentis
illuminatio.*[11]

From this point, in fact, our conceptual thought will
move around the divine simplicity regarding it from every
side in order to express its inexhaustible riches in a multiplicity of complementary views. When we try to express
God as He is in Himself we can only repeat with St. Augustine the divine name that God Himself announced : *non
aliquo modo est, sed est, est.** If we would go further we can
but try to make explicit in multiple judgments, none of
which will in itself suffice, the whole content that we can see
in *est.* Now it appears that Christian speculation has pursued
this work along two converging paths, one of which leads
us to assert God as perfect, the other to assert Him as infinite,
this perfection and this infinity being reciprocally implied
in each other as two equally necessary aspects of the Being
they qualify.

Considered from the first point of view, the pure being is
endowed with absolute sufficiency in virtue of its very
actuality. It would be contradictory to say that what is

* This text is quoted from the edition of P. Knöll, Leipzig, Teubner, 1919,
p. 329. The reading *quod est, est,* might be justifiable, but, as it seems to me,
with greater difficulty.

being by definition could receive anything whatever from outside, for it could receive only what was lacking to its actuality. Thus to say that God is Being is equivalent to asserting His aseity. We must, once more, be quite clear as to the meaning of this last term. God exists in virtue of Himself (*per se*) in an absolute sense, that is to say as Being He enjoys complete independence not only as regards everything without but also as regards everything within Himself. Just as His existence is not derived from any.other than Himself, so neither does He depend on any kind of internal essence, which would have in itself the power to bring itself to existence. If He is *essentia* this is because the word signifies the positive act itself by which Being is, as if *esse* could generate the present participle active *essens*, whence *essentia* would be derived.[12] When St. Jerome says that God is His own origin and the cause of His own substance, he does not mean, as Descartes does, that God in a certain way posits Himself in being by His almighty power as by a cause, but simply that we must not look outside God for a cause of the existence of God. Now this complete aseity of God involves His absolute perfection as an immediate corollary.

Since in fact God is being *per se*, and since our conception of God absolutely excludes all non-being, and all that dependence that would result from non-being, it follows that in Him the fullness of existence must be completely realized. God is thus pure being in its state of complete fulfilment and realization, as that being alone can be which can receive no addition either from within or from without. Moreover His perfection is not a perfection received, but a perfection, so to speak, *existed*, and it is just that which will always keep Christian philosophy distinct from Platonism, in spite of all the efforts that may be made to identify them. Even when it is granted that Plato's true God is the Idea of

the Good as set forth in the *Republic* (509 B), the supreme
term thereby attained would be only an intelligible, the
source of all being because the source of all intelligibility.
Now the primacy of the Good as Greek thought conceived
it, compels the subordination of existence to the good, while
on the other hand the primacy of being as Christian thought,
under the inspiration of Exodus, conceived it, compels
the subordination of good to existence. Thus the perfection of the Christian God is that perfection which is proper
to being as being, that which being posits along with itself ;
we do not say that He is because He is perfect, but on the
contrary, He is perfect because He is. And it is just that
difference, so nearly imperceptible at its point of origin,
and yet so fundamental, that carries with it such startling
consequences, when at last it brings forth from the very
perfection of God, His total freedom from all limits and His
infinity.

That which is in virtue of itself and is not made, presents
itself to thought in fact as the very type of the immutable
and fully realized. The divine being is necessarily eternal,
because existence is His very essence ; He is not the less
necessarily immutable, since nothing can be added or withdrawn from Him without destroying His essence simultaneously with His perfection ; He is therefore finally
repose, as a tranquil ocean of substance integrally present
to Himself and for Whom the very conception of an event
would be altogether meaningless. But, at the same time,
because it is of being that God is the perfection, He is not
merely its complete fulfilment and realization, He is also
its absolute expansion, that is to say its infinity. If we hold
to the primacy of the good the idea of perfection implies
that of limitation, and that is why the Greeks prior to the
Christian era never conceived infinity save as an imperfec-
tion. But when, on the contrary, we assert the primacy of

being, it is very true that nothing can be lacking in Being, and that, consequently, He is perfect ; but, since we now have to do with a perfection in the order of being, the internal exigencies of the concept of the good are subordinated to those of the concept of being, for goodness is nothing but an aspect of being. The perfection of being not only calls for all realizations, it also excludes all limits, generating thereby a positive infinity which refuses all determination.

Regarded in this light the divine Being more than ever eludes the grasp of our concepts. There is no single idea at our disposal which does not break down in some way when we attempt to apply it to Him. Every denomination is a limitation, but God is above all limitation, and therefore above all denomination no matter how exalted it may be. In other words, an adequate expression of God would be God ; and that is why Christian theology admits one, and only one, Who is the Word ; but our poor human words, however ample their extension, express only a part of that which has no parts, and would endeavour to circumscribe within an essence that which, in Dionysius' phrase, is super-essential. Even the divine ideas express God only *quatenus*, as so many possible participations, therefore partial and limited participations, of that which participates nothing and overflows all limits. Infinity, in this sense, becomes one of the primary characteristics of the Christian God, and the one which, after Being, most clearly distinguishes Him from all other conceptions of God.

Nothing is more remarkable than the agreement of the mediæval thinkers on this point. Perhaps it is in the doctrine of Duns Scotus that this particular aspect of the Christian God is most easily recognizable. For Duns Scotus, in fact, it is altogether one and the same thing to prove the existence of God and to prove the existence of an infinite being, and this undoubtedly means that until the existence of an infinite being has been established it is not God Whose existence has been proved. He asks himself : *utrum in entibus sit aliquid actu existens infinitum,* and in that there is nothing out of harmony with the thought of St. Thomas Aquinas and other Christian philosophers of the Middle Ages, although this very special way of stating the problem brings out the aspect before us in a sufficiently striking light. Duns Scotus starts, in fact, from the idea of being in order to prove that we must necessarily admit a first being ; from the fact that it is first he deduces that it is uncausable ; from the fact that it is uncausable he deduces that this first being exists necessarily. Passing on to the properties of this first and necessary being he shows that it is efficient cause, endowed with intelligence and will, that its intelligence embraces the infinite, and that since this intelligence is identical with its essence, its essence also envelops the infinite : *Primum est infinitum in cognoscibilitate, sic ergo et in entitate.* To demonstrate such a conclusion is, according to the Franciscan Doctor, to establish the most perfect concept conceivable by us, that is to say the most perfect concept we can possibly possess on the subject of God : *conceptum perfectissimum conceptibilem, vel possibilem a nobis haberi de Deo.*[13]

We ought however to add that St. Bonaventure and St. Thomas are perfectly at one with Duns Scotus in affirming the subsistence of a being in face of Whom both absolute Eleatism and absolute Heracliteanism are equally vain, because at one and the same time this being transcends the most intense actual dynamism and the most fully realized formal staticism. Even in the thought of those who are most attracted to the aspect of realization and perfection which characterizes the Pure Being, we may easily discern the presence of the element of " energy " which, as we know, is inseparable from the conception of act. In this sense St. Thomas himself, who speaks of God in the pure language of Aristotle, is nevertheless far enough from the thought of Aristotle. The " Pure Act " of the Peripatetics is pure act only in the order of thought ; that of St. Thomas is pure act in the order of being, and therefore, as we have seen, it is at once infinite and perfect. Whether, in fact, we refuse to withdraw the reality of such an act from all limit, or whether we refuse to enclose it within the perfection of its own realization, in either case we re-introduce virtuality and at the same blow destroy its essence. " The infinite," said Aristotle, " is not that outside of which there is nothing, but, on the contrary, that outside of which there is always something." [14] The infinity of the Thomist God is precisely that outside of which there is nothing, and that is why, having just said that the true name of God is *being* because this name does not signify any determinate form—*non significat formam aliquam*—St. Thomas tranquilly writes, in Aristotelian formulæ, something which it is questionable whether Aristotle would have understood, namely : it is because God is Form, that He is Infinite Being—*cum igitur Deus ex hoc infinitus sit, quod tantum forma vel actus est.*[15] St. Thomas is well aware that form, as such, is the principle of perfection and completion ; *perfectio autem omnis ex forma est,* and this is precisely why he has just said that God is called being, because this word does not designate any form ; but he also knows that in the unique case when the pure act in question is that of being itself, the plenitude of its actuality of being confers on it, as of full right, a positive infinity unknown to Aristotle, an infinity outside of which there is nothing. By a paradox which would have no meaning save as applied to God : *sua infinitas ad summam perfectionem ipsius pertinet.*[16] For St. Thomas, as for Duns Scotus, it is of the very essence of God, as the pure form of being, to be infinite.

When we reflect on the meaning of this conception, it

becomes clear that, sooner or later, it was bound to give rise to a new proof of the existence of God, that namely which, since Kant, has gone by the name of the ontological argument, and which St. Anselm has the signal honour of having first put into definite shape. Even those who deny all creative originality to Christian thought usually make some reserves in favour of Anselm's argument, which, since the Middle Ages, has reappeared again and again in the most diverse forms in the systems of Descartes, Malebranche, Leibniz, Spinoza, and even that of Hegel. That no trace of it exists in Greek thought is quite undisputed, but it does not seem to have occurred to anyone to ask either why the Greeks never dreamt of it, or why, on the contrary, it was perfectly natural that Christians should be the first to conceive it.

Once the question is asked the answer is obvious. Thinkers like Plato and Aristotle, who do not identify God and being, could never dream of deducing God's existence from His idea ; but when a Christian thinker like St. Anselm asks himself whether God exists he asks, in fact, whether Being exists, and to deny God is to affirm that Being does not exist. That is why the mind of St. Anselm was so long filled with the desire of finding a direct proof of the existence of God which should depend on nothing but the principle of contradiction. The argument is sufficiently well known, and there is no need to set it out in detail, but those who do so are not always as clear as they might be as to its significance. The inconceivability of the non-existence of God could have no meaning at all save in a Christian outlook where God is identified with being, and where, consequently, it becomes contradictory to suppose that we think of Him and think of Him as non-existent.

Leaving on one side the technical mechanism of the proof in the *Proslogion*, for which I profess no excessive admiration, it comes essentially to this : that there exists a being whose intrinsic necessity is such as to be reflected in the very idea we form of Him. God exists so necessarily in Himself that even in our thought He cannot not exist : *quod qui bene intelligit, utique intelligit idipsum sic esse, ut nec cogitatione queat non esse.*[17] Where St. Anselm went wrong, as his successors very well saw, was in failing to notice that the necessity of affirming God, instead of constituting in itself a deductive proof of His existence, is really no more than the basis for an induction. In other words, the analytical process, by which from the idea of God is drawn the necessity of His existence, is not in fact the proof that God exists, but might very well become the initial datum of this proof, for we might try to show that the very necessity of affirming God postulates God's existence as its sole sufficient reason. What St. Anselm only half divined was left for others to put in a clear light. St. Bonaventure, for example, very well saw that the necessity of God's being *quoad se* is the sole conceivable sufficient reason of the necessity of His existence *quoad nos*. Let him who would contemplate the unity of the Divine Essence, he says, first fix his eyes on being itself : *in ipsum esse,* and there he will see that being itself is in itself so absolutely self-evident that it cannot be thought of as not being : *et videat ipsum esse adeo in se certissimum, quod non potest cogitari non esse.*[18] The whole Bonaventurian metaphysic of illumination lies behind this text, in readiness to explain our certitude of His existence by an irradiation of the divine being on our thought. Another theory of knowledge, but one not less carefully elaborated, justifies the same conclusion in Duns Scotus. According to him the proper object of our intellect is *being ;* how then could we doubt of that which the intellect affirms of being with such fullness of light, that is to say its infinity and its existence ? [19]

Finally, if we leave the Middle Ages and pass to the beginnings of modern philosophy in Descartes and Malebranche,
we still see St. Anselm's discovery bearing fruit. In Descartes
in particular it is interesting to observe that the two possible
ways of proving God from our conception of God are successively tried out. In the *Fifth Meditation*, he attempts once
more, following St. Anselm, to pass direct from the idea of
God to the affirmation of His existence ; but already in the
Third Meditation he has attempted to prove the existence of
God as the necessary cause of the idea we have of Him.
And that also is the route followed by Malebranche, for
whom the idea of God is as an imprint left by God Himself
upon our soul. In the remarkable texts in which the
Oratorian thinker, analysing our general, abstract and
confused idea of being, shows that it is the sign of the
presence of Being Itself to our thought, he authentically
prolongs one of the ways followed by the Christian
philosophic tradition in order to rejoin God : if God
is possible, He is real ; if we think of God, He must necessarily be.

But whatever we may think of its modern prolongations,
all Christian and mediæval philosophy must be regarded as
one in affirming the metaphysical primacy of being, and its
sequel, the identity of essence and existence in God. This
unanimity, which for the rest is of capital importance, does
not extend only to the principle, but also to all its necessary
consequences in the field of ontology. We shall soon have an
opportunity of watching the evolution of some of the most
important of these, especially in all that concerns the
relation in which the world stands to God. But as far
as the legitimacy of a proof of the existence of Being from
our idea of Him is concerned there is no great measure
of agreement to this day. Those Christian philosophers
who follow the tradition of St. Anselm always tend to

consider this proof as the best, even sometimes as the only one possible. But they, too, seem to labour under a double preoccupation, to be attracted so to speak by a double virtuality : either to base everything on the ontological value of rational evidence, and then, with the Anselm of the *Proslogion* and the Descartes of the *Fifth Meditation* they maintain that a real existence necessarily corresponds to the necessary affirmation of an existence ; or else to construct an ontology based on the objective content of the concepts, and then they prove the existence of God inductively as the sole conceivable cause of the idea of God in us ; a way opened up by St. Augustine and the St. Anselm of the *De Veritate*, and pursued in turn by St. Bonaventure, Descartes, and Malebranche. This is not the place to discuss the respective values of these two methods, especially since we are about to compare them with a third ; but I may perhaps be allowed to indicate that, for reasons which will appear more clearly later on, the way followed by St. Augustine and St. Bonaventure seems to me to be much the better. To show that the affirmation of necessary existence is analytically implied in the idea of God, would be, as Gaunilo remarked, to show that God is necessary if He exists, but would not prove that He does exist.[20] If, on the contrary, we seek the sufficient reason of a being capable of conceiving the idea of being, and there read the inclusion of essence in existence, we are dealing with a question which must remain an open one in any epistemology. It will always be legitimate to attempt the construction of a metaphysic on the basis of the presence in our minds of the idea of God, provided, however, that we do not attempt a deduction *a priori* with its starting-point in God, but an induction *a posteriori* with its starting-point in the content of our conception of God. Perhaps it would not be impossible to show that, in this sense, the Thomist method is necessary to bring the Augustinian to a full consciousness of its own nature and legitimate conditions of exercise ; but this is a point which will arise of itself when we have considered the route to God that was followed by St. Thomas Aquinas.

CHAPTER IV

BEINGS AND THEIR CONTINGENCE

IF all we have said is correct, the Christian revelation exerted a decisive influence on the development of metaphysics by introducing the identification of God and Being. Now this involved a correlative modification in the Christian conception of the universe. If God is Being, He is not only total being : *totum esse*, but, as we have seen, He is more especially true being : *verum esse*, and that means that everything else is only partial being, hardly deserves the name of being at all. And thus all that seems to us most obviously real, the world of extension and change around us, is banished at one stroke into the penumbra of mere appearance, relegated to the inferior status of a quasi-unreality. It is impossible to insist too much on the importance of this corollary, and its essential meaning at least must now be made clear.

That sensible reality is not the true reality was certainly not revealed to the world for the first time by Christianity. We all think at once of Plato, and the way in which he subordinates things themselves to their Ideas. The latter are eternal, immutable, necessary, and thus they really are ; the former are mutable, perishable, contingent ; they are as if they were not. Such being as they possess is due merely to their participation in the Ideas ; but they do not participate in the ideas alone, since the transitory forms of things are merely reflections cast by the Ideas on a passive recipient, a kind of indetermination hovering between being and non-being, a miserable and precarious existence which, in flux and reflux like the waves of some vast Euripus, communicates its own indetermination to the shadows of the Ideas it bears on its bosom. What Plato here suggests is true for a Christian, but true with a much deeper truth than Plato ever guessed, perhaps we ought to say, in a sense, with another truth. The main distinction between Hellenism and the Christian philosophies lies precisely in the fact that the latter are based on a conception of the divine Being which neither Plato nor Aristotle ever attained to.

As soon as we identify God with Being it becomes clear that there is a sense in which God alone is. If we refuse to admit this we shall have to assert that all things are God, and that is precisely what a Christian can never do, and this not merely for religious but also for philosophical reasons ; of which the chief one is that if all things are God, then there is no God. None of the things we know directly possesses all the characters of being. Bodies, in the first place, are not infinite, since each is determined by an essence defining and therefore limiting it. It is always with this or that being that we come into contact, never with Being : and even if we suppose the whole totality of the real and the possible to be realised, no such summation of particular beings would re-constitute the unity of that which purely and simply *is*. But that is not all. To the *Ego sum qui sum* of Exodus, there exactly corresponds this other word of Malachi (iii. 6) : *Ego Dominus et non mutor*. And, indeed, all that we know is subject to becoming, that is, to change ; and thus no single one of these things is perfect and immutable as must of necessity be the case with Being Itself. In this sense, then, there is no fact or problem more vital to Christian thought than that of movement, and it is precisely because the philosophy of Aristotle is essentially an analysis of becoming and of its metaphysical conditions, that it has itself become an integral part of Christian metaphysics, and will always remain so.

It is not a little surprising at times to watch St. Thomas Aquinas commenting on Aristotle's physics often almost word for word, subtilizing on the notions of potency and act as though the whole fate of natural theology were at stake. And so indeed, in a way, it is. Aristotle's language almost resembles a system of algebraic notation, which is just why the ideas it expresses go to constitute a science ; but we can always penetrate beneath his technicalities to the reality he has in mind and this usually turns out to be movement. No one has ever better discerned the mystery that the very familiarity of movement hides from our eyes. All movement implies some being, for if there were no being there, there would be nothing that could move ; wherever, then, there is movement, there is something that moves. On the other hand this something that moves never fully *is ;* if it were it could not be in movement, since to change is either to acquire being or to lose it. Clearly, if a thing becomes it could not at the outset be what it becomes, and indeed to become anything it often has to cease to be something else, so that movement is a state of that which, while by no means merely nothing, is yet not fully being. M. Bergson accuses Aristotle and his successors of having *reified* movement, and of having cut it up into a series of successive immobilities ; but nothing could be more unjust, it is to saddle Aristotle with the errors of Descartes, who on this precise point is his very negation. The whole of mediæval Aristotelianism, looking beyond even the series of states of the moving thing, saw in movement a certain *mode of being,* that is to say, in the fullest sense of the term, a way of existing, metaphysically inherent in the essence of the thing which thus exists, and, consequently, bound up with its nature. In order that things should change, as we see they do, it is not sufficient that, stable in themselves, they should simply pass from one state to another as, in the

Cartesian physics, a body passes from one place to another without ceasing to be what it is ; on the contrary, as in the Aristotelian physics, even the local displacement of a body marks the intrinsic mutability of the body which changes place, so that, in a way, the possibility of ceasing to be where it is attests a possibility of its ceasing to be what it is.

It is this fundamental insight which Aristotle is trying to express when he says that movement is the act of a thing in potency in so far as it is in potency. Since the time of Descartes it has become fashionable to scoff at this definition, and certainly Descartes' own seems a good deal clearer ; probably, as Leibniz saw, because it altogether fails to define movement. The obscurity does not lie in Aristotle's definition, it lies rather in the thing defined : something, namely, which is in act because it is, but is not pure actuality because it becomes, yet has a potency which tends to actualize itself progressively because it changes. If we look at things instead of words, we cannot fail to see that the presence of movement in a being reveals a certain lack of actuality.

It is already apparent, no doubt, what a profound interest this analysis of becoming has for Christian thought, and why such importance was attached to it by the mediæval philosophers. However, and this is well worth noting, it is also precisely one of those points where we most clearly perceive how Christian thought, taking up ideas of Greek origin, saw so much more deeply into them than the Greeks did. The Christian philosophers gathered from the Bible the identity of essence and existence in God ; and then could hardly fail to see that such identity exists nowhere save in God. Henceforth movement is seen to involve something more than the contingency of modes of being, something more even than the contingency of the substantiality of beings that arise and vanish according to their changing participations in the intelligibility of the form or idea : it means now the radical contingency of the very existence of beings subject to becoming. Into Aristotle's eternal world, existing outside God and without God, the Christian philosopher introduces the distinction of essence and existence. Not only does it remain true to say that all that is, save God, might be other than it is, but it now becomes true to say that all save God might possibly not exist.[1] This radical contingency stamps the world with a character of metaphysical novelty of immense significance, the nature of which will fully appear when we open up the question of its origin.

Nothing could be more familiar than the first verse of the Bible : " In the beginning God created the Heavens and the Earth " (Gen. i. 1). Here once more we have no trace of philosophy. God asserts His creative action, just as He asserts the definition of His being, without any kind of metaphysical justification. And yet between these two unproved assertions, how profound, how inevitable is the metaphysical accord ! If God is Being, if He alone is Being, then all that is not God must of necessity hold its existence from God. At one bound, and with no help from philosophy, the whole Greek contingency is left behind and rejoined at its ultimate metaphysical root.[2] In uttering so simply the secret of His creative action, it seems that God puts us in possession of one of those enigmatic key-words, which we knew all along must exist but could never discover for ourselves, and the truth of which comes home to us with irresistible force as soon as it is gratuitously given. The Demiurge of the *Timaeus* so closely resembles the Christian God that the whole Middle Ages saw his activity as a kind of foreshadowing of creation ; and yet he endows the universe with everything except, precisely, existence.[3] The first unmoved mover of Aristotle is also in a certain sense the cause and father of all that is, so that St. Thomas will go so far as to write : *Plato et Aristoteles pervenerunt ad cognescendum principium totius esse.* But St. Thomas never credits the Philosopher with the notion of creation, never once does he qualify as creationism his doctrine of the origin of the world ; and if in fact he does not do so it is because the first principle of all being, as Plato and Aristotle conceived it, integrally explains indeed why the universe is what it is, but does not explain why it exists.[4]

Less conciliatory in form than St. Thomas, the mediæval Augustinians took a certain pleasure in emphasizing this lacuna in Greek philosophy, sometimes reproaching it rather bitterly on that score. Other interpreters, especially among the moderns, do not go so far as to attribute the omission to any congenital vice in Aristotelianism, but realizing that he remained a complete stranger to the conception of creation, they put it down to a serious logical deficiency in Aristotle, putting him in contradiction with his own principles. But perhaps the truth is still simpler, for what Aristotle lacked in order to conceive creation was precisely the essential principle and starting-point. Had he known that God is Being, and that in God alone essence and existence are identical, then indeed it would have been inexcusable in him to have missed it. A first cause which is in fact Being and yet is not the cause of being for all the rest, would be evidently absurd. It did not need the metaphysical genius of Plato and Aristotle to see that, and however little inclined to speculative thought the first Christians may be supposed to have been, they were quite sufficiently speculative to take it into account. From the time of the Epistle of Clement, that is to say from the first century after Jesus Christ, the Christian universe begins to appear in all the contingency of existence that properly belongs to it ; for God " has constituted all things by the word of His majesty and by that same word He could destroy them all " (*Epist. ad Corinth.*, XXVII, 4). However modest a metaphysician may have been the author of the *Pastor of Hermas*, he was, nevertheless, speculative enough to understand that the first commandment of the Law implies the conception of creation : " Before all else believe that there exists one only God, Who created and finished all things, and brought all things into being out of nothing : He comprehends all and nothing can comprehend Him " (*Mand.*, I, 1). And observe, we are as yet only at the beginning of the second century. The *Apology* of Aristides, belonging to the same period, draws a proof of creation from the very fact of movement, thus anticipating the doctrine which Thomism will develop in the thirteenth century with a more rigorous technique no doubt, but exactly in the same spirit. And to come down to the end of the second century, we find in the *Cohortatio ad Graecos* (XXII–XXIII) a direct criticism of Platonism, with its artificer but non-creative god, whose power stops short at the very being of the material principle. It was all perfectly simple for these Christians ; but then, if they understood things that had remained hidden from the philosophers, it was simply because, as Theophilus of Antioch recognized without difficulty (*Ad Autolyc.*, II, 10), they had had the advantage of reading the first lines of Genesis. Neither Plato nor Aristotle knew anything of Genesis ; had they done so the whole history of philosophy might have been different. It would be easy enough, of course, to collect texts in which Plato posits the One as the source of the Many, and Aristotle posits the necessary as the source of the contingent, but in any case the metaphysical contingency of which they are speaking cannot possibly go deeper than the unity and being of which they are thinking. That the multiplicity of Plato's world is contingent with respect to the unity of the Idea is entirely obvious ; that the beings of Aristotle's world, involved in a long series of generations and corruptions by the incessant flux of becoming, are contingent with respect to the necessity of the first unmoved mover, that too is no less clear ; but that this Greek contingency in the order of intelligibility and becoming ever touched the depths of the Christian contingency in the order of existence—that is something of which we have no sign at all and nobody could even conceive it without having first conceived the Christian God. The pure and simple production of being is the action proper to Being Itself. No one can possibly attain to the idea of creation, or to the real distinction of essence and existence in all that is not God, as long as he admits forty-nine " beings as being." What Plato and Aristotle both lacked was the *Ego sum qui sum.*

This metaphysical achievement evidently marked a considerable advance in the idea of God ; but the current conception of the universe was modified correlatively at the same time and in no less profound a way. As soon as the sensible world is regarded as the result of a creative act, which not only gives it existence but conserves it in existence through all successive moments of its duration, it becomes so utterly dependent as to be struck through with contingency down to the very roots of its being. The universe is no longer suspended from the necessity of a thought that thinks itself, it is suspended now from the freedom of a will that wills it. This metaphysical outlook is familiar enough to-day, for the Christian world is not only the world of St. Thomas, St. Bonaventure and Duns Scotus, but also the world of Descartes, Leibniz and Malebranche ; it is only with difficulty that we realize the change of view it presupposes with respect to the Greek conception of nature. Nevertheless, familiar as it is, it is impossible to think of it seriously without a kind of dismay. Henceforth existences themselves, not merely forms, harmonies, and numbers, no longer suffice to themselves. This created universe, of which St. Augustine said that it unceasingly leans over towards the abyss of nothingness, is saved at each moment from collapse into nothingness by the continuous giving of a being which, of itself, it could neither give nor preserve. Nothing exists, nothing develops, nothing acts, but it receives existence, development and efficiency from the motionless subsistence of the Infinite Being. The Christian world manifests the glory of God not only by the spectacle of its splendour, but also by the very fact that it exists : " To all those things that surrounded my bodily sense I said, Speak to me of God, you who are not He, speak to me of Him ! And they all cried out with a loud voice : He made us. I questioned them merely by looking at them, and seeing them I had their answer." [5] *Ipse fecit nos :* the words of the ancient psalm had never sounded in the ears of Aristotle, but St. Augustine heard them, and the cosmological proofs of the existence of God were altogether transformed.

Since, in short, the relation of the world to God takes on a new aspect in Christian philosophy, the proofs of the existence of God must of necessity undergo a change of meaning. Everyone knows that the whole speculative effort of the Fathers of the Church and the thinkers of the Middle Ages concerning the possibility of proving God from His works, hangs directly from the famous words of St. Paul in the Epistle to the Romans (i, 20) : *Invisibilia Dei per ea quae facta sunt, intellecta conspiciuntur.* On the other hand, however, it seems that insufficient attention has been paid to a fact that is none the less of capital importance ; namely, that in the act of attaching themselves to St. Paul all the Christian thinkers, *ipso facto*, cut themselves loose from Greek philosophy. Whoever undertakes to prove the existence of God *per ea quae facta sunt* undertakes in advance to prove His existence as Creator of the Universe ; in other words he is committed to the view that the efficient cause to which the world testifies can be none other than a creative cause ; and thus also that the idea of creation is necessarily implied in every demonstration of the existence of the Christian God.

That such also was St. Augustine's thought we cannot doubt for an instant, since the celebrated ascent of the soul towards God in the tenth book of the *Confessions* supposes that the soul successively passes over all things that proclaim they did not make themselves, in order to lift itself up to the Creator Who made them. On the other hand, the Aristotelian language adopted by St. Thomas, both here and elsewhere, seems to have misled some excellent historians as to the true significance of the cosmological proofs, or " ways," as he calls them, of establishing God's existence.

Note in the first place that for St. Thomas, as for every Christian thinker, the relation of effect to cause that links up nature with God, lies in the order and on the plane of existence itself. There can be no possible doubt on this point. " Absolutely everything that exists owes its existence to God. Generally, and in all orders, we see that what is first in any particular order is the cause of all that is posterior in that order. Fire, for example, which of all bodies is the hottest, is the cause of the heat in other hot bodies, for the imperfect is always derived from the perfect, as seed from animals and plants. Now we have already shown that God is the first and absolutely perfect being. Therefore He is necessarily the cause of the being of all else." [6] The sensible examples that St. Thomas here uses should occasion no difficulty, for it is clear that, far from demanding any pre-existing matter on which to work, the creative action positively excludes anything of the kind. It is as the first act of being that God is the cause of beings ; matter is but being in potency, and how then could it condition the activity of the Pure Act ? [7] Everything, down to matter itself, falls under the creative action ; and therefore we must admit, before all causality put forth by God within nature that by which He causes the very being of nature, and it is for this reason that all Christian demonstrations of the existence of God by way of efficient causality are, in reality, so many proofs of creation. It is quite possible to miss this at first sight ; but even the proof by the first mover itself, the most Aristotelian of them all, is open to no other inter-pretation. *Movere praesupponit esse ;* [8] what does Aristotle's proof become in the light of this principle ?

Our senses testify that there is movement in the world. Now, nothing is moved save in so far as it is in potency, and nothing moves save in so far as it is in act ; and since nothing can be both in potency and act at the same time and in the same respect, it follows that all that is in move-ment is moved by another. But we cannot have an infinite series of motive causes and things moved, since then there would be no first mover, and, consequently, no movement. There is therefore a first mover not itself moved by any other ; and this is God. At first sight nothing could be more purely Greek than such an argumentation : a universe in movement, a hierarchical series of movers and things moved, a first mover, which, remaining itself unmoved, communicates movement to the entire series—have we not here a complete picture of Aristotle's world, and do we not know, moreover, that the proof is taken from Aristotle ?

Undoubtedly we have here the very cosmography of Aristotle, for as far as concerns its physical structure St. Thomas' world is indistinguishable from the Greek ; but beneath the physical analogy how profound is the meta-physical difference ! We might have divined it indeed from the simple fact that the five Thomist proofs are hung expressly from the text of Exodus.[9] From the very outset we are on the plane of Being. With Aristotle it is as final cause that the Thought that thinks itself puts all the rest in motion. That the Pure Act is, in a certain sense, the source of all the efficient and motive causality in the world is certainly true enough, since if the secondary motive causes had no last end there would be no reason why any of them should either move or be moved, that is to say exert their motive power. But if the First Mover makes causes to be causes, it is not by any kind of transitive action which would make these second causes at once to be, and to be causes. It moves only by the love it excites—which it excites, observe, but does not breathe in. When we read in the commentaries on the *Divina Commedia* that the last verse of the great poem merely echoes a thought of Aristotle's, we are very wide of the mark : *l'amor che muove il Sole e l'altre stelle* has nothing but the name in common with the first unmoved mover. The God of St. Thomas and Dante is a God Who loves, the god of Aristotle is a god who does not refuse to be loved ; the love that moves the heavens and the stars in Aristotle is the love of the heavens and the stars for god, but the love that moves them in St. Thomas and Dante is the love of God for the world ; between these two motive causes there is all the difference between an efficient cause on the one hand and a final cause on the other. And that does not end the matter.

Even if we suppose that Aristotle's god were a moving and efficient cause properly so called, which is by no means certain, his causality nevertheless would fall upon a universe which does not owe its existence to him, on beings whose being does not depend on his. In this sense he would merely be the first unmoved mover, that is to say the originating point in the communication of movements, but he would not always be the creator of the movement itself. To make the bearing of the question clear we must remember that movement lies at the origin of the generation of beings, and that, consequently, the cause of the generative movement is the cause of the beings generated. In a world like Aristotle's all is given, the First Mover, the intermediary movers, the movement, and the beings generated by the movement. Even then if we admit that the First Mover is the first of the motive causes which move by transitive causality, the very being of the movement would still escape his causality. But the case is very different in a Christian philosophy, and that is why St. Thomas, when he would demonstrate creation, needs only to recall the conclusion of his proof of God by movement. " It has been shown," he says, " by Aristotle's arguments, that there exists a first unmoved mover whom we call God. Now the first mover in any order is the cause of all the movements in that order. Since, then, many things come to existence in consequence of the movements of the heavens, and since God has been shown to be the first mover in the order of movements, it follows that God is the cause of the existence of all these things." [10] It is obvious that if God creates things solely because He moves the causes which produce these things by their movement, God must be a Mover as Creator of movement. In other words, if the proof by the first mover suffices to prove creation, then this proof must of necessity imply the idea of creation. Now the idea of creation is wanting in Aristotle, and so the Thomist proof of the existence of God, even if it merely literally reproduces an argumentation of Aristotle's, has a meaning altogether of its own, a meaning that the Greek philosopher never intended to give it.

With all the more reason then must this be true of the proof from efficient causality ; here too the same difference appears between the Greek and Christian worlds. In both we encounter the same hierarchy of second causes subordinated to a first cause, but for lack of passing beyond the

plane of efficiency to the plane of being, Greek philosophy
fails to emerge from the order of becoming. And that,
moreover, if we look a little more closely, is why Aristotle
is able to subordinate to the first cause a plurality of second
causes, all unmoved like the first, for if these causes received
the efficiency which they exert, how could they be unmoved ?
But they can and must be unmoved if, not dependent on
any other being in their being, they find in the first cause
rather the cause of the exercise of their causality than of
their causality itself. But a glance at St. Thomas is enough
to show that his proof moves on quite other lines, for with
him the proof of the existence of God by efficient causality
is the typical proof of creation. " We have shown by Aristotle's arguments that there exists a first efficient cause whom
we call God. But the efficient cause produces the being of
its effects. God, therefore, is the cause of the being of all
other things." [11] Impossible to say more clearly that in the
case of God, efficient causality means creative causality,
and that to prove the existence of a first efficient cause is
to prove the existence of a first creative cause. St. Thomas
is pleased to invoke Aristotle in the matter ; well and good !—
but since the efficiency in question does not bear upon the
same aspect of reality in the two systems, we must make up
our minds to admit that the Thomist proof of God by
efficient causality means something other than the Aristotelian.[12] The problem that faces us henceforth, the
problem that faced the whole of the classical metaphysics,
is a problem which would have been unintelligible to the
Greeks, the problem *de rerum originatione radicali*. Leibniz
will put it quite simply : why should there be anything
rather than nothing ? And exactly the same question recurs,
in Christian philosophy, on the plane of finality.

It is generally supposed to-day that modern science has
once and for all eliminated finality from our system of

thought. Whether the case is quite so definitely concluded as all that is possibly open to question, but for the moment we propose to do no more than indicate the precise point on which the proofs of God from finality turn. We grant, then, the existence of order in the world and we ask : what is the cause of this order ? Two remarks are called for at once. In the first place we ask no one to suppose that the order of the world is perfect—far from that !—the amount of disorder might even preponderate, but no matter how little order there might be we should still have to ask its cause. Secondly, we ask no one to go into raptures about the wonderful adaptation of means to ends in nature and to subtilize on their finer points in the naïve manner of a Bernadin de Saint-Pierre. The well-intentioned but rather silly zeal of some of its representatives has indisputably done much to discredit finalism in the eyes of science ; that is only too true, but fortunately the proof from finality is not bound up with these erroneous notions. It requires us only to grant that physico-biological mechanism is a mechanism with an orientation ; and we ask at once, whence comes this orientation ? Philosophers who put the question often fail to notice that it really covers two. The first, which leads nowhere, consists in asking the cause of these " wonders of nature " ; but even supposing that we make no mistake about these wonders—and mistakes of this kind will happen at times—they never introduce us to anything better than a kind of chief engineer of the universe, whose power, as astonishing to us as our own is to a savage, remains, nevertheless, within the human order. Over against such finalism Descartes set up his mechanism, and, moreover, justified it. It would doubtless be difficult to manufacture an animal, but there is no *a priori* proof that it could not be done, that it is not in the nature of an animal to be manufactured. Descartes himself, the prophet of mechanism, considered that an angel at least would be needed to make a flying-machine ; in the twentieth century, however, he might observe mere men turning them out wholesale and with an ease and efficiency that increases every day. It is useless therefore to press this question ; and we must pass to the second. Just as the proof from movement does not consider God as the Central Generating Station for the energies of nature, so neither does the proof from finality consider Him as the Chief Engineer of the whole vast enterprise. The precise question is this : if there is order what is the cause of the being of this order ? The celebrated example of the watch-maker misses the point, unless we leave the plane of making for the plane of creating. Just as, whenever we observe an artificial arrangement, we infer the existence of an artificer as the sole conceivable sufficient reason of the arrangement, so also when we observe, over and above the being of things, an order between the things, we infer the existence of a supreme orderer. But what we have to consider in this orderer is not so much the ingenuity displayed in the work, the precise nature of which too often, possibly always, escapes us, but the causality whereby he confers being on order. Descartes was quite within his rights in rallying those who, pretending to penetrate the inner counsels of God, set out to legislate in His name ; but there is no need to surprise the secrets of the divine legislation in order to be assured of its existence ; it is enough to know that a legislation exists, for if so, it appertains to being, that is to say either to contingent being which cannot explain itself, or to necessary being which, while sufficient to itself, suffices also as the reason of the contingent that derives from it.

Once we grasp this, the interpretation of the cosmological proofs of the existence of God becomes clear ; and clear also why we were able to say that even when the Christian philosophers cite the letter of Aristotle they move on a different plane. The better to understand this truth it suffices to recall the well-known mediæval controversy between those who admitted only purely physical proofs of God's existence like Averroes, and those who would admit none but metaphysical proofs, like Avicenna. The tradition here represented by Averroes is much more nearly allied to the Greek ; for in a universe like those of Plato and of Aristotle, where God and the world stand eternally over against each other, God is but the keystone of the cosmos and its animator, He is not put forward as the first term of a series, which is at the same time transcendent to the series. Avicenna, on the other hand, represents the Jewish tradition, and the Jewish tradition most fully conscious of itself ; for his God, whom he calls strictly and absolutely the First, is no longer merely the first being of the universe ; He is first with respect to the being of the universe, prior to that being, and consequently also outside it. That, to speak precisely, is why we ought to say that Christian philosophy essentially excludes all merely physical proofs of the existence of God, and admits only physico-metaphysical proofs, that is to say proofs suspended from Being as being. The fact that in these matters St. Thomas avails himself of Aristotle's physics as a starting-point, proves nothing, if, as we have just said, he always ends as a metaphysician ; we might show rather that even his general interpretation of Aristotle's metaphysics transcends the authentic Aristotelianism, for in raising our thoughts to the consideration of Him Who Is, Christianity revealed to metaphysics the true nature of its proper object. When, with Aristotle, a Christian defines metaphysics as the science of being as being, we may rest assured that he understands it always as the science of Being as Being : *id cujus actus est esse*, that is to say, God.

It seems, then, to borrow an expression from William James, that the Christian mental universe is distinguished from the Greek mental universe, by ever more and more profound structural differences. On the one side we have a god defined by a perfection in the order of quality : Plato's Good ; or by a perfection in one of the orders of being : Aristotle's Thought ; on the other side stands the Christian God Who is first in the order of being, and Whose transcendence is such that, in the vigorous phrase of Duns Scotus, when we have a first mover of this kind it needs more of a metaphysician to prove that He is first than it does of a physicist to prove that He is a mover. On the Greek side stands a god who is doubtless the cause of all being, including its intelligibility, efficiency and finality—all, save existence itself ; on the Christian side a God Who causes the very existence of being. On the Greek side we have a universe eternally informed or eternally moved ; on the Christian side a universe which begins to be by a creation. On the Greek side, stands a universe contingent in the order of intelligibility or in the order of becoming ; on the Christian side a universe contingent in the order of existence. On the Greek side, there is the immanent finality of an order interior to beings ; on the Christian side the transcendent finality of a Providence who creates the very being of order along with that of the things ordered.[13]

Having said so much we may say something on a difficult question, which, to say the truth, is neither to be fully elucidated nor altogether avoided. When Christian thought thus passed beyond the limits of Greek thought must we say that it placed itself in opposition to Greek thought, or simply that it carried it forward and completed it ? For my part I see no contradiction between the principles laid down by the Greek thinkers of the classical period and the conclusions which the Christian thinkers drew out of them.[14]

It would seem, on the contrary, that from the moment these conclusions were deduced they presented themselves as evidently contained in the principles, so that it then becomes a question how those who discovered the principles could so wholly fail to appreciate the necessary consequences there implied. My own view of the matter is this : that Plato and Aristotle missed the full meaning of the ideas they were the first to define, because they failed to explore the problem of being to that point where, transcending the plane of intelligibility, it touches that of existence. The questions they put were the right ones, for the problem they had in hand was certainly the problem of being ; and for that reason their formulæ remain good. The thinkers of the thirteenth century, seeing there the reflection of their own minds, welcomed them not merely without difficulty but with joy, for they found themselves able to read the truths they contained, although neither Plato, nor even Aristotle, had ever been able to decipher them. And so it came about, at one and the same time, that Greek metaphysics made decisive progress, and that the progress was realized under the impulsion of the Christian revelation. " The religious side of Plato's thought was not revealed in its full power till the time of Plotinus in the third century A.D. ; that of Aristotle's thought one might say without undue paradox, not till its exposition by Aquinas in the thirteenth." [15] Substituting rather the name of Augustine for that of Plotinus, and bearing in mind in any case that Plotinus himself was not altogether ignorant of Christianity, we can conclude that if mediæval thought succeeded in bringing Greek thought to its point of perfection it was at once because Greek thought was already true, and because Christian thought, in virtue of its very Christianity, had the power of making it still more so. When they raised the problem of the origin of being Plato and Aristotle were on the right road and it is precisely because they were on the right road that to go further along it was a progress. In their march towards the truth they stopped short at the threshold of the doctrine of essence and existence conceived as really identical in God and really distinct in everything else. There we have the fundamental verity of the Thomist philosophy and also, we may say, of all Christian philosophy whatsoever—for those of its representatives who think it proper to contest the formula agree at bottom in recognizing the truth.[16] Plato and Aristotle were building a magnificent arch all the stones of which converged upon this keystone; but it was due to the Bible that the keystone was put in position and it was the Christians who actually put it there. History ought never to forget either what Christian philosophy owes to the Greek tradition on the one side or what it owes to the Divine Pedagogue on the other. His lessons carry with them a luminous evidence such as we do not remember always to have been vouchsafed.

CHAPTER V

ANALOGY, CAUSALITY AND FINALITY

THE relation of contingent beings to the necessary being, as above explained, attains its full significance for thought only if we start from the Christian idea of God conceived as Being. It may be objected that this very idea, taken strictly, excludes the very possibility of any relation between things and God—for the simple reason that it makes the existence of things impossible. Let us grant hypothetically that the given universe, characterized as it is by change, does not contain within itself its own sufficient reason, and that its existence postulates that of Being ; then, once Being Itself is posited in its pure actuality, does it not become impossible to imagine the existence of anything that would not be Being ? If God is not Being, how is the world to be explained ? But if God is Being, how can there be anything other than Himself? There is only one God, said Leibniz, and this God suffices. Doubtless : but then not only does He suffice, but He suffices to Himself. Is there any possible escape from this dilemma ?

Observe, in the first place, that it is a Christian dilemma ; I mean a dilemma characteristic of Christian metaphysics, and only arising as a result of rational reflection on the data of revelation. The Greek universe and its interpretation occasioned no such difficulty. For Plato and Aristotle the world and its gods were given together ; neither the one nor the others laid claim to the exclusive possession of being, nothing prevented the latter being posited within the former, and the problem of their compossibility did not arise. It is quite otherwise in the Christian universe, and one may say that the fact is recognized even by philosophers who regard such an antinomy as insoluble. To balance between affirmation and denial of the necessary being and cause of the world, or to feel constrained to affirm and deny this being simultaneously [1]—these were embarrassments undreamt of by the Greeks, and felt by modern thought only because it lives and moves in a Christian scheme of things.

We must next notice the abstract, non-realist, and therefore non-Christian character of the difficulty. However metaphysical were the considerations developed in the preceding lectures, at least they never lost touch with reality. Whether we start from the idea of God conceived in the human mind with St. Anselm, or from man and the world with St. Thomas, never, at any rate, do we start from God Himself—He is invariably the goal. What therefore has to be said is not this : that the idea of beings postulates the idea of Being, while that of Being excludes that of beings ; but rather this : that beings, which are given as facts, have no sufficient reason save in Being. If there is any difficulty in the simultaneous admission of both, we can be quite sure in advance that it is only apparent ; for we can neither deny the fact which is a matter of experience, nor avoid the assertion of the sufficient reason which is a matter of necessity. It is only in its setting of critical idealism that Kant's fourth antinomy is insoluble ; in a realist rationalism it is evident *a priori* that a solution exists, and that we ought to be able to find in the idea of God, a justification for the co-existence of creatures and God. Any such justification will presuppose in the first place that we discover some conceivable reason for the production of beings by Being, and then some intelligible mode of setting forth the relations of beings to Being.

To overcome the first difficulty we must return to the

central point of the whole debate, that is to say to the Christian idea of God, and show what a new light it throws on the conception of cause. This is no easy matter, because the criticism of the idea of transitive causality as developed in the philosophies of Malebranche and Hume, has made it well-nigh unintelligible. The world is made up of a vast series of necessary connections; the actual arrangement is given as a fact, but the *why* is altogether beyond our power to discern : that is the modern view of the matter, and it has become so thoroughly familiar as to seem self-evident. To recover the mediæval conception of cause we must return to a realism which may seem a little naïve, and which St. Thomas has put into a perfectly clear formula : *causa importat influxum quemdam ad esse causati.*[2] That we may have causality in the strict sense of the term means that we must have two beings and that something of the being of the cause passes into the being of that which undergoes the effect.

The meaning of this conception is not to be grasped unless first we understand the profound relation which the mediæval thinkers considered to exist between being and causality. Before there can be any making there must first of all be being, for if causal action is to be conceived as a giving of itself by the cause to a subject, or even as the invasion of this subject by a cause, it is clear that the cause can give only what it has, and establish itself in another only in virtue of what it is. Thus being is the ultimate root of causality. Moreover, being not only makes causality possible, but even, in a way, demands it, and one of the most difficult problems which the classical metaphysicians had to face was the determination of the relation of being to its causal activity. I shall make no attempt to settle it, nor do I even pretend to define its terms with technical rigour ; I would merely try to suggest its meaning with the help of a comparison and at the cost of a brief digression.

The mediævals have often been taken to task for their naïve anthropomorphism. The reproach is natural enough when it comes from minds altogether formed in the scientific disciplines, and everywhere eager to substitute science for philosophy ; but although the scholastics were not all quite so ignorant of the sciences as we sometimes suppose, it was not their primary ambition to be regarded as *savants*, but rather as theologians and philosophers. They set out to discover first principles, and to achieve a rational interpretation of those elementary data of reality, which, though accepted by the scientist as data pure and simple, nevertheless demand an explanation at the hands of the philosopher. From the standpoint of science being and movement can be taken for granted, but philosophically they stand in need of justification, and the same has to be said for causality. If a scientist disclaims interest in such questions he is quite in the right as a scientist ; even if he maintains that they do not arise he is still in the right in the sense at least that they do not arise for science. And if he adds that anthropomorphism is the death of science, he is in the right once more ; it is certainly fatal when we try to apply it to scientific problems. But ought we therefore to conclude that it is fatal to philosophy ?

The very contrary is the truth. To suppose that all the phenomena of nature stand on a human footing would be manifestly ridiculous. To suppose that the scheme of finality could be reconstructed *a priori* by laying down that all things are what it seems to us they should be in order to serve what seems the greatest good of humanity, that would be something still worse—for the detection of such relations is infinitely risky in any event, and even were it possible it would not constitute a scientific explanation. It is none the less true that the universe is a system of beings and intertwined relations and that man enters into this system as a part. Now if man is a part of nature there seems to be no reason why a philosopher should not consider man in order the better to understand nature. There is nothing to prove *a priori* that what is true of human beings is false of other beings ; especially when what we are considering in either is precisely their being itself and its immediate properties as being. Here, then, and in this sense, this much decried anthropomorphism, so much in evidence in mediæval times, may prove perhaps to be an indispensable element in method. Since I am a part of nature, and my experience of self, in virtue of its very immediacy, is a privileged case, why should I not interpret what I know only from without in function of the sole reality that I know from within ? In man, and in man alone, nature attains to consciousness of itself. That is the foundation of all legitimate anthropomorphism, and in this we may find the ultimate justification of the mediæval conception of causality.[3]

Man may exercise several different modes of causality ; first that of a physical body, because he *is* such a body, then that of a living and organized body, since he is that too, and finally that of a rational being, because he is a living being gifted with reason. And since it is precisely rationality which specifically distinguishes man from other animals, the only causality which is specifically human is rational causality, that is to say that kind of causal activity in which reason plays the part of directive principle. Now all such causality is characterized by the presence in the mind of him who acts or makes of a certain preconceived idea of the act to be accomplished or the product of its action ; and that amounts to saying that our actions, or the products of our actions, must of necessity be first of all in us before they can be in themselves what they will be when we have produced them. Our effects, in other words, before existing in themselves as effects, exist in us as causes, and partake of the being of their cause. The possibility of the typically human mode of causality, that of the *homo faber*, rests precisely on the fact that man, being gifted with reason, is capable of containing within himself, by way of representation, the being of possible effects which shall be distinct from himself. And that, moreover, is why what we do or produce is ours ; for if we are responsible for our acts and legitimate owners of the works of our hands, it is because, as these effects were at first but ourselves as cause, so it is still we ourselves who exist in them in their being as effects. The plays of Shakespeare, the comedies of Molière, the symphonies of Beethoven *are* Shakespeare, Molière, Beethoven ; so much so that we might reasonably ask whether they do not constitute the best part of their authors' being, the very summit of their personality.

But if we push the analysis a little further in this direction, it soon appears that this result is merely provisional. Man exerts no causality save in so far as he *is*, and it is very true that since nothing is prior to being, we cannot pass beyond that point. But what is the meaning of the verb *to be?* When I say *I am*, my thought does not, as a rule, pass beyond the empirical apprehension of a fact given by internal observation. It was otherwise, however, in the eyes of the mediæval thinkers. For them the verb *to be* was essentially an active verb signifying the very act of existing ; to affirm their own actual existence was much more to them than to affirm their present existence, it was an affirmation of the actuality, that is to say the very energy, by which their being existed. If, then, we would arrive at an exact understanding of the mediæval conception of causality we must ascend to this very act of existence, for it is clear that if being is act, the causal act must necessarily be rooted in very being of the cause. This relation is expressed in the technical distinction, somewhat alarming at first sight, but very clear in the upshot, between *first* act and *second* act. The first act is the being of the thing, of that which is called being in virtue of the very act of existing exerted : *ens dicitur ab actu essendi :* the second act is the causal operation of this being, the intrinsic or extrinsic manifestation of its first actuality by the effects it produces within or without itself.[4] That is why causal action, which is nothing but an aspect of the actuality of the being as such, finally resolves itself into a transmission or communication of being : *influxum quemdam ad esse causati.*

When this doctrine is once understood, we are thereby prepared for a philosophically precise statement of the idea of creation. To create is to cause being. If then each thing is capable of being a cause in the exact measure in which it is a being, God, Who is Being, must be able to cause being, and He must be the only one able to do it.[5] Every contingent being owes its contingence to the fact that it only participates being ; it has its being, but it is not its own being ; in the unique sense, that is, in which God is His own being. That is why contingent beings are never more than second causes ; they are no more than second beings. Their causal activity is strictly limited to the transmission of modes of being and to the alteration of the dispositions of the subjects on which they act ; it can never go so far as to cause the very existence of the effects produced, and, in a word, *homo faber* can never become *homo creator* because, having only a received being he cannot produce what he himself is not, nor overstep in the order of causality the rank he holds in the order of being. Creation, therefore, is a causal action proper to God, it is possible for Him, and it is not possible for any save Him. But does it therefore follow that creation can be conceived, or rationally grasped in its nature or cause ? These are other questions and they demand a little separate attention.

The first raises no very serious difficulty. All the Christian philosophers recognize that if the creative act is conceivable, it is not representable. We never create, we are incapable of creating, also we are incapable of representing to ourselves any truly creative action. Whatever we make we make out of something else, and therefore we cannot conceive an act the term of which would be the very being of the effect produced. Nothing is easier than to repeat that God has created things and created them *ex nihilo*, but how can we prevent ourselves imagining, even while we deny it, that this *nothing* is a kind of matter from which the creative act draws its effects? We can think only of change, of transmutation, of alteration; in order to think creation, we should need the power of transcending both our degree in being and our degree of causality.[6]

We are faced with the same kind of difficulty, in a rather less radical form perhaps, when we try to conceive the why of creation. We must, to begin with, dissipate an illusion which threatens to falsify the whole meaning of the question. To seek the sufficient reason of creation does not mean to seek the cause of the creative act, for the creative act is God Himself; He has no cause, He Himself is cause. St. Augustine settled this point long ago in his own way, that is to say with a just intuition not always accompanied with the necessary technical justifications. To ask the cause of the will of God is implicitly to suppose that there could be something prior to His will, whereas the fact is that His will is prior to all the rest.[7] Evidently St. Augustine is thinking here of the contradictory hypothesis of a cause of creation which would be a part of creation itself. Passing beyond that standpoint, which for the rest he adopts while going deeper, St. Thomas demonstrates the impossibility of supposing any cause at all, whether external or internal, of the creative act. That there should be within God Himself any

cause whatever of His own will, would demand a real distinction of powers and attributes within God. It would
hardly in fact be possible to imagine any cause of His will
except perhaps His understanding. Now we have already
seen that Being Itself, in its pure actuality, is a perfect
unity, and that this unity excludes by definition all such
internal division as would permit the opposition of one
divine attribute to another, or allow of any causal action
whatever of one upon another. Doubtless from the standpoint of our discursive mode of apprehension—a perfectly,
well-founded one for the rest—we are constrained to say
that God's understanding acts upon His will; but the
intellect of God is God just as the will of God is God, and
since, evidently, *idem non est causa sui ipsius*, it is quite impossible to see how any relations of causality, properly so called,
can be set up within the bosom of God.

But in thus identifying the will of God with God we are
forced to ask ourselves whether the conception of will has
any meaning left in it, since, however it may be for the Pure
Being Himself, a will without an end is hardly conceivable
by us; and can we suppose a divine will acting for an end
unless this end is the final cause determining this will?
Well, we can; but on one condition, that is to say when
God is in question and provided we remember that God is
Being Itself. To deny that the will of God has an end would
be tantamount to subjecting it either to blind necessity or
to irrational contingence, and in either case this would be
to admit imperfections in God incompatible with the
actuality of the pure Being. On the other hand, and precisely because God is Being Itself, it would be contradictory
to suppose that He could have any other end than Himself.
The sole conceivable end of the divine will is therefore the
divine Being, and since this being, as end of the will, is
identical with the good, we may say that the only possible

end for God is His own perfection. In other words, God wills Himself necessarily but does not will anything other than Himself necessarily, and all that He does will He wills with respect to Himself.

We can say then, and all Christian philosophers have said it, that the reason for creation lies in the goodness of God. In St. Augustine's phrase, adopted later by St. Thomas, it is because God is good that we exist : *quia Deus bonus est, sumus ;*[8] or as St. Thomas once more puts it in a formula deduced from Dionysius and commented again and again by the Middle Ages : *bonum est diffusivum sui et communicativum.*[9] The Platonic origin of this idea is beyond doubt. Already in the *Timaeus* the ordering activity of the divine Demiurge is attributed to his liberality, his freedom from envy.[10] But if the good is the last reason for creation[11] how are we to account for this good itself?

To put such a question to Plato would be to run a serious risk of getting no answer, since Plato conceives the Good as the supreme reality. To put it to Dionysius would have much the same result. Penetrated through and through with Platonism as he was, this Christian never rose above the idea of the primacy of the Good, never grasped the primacy of Being. The fact emerges quite clearly from that capital text of the *Divine Names* where Dionysius comments on the text of Exodus : *Ego sum qui sum.* When God declares Himself as Being we ought to understand, according to his view, that of all possible participations of goodness, which is His essence, being is the first. Commenting in his turn on the commentary of the Areopagite, St. Thomas signifies his agreement, but it has been justly remarked that he does not agree,[12] for instead of seeing in being a participation of the good, as Dionysius' text supposes, he sees goodness as an aspect of being. It is just this that makes the Thomist interpretation of Dionysius a matter of no little philosophic interest ; here we can watch the process by which Christian thought achieves clear consciousness of its own metaphysical principles, rises out of the level of Hellenism, and elaborates at last in definitive form what we may call the metaphysics of Exodus. Let us see how it proceeds.

St. Thomas grants without reserve, without even any mental reserve, that goodness is diffusive and communicative of itself. But whence has it this character ? From the fact that goodness is nothing else than one of the transcendental aspects of being. Considered in its metaphysical root the good is being itself as desirable, that is to say it is being considered as the possible object of a will ; and if then we would understand why it spontaneously tends to diffuse and communicate itself, we must turn to the immanent actuality of being for an answer. To say that being is at once act and good, is not merely to indicate that it may act as cause, but it is also to suggest at the same time that it contains a reason for the exercise of this causal power. The perfection of its actuality, conceived as good, invites it to communicate that actuality freely to the being of its possible effects.[13] To return to the anthropomorphic standpoint I have tried to defend above, we might avail ourselves here too of a human analogy. What do we admire in the hero, the artist and the sage ? Is it not precisely the overflowing actuality of their being, which, issuing into acts and works, thus passes into a world of inferior beings who stand amazed at the sight. But even if we put aside such extreme cases, do we not all feel in ourselves the truth of the metaphysical principle *operatio sequitur esse ?* For to be is to act, and to act is to be. The generosity with which goodness gives of itself is, in the case of an intelligible being, a free manifestation of the energy by which that being exists.

It is very true that man is not all generosity, but that is

because he is not all being ; before giving of what he is, he has to take of what he is not. Often enough he wills the good of another in order to supply his own deficiencies, to preserve or aggrandize his own being ; but avid as he may be for what he lacks, he is generous with what he is, because, in so far as he is, he is good : *ens est diffusivum sui et communicativum*. Having arrived at this point, however, we shall have to go further.

As soon as causality is interpreted as a gift of being, we are necessarily led to set up a new relation between effect and cause ; the relation, namely, of analogy. This consequence seemed so obvious to the mediæval philosophers that they felt no need at all to justify it ; to them it seemed to be a fact given in the most ordinary and everyday experience. An animal or plant gives birth to an animal or plant of the same species, fire breeds fire, and movement movement ; the thing is evident everywhere. But the metaphysical reason for this fact is as evident to the mind as the fact itself is to the senses ; for if the being that causes does nothing but communicate itself to the effect, diffuse itself into the effect, then it is still the cause that is found in the effect, under a new mode of being, doubtless, and with differences due to the conditions imposed by the matter on which it exerts its efficacy. Few formulæ recur as often in the writings of St. Thomas as the one that expresses this relation : since all that causes acts according as it is in act, every cause produces an effect that resembles it : *omne agens agit sibi simile*. And this resemblance is no additional or contingent quality supervening no one knows why or whence to crown the efficacy, it is co-essential with the very nature of efficiency, it is merely its external sign and sensible manifestation.

If then, as the idea of creation implies, the Christian universe is an effect of God, it must of necessity be an analogue of God. No more, however, than an analogue, for when we compare being *per se* with the being caused even in its very existence, we are dealing with two orders of being not to be added together nor subtracted ; they are, in all rigour, incommensurable, and that is also why they are compossible. God added nothing to Himself by the creation of the world, nor would anything be taken away from Him by its annihilation—events which would be of capital importance for the created things concerned, but null for Being Who would be in no wise concerned *qua* being.[14] But although creation is no more than an analogue, it is an analogue—that is to say much more than an effect to which has been accidentally added a certain resemblance to its cause. And since the effect in question here is one which all others presuppose—that is, being— it is precisely in its existence and substantially that the creature is an analogue of its Creator.

That, then, is why every Christian metaphysic involves the conceptions of participation and similitude, but gives them nevertheless a much more profound meaning than did the Platonism whence it borrowed them ; for the matter in which the Demiurge of the *Timaeus* works is simply *informed* by the Ideas in which it participates, whereas the matter of the Christian world receives its *existence* from God simultaneously with the existence of its forms. We are well aware of all the difficulties that may be brought to bear on this point, but it seems that every one of them defeats itself in the end. That the idea of participation is repugnant to logical thought is very possible, since every participation supposes that the participator both is, and is not, that in which it participates. But does not any exercise of logical thought demand that certain real unifications be given it for analysis ? Is not this precisely its proper work, and would it otherwise have anything left to do ? And does not the very enunciation of the principle of contradiction imply the presence to the mind of the idea of participation, and consequently also its relative intelligibility? When I say that the same thing cannot, at one and the same time and in one and the same respect, be both itself and its contrary, I integrate with the definition of the thing the presence of its " respects," and I suppose that what is such in a certain respect might perhaps be such in another. Besides, what would be the use of discursive thought at all, in face of a reality which simply derided and defied its formulæ? The problem of the same and the other is written in things themselves, and if the famous question *de eodem et diverso* discouraged more than one mediæval logician, that is no reason why the metaphysician should run away from it. Now resemblance is a fact. No one will deny for a moment that the world in which we live lends itself to classification, demands it even, and once that is granted the whole doctrine of participation inevitably re-enters philosophy. Nothing is merely itself, and the μίξις εἰδων is written in the definition of every essence. It would be suicidal for logical thought, when it comes to analyse the syntheses thus given it to deal with, to pretend to dissolve them instead. Not only would it have nothing left to do, but it would also destroy itself, since merely to say that A is A is to admit that the same, without ceasing to be the same, might become other in a certain respect.

It will be objected, next, that the metaphysics of analogy involve a naïve materialization of causality. To believe that the world represents God as a portrait represents its original or an animal its progenitors—surely all that would be to fall back into a pre-scientific state of mind and reason in the style of a primitive? The fact is that a good deal too much has been made of the age of ideas—even were it shown that such and such a conception is primitive the problem of its legitimacy would still remain intact. It would be rather surprising, on the other hand, if primitive modes of explanation imagined by human beings gave expression to no imperious necessities of human thought. It may very well be—no one is at all likely to dispute it—that participations and analogies flourished rather over-luxuriantly in primitive or even in mediæval cosmogonies ; let us admit at once that they sprawl crudely everywhere, without criticism, without method, without any kind of rational justification. It does not in the least follow that they correspond to no authentic aspect of reality, and may in consequence be altogether dispensed with in our efforts to explain or even simply to describe it : and this the less so since the idea of analogy is not quite so naïve as it is commonly imagined.

The Christian thinkers always took good care to distinguish several species or degrees of analogy. Resemblance is its most striking form for the imagination, but not its only form ; there may well be analogy where there is no resemblance. A portrait interests us, perhaps, on account of the resemblance—which occasionally exists—between the image and the original ; but if we compare several portraits of different people executed by the same painter, especially if they belong to the same period of his life, we may notice that they bear an unmistakable resemblance to each other, and if they thus resemble each other, this is because they all resemble the artist. And this holds not only of the works of the same artist, but also of the works of his pupils, or, as we say, of his school, because he is its cause, and because here, as in all other cases of the same kind, it is something of his being which, directly or indirectly, has been communicated to his effects. In this sense we might say that all Rembrandt's pictures are pictures of Rembrandt painted by himself, no matter whether they depict the *Ecce Homo*, or the *Philosopher in Meditation*, or the *Pilgrims of Emmaus*.

Thus, it seems, should be interpreted the familiar mediæval theme of a universe in which all things bear the traces of the divinity. It is not to be denied that the imagination of the Christian thinkers was given free rein in this field. With an extraordinary abundance, with a kind of poetical joy, Hugh of St. Victor, St. Bonaventure, Raymond Lully strive to make out the ternary order in the structure of things which symbolizes for them the Trinity of the Christian God. The *Itinerary of the Soul to God* is filled from cover to cover with this idea, and it has been maintained that even the *terza rima* of the *Divine Comedy*, that admirable mirror of the mediæval world, was chosen by the poet so that the poem, like the world it describes, should be stamped in its very matter with the likeness of God. It is difficult to judge such a state of mind without taking account of the principle that inspired it. Where is the search for analogies to stop once it is granted that analogies exist? The more sober minds, that of St. Thomas Aquinas for example, do not for the most part take the resemblances accumulated by the great meditatives, as if they were proofs, and yet they do not refuse to see in the substance, form, and order co-essential with things the mark of the Triune God Who is their Author.[15]

Hence what we may call, with Newman, the sacramental character of the Christian world. We may call a halt to our explanations at any point we will, but nature has depths, nevertheless, which transcend the physical order, involve the metaphysical, and finally prepare the way for the mystic. I do not mean that the Christian conception of nature ignores or despises purely scientific or philosophical explanations, on the contrary it welcomes them, calls for them insistently, for every truth of science or philosophy teaches us something about God. But what I would say is that a Christian philosopher, besides seeing the universe as all the world also sees it, admits the need of another stand-point peculiar to himself. Just as it is by His goodness that God gives being to beings, so also it is by His goodness that He makes causes to be causes, thus delegating to them a certain participation in His power, along with a participation in His actuality. Or rather, since causality flows from actuality, let us say that He confers the one in conferring the other, so that to the Christian mind the physical world in which we live offers a face which is the reverse of its physicism itself, a face where all that was read on the one side in terms of force, energy and law, is now read on the other in terms of participations and analogies of the divine Being. For whoever understands this, the Christian world takes on the character of a sacred world with a relation to God inscribed in its very being and in every law that rules its functioning.

Perhaps it will not be out of place here to dissipate a mis-conception recently become current and likely to become more so with the general growth of interest in mediæval philosophy. The Middle Ages that first appeared over the historical horizon was the Middle Ages of the romantics, a stirring, picturesque and brightly coloured world where saints and sinners jostled familiarly in the crowd, a world which expressed its deepest aspirations in architecture, sculpture and poetry. And that, too, is the Middle Ages of symbolism, where realities dissolved into the mystical meanings with which they were charged by artists and thinkers, so that the book of nature became a sort of Bible with things for words. Bestiaries, Mirrors of the World, stained glass, cathedral porches, each in its own way expressed a symbolic universe in which things, taken in their very essences, are merely so many expressions of God. But by a very natural reaction the study of the classical systems of the thirteenth century led historians to oppose to this poetical vision of the mediæval world, the scientific and rational conception that presented itself in the writings of Robert Grosseteste, Roger Bacon and St. Thomas Aquinas. And this was entirely justifiable, in this sense at least, that from the thirteenth century onwards the universe of science begins to interpose between ourselves and the symbolic universe of the early Middle Ages ; but it would be wrong to suppose that it suppressed it or even tended to suppress it. What really then took place was this : first, things, instead of being nothing more than symbols, became concrete beings which, above and beyond their own proper nature, were still charged with symbolic significances ; and then, next, the analogy of the world to God, instead of being expressed only on the plane of imagery and feeling, was now formulated in precise laws and definite metaphysical conceptions. God in fact penetrated more deeply into nature as the depths of nature became better known. For a Bonaventure, for instance, there is no joy like the joy of the contemplation of God as mirrored in the analogical structure of beings ; and even the more sober mind of St. Thomas expresses, nevertheless, the same philosophy of nature when he reduces the efficacy of second causes to nothing but an analogical participation in the divine efficiency. Physical causality is to the act of creation what beings are to Being, and time to eternity. Thus, under whatever aspect we consider it, there exists in reality but one mediæval vision of the world, whether it expresses itself now in works of art or now in defined philosophical concepts : that, namely, which St. Augustine drew with a master-hand in his *De Trinitate*, and which is directly referable to the words of the Book of Wisdom (xi. 21) : *omnia in mensura, et numero, et pondere disposuisti.*

It is true that as soon as we emerge from this difficulty we encounter another no less formidable. If we grant the

compossibility of beings and Being, and even the metaphysical possibility of a gift of being by Being, the moral reason
for this gift has still to be made intelligible. Creation, we
have said, is an act of generosity. So be it : but of generosity
to whom ? Since God is the Sovereign Good, what gift
could He give Himself ? Since the creature is nothing, what
could He give it ? In other words, even were it admitted
that an efficient cause of the creative act is conceivable,
it would seem difficult to suggest a final cause : to the
problem of the relation of beings to Being there is added the
problem of the relation of good things to Goodness.

However, even in the face of this formidable difficulty,
Christian thought is not left without resources. Turning
once more to Scripture, it there finds the solution of the
problem, and all that reason will have to do now is to recon-
struct it, very much as a geometrician analyses the conditions of possibility of some problem supposed to be already
resolved. *Universa propter semetipsum operatus est Dominus.*
It is for Himself, says the Book of Proverbs (xvi. 4), that
God made all things, and as soon as we read this we realize
that in the unique case of God, where cause and end are
one, He Who made all things could have made them only
for Himself. But then, again, if He made them for Himself,
what becomes of the pure liberality of the creative action ?
On what transcendent plane can be realized this manifest
contradiction in terms : an interested generosity ?

It is entirely true to say that every Christian conception
of the universe, whatever else it be, is theocentric. The
trait may be emphasized in one doctrine more insistently
than in another, but none can efface it without losing its
Christian character, and, moreover, without at once becoming metaphysically contradictory by that very fact. If good
is understood in terms of being, then finality, which is
directed to the achievement of the good, is also, by way of

the good, reduced to being. In other words the good is nothing else than the desirability of being, so that the sovereignly desirable, from the very fact that it is the sovereign good, is one by definition with the sovereign being. If, then, we lay down a creative cause as required for the intelligibility of the universe, the final cause of this creative cause can be none other than itself. To suppose that God could find an end for His acts outside Himself would be to limit His actuality, and since creation is the action proper to the pure Being creation would then be impossible. Thus the good in view of which God creates can be nothing but Being itself, creative in virtue of His perfect actuality : *universa propter semetipsum operatus est Dominus.*

Only we must be careful to note what is implied in the idea of an act of the supreme good. From our natural human standpoint we are apt to reason as though the end proposed by Infinite Goodness could be that proposed by a finite good. Now in the case of finite goods like ourselves, for whom perfection stands at the term of action as an end to be acquired, operation for the most part is directed to some utility. It is just this that makes the divine action so all but incomprehensible. For a being who has always some further being to acquire the act of a good which has no good to acquire remains a mystery ; but, nevertheless, it is enough to reflect on the idea of a sovereign good to see the necessity of positing this incomprehensible act at the origin of things, for when it is the sovereign good that acts then we have the unique case in which the sole possible end of the act is self-communication. Beings always strive more or less to realize themselves ; Being, since He is already fully realized, can act only to give. It is therefore only by a very deficient metaphor that it is possible sometimes to speak of the divine egoism as the only legitimate egoism, for no egoism is conceivable except where there is something to be gained. Quite other is the action of the sovereign Being, Who, knowing Himself as sovereignly desirable, wills that there shall exist some analogues of His being so that there may exist some analogues of His desire. God creates, not that there may be witnesses to render Him His due glory, but beings who shall rejoice in it as He rejoices in it Himself and who, participating in His being, participate at the same time in His beatitude. It is not therefore for Himself, but for us, that God seeks His glory ; it is not to gain it, for He possesses it already, nor to increase it, for already it is perfect, but to communicate it to us.

These considerations lead to consequences of a metaphysical importance impossible to exaggerate. Born of a final cause, the universe is necessarily saturated with finality, that is to say, we can never in any case disassociate the explanation of things from the consideration of their *raison d'être*. That is the reason why, in spite of all the resistance, and occasionally even violent opposition, of modern science and philosophy to the idea of finality, Christian thought has never yet renounced it, and never will. No doubt, it may be maintained with Bacon and Descartes that even if final causes exist, the finalist standpoint is scientifically sterile : *causarum finalium inquisitio sterilis est, et tanquam virgo Deo consecrata nihil parit*.[16] Possibly, however, as we shall see later on, Bacon is in fundamental disagreement with the mediævals on the very idea of philosophy itself ; and, moreover, we ought above all to note that it is to God that this virgin is consecrated—even if she bear none of those practical results so dear to Bacon's heart, she nevertheless keeps vigil over the intelligibility of the world, and that is why Christian thought guards her so jealously. To reproach finalism for its scientific sterility—

even supposing it to be as complete as alleged—is to disregard what later on we shall call the primacy of contemplation, and to confuse two orders which, in our view, it is absolutely necessary to distinguish.

Perhaps it involves something still worse. No one, as we have said, dreams of defending the endless follies of naïve finalism nor even of maintaining that an explanation by final causes could be a scientific explanation ; knowledge of the why, even were it possible, could in no case dispense us from seeking the how—and that is all that concerns science. Descartes has already said sufficiently strong things on that head, and we need not return to them here. On the other hand there is and always will be a need to return to the question whether or no there is any why at all in nature. Now we are very certain that there is in man, who is, indisputably, a part of nature ; and from this standpoint all that seemed to us to be true in the case of analogy seems to be even much more evidently true as regards finality. In every one of his actions man is a living witness to the presence of finality in the universe, and if it would be a very naïve piece of anthropomorphism to regard all natural events as the work of a hidden superman, it would be no less naïve in another way to enter a universal denial of the causality of ends in the name of a method which holds itself bound to take no notice of it even where it exists. The discovery of the why does not absolve us from looking for the how, but, if anyone looks only for the how, can he be surprised if he fails to find the why ? Can he, above all, be surprised when at last he meets with it in himself, that he fails to invest it with that type of intelligibility which only the how carries with it ? Let that be as it may, on this point Christian thought has never wavered. Faithful to its principles it finds itself at home in the world of Plato and Aristotle, and that

world, for the first time, received at its hands its own
full rationality. Below man, who acts in view of ends
known, mediæval philosophy always placed the animal
determined by ends perceived, and below the animal
the mineral world actuated by its ends, that is to say
submitting to them blindly without knowledge.[17] But
above man it always placed God, Whose action, transcending ours, is not ruled by ends, but gives being
simultaneously both to means and ends, and this for no
other reason than to communicate His beatitude along
with His intelligibility.

That, then, is why the universe was conceived by the
Christian Middle Ages not only as sacramental, but also as
orientated ; and whenever Christian thought has returned
to full consciousness of its own true nature it has always
been conceived in both these ways. In the face of that
fundamentally stable outlook the recurrent revolutions in
our scientific views of the universe appear as accidents of
no overwhelming importance. Whether the earth revolves
or not, whether it stands at the centre or not, whether it is
ruled by physico-chemical agencies to a greater or lesser
extent, all things at bottom remain, for the Christian, so
many vestiges left by God's creative touch as it passes.
Pascal was not altogether ignorant of the nature of a scientific
explanation, yet nevertheless, after Descartes, in the full
seventeenth century, he was bold enough to write : " All
things cover a mystery, all are veils that hide God." [18]
In this sense Christian finalism itself is an immediate corollary from the idea of creation, so much so that we may go
so far as to say that it is only in a universe that depends on
the free act of the God of the Bible and the Gospel that the
idea of final cause attains its full significance. For there is
no true finality unless intelligence is at the source of things,
and unless that intelligence is that of a creative person.[19]

But this idea of personality will come up for consideration only later on : our immediate task is to examine the nature of a universe born of the omnipotence of Being, and more particularly to consider that essential goodness which belongs to it of full right.

CHAPTER VI

CHRISTIAN OPTIMISM

THE opinion is widespread that Christianity is a radical pessimism, inasmuch as it inculcates despair of the only world of whose existence we are assured, and asks us to pin our hopes to another which may be nothing but a dream after all. Jesus Christ never ceased to preach renunciation of worldly goods. St. Paul condemns the flesh and exalts virginity. The Fathers of the Desert, driven mad by an insensate hatred of nature, embark on a life which is a radical negation of all social or even simply human values. And finally the Middle Ages, codifying so to speak the rules of this *contemptus saeculi*, go on to justify it metaphysically. The world is infected by sin, corrupted in its very roots, essentially evil, a thing to be fled, denied, destroyed. St. Peter Damian and St. Bernard condemn every natural impulse : at their bidding thousands of young men and women betake themselves into solitude, or follow St. Bruno into the desert of the Chartreuse : sometimes fully constituted families are broken up, and their scattered and emancipated members find nothing better to do with their new-found liberty than to mortify the flesh, deaden the senses, and repress the exercise of that very faculty of reason that precisely makes them men. *Ubi solitudinem fecerunt, pacem appellant.* Is not this senseless aspiration of whole generations towards nothingness the normal fruit of Christian preaching ? But then, on the other hand, is not the negation of this negation, the refusal of this refusal, one of the essential affirmations of the modern conscience ? Acceptance of nature, confidence in the intrinsic worth of all its manifestations, hope for its indefinite progress, if only we know how to perceive what is good in it and work to make it better—there we have modern optimism, there we have the gage flung down to Christian pessimism. The Renaissance called up the gods of Greece once more from the place where they had laid them, at least it called up the spirit that gave them birth : and to set over against Pascal we have Voltaire, against St. Bernard, Condorcet.

We might have a word or two to say about the perfect serenity of the Greek world were this the place to say it ; but that its optimism had its limitations will doubtless best appear on a philosophical plane. It will be pertinent to remark on the other hand that if we wish to determine the true line of Christian thought in this matter we should by no means be content to consult only the heroes of the interior life ; rather it would be most dangerous to rely only on these without turning to the background of Christian dogma to which they all appealed, and which alone sets their activities in a comprehensible light. However great was St. Bernard, however indispensable is Pascal, they cannot replace the long tradition of the Fathers of the Church and the thinkers of the Middle Ages. Here, as elsewhere, our best witnesses will be St. Augustine, St. Bonaventure, St. Thomas Aquinas and Duns Scotus, not forgetting, moreover, the Bible whence their inspiration came.

For if we would discover the basic principle on which Christian optimism, as I propose to call it, has always rested, we have only to open the first chapter of Genesis. At once we find ourselves face to face with the capital fact of creation, and the Creator Himself, contemplating His work on the evening of each day, declares not merely that He made it, but that because He made it, it is good : *et vidit Deus quod esset bonum ;* and then, on the evening of the sixth day, casting a comprehensive glance over all His work, He gives for the last time a similar testimony, proclaiming His creation very good : *viditque Deus cuncta quae fecerat, et erant valde bona* (Gen. i. 31). There, from the time of Irenaeus, we have the unshakable foundation of Christian optimism. Once more, no metaphysics ; good ground, however, for rejecting any number of metaphysical systems until a more satisfactory one shall be forthcoming. All those gnostic sects which would throw responsibility for creation upon some inferior Demiurge in order the better to absolve God from the crime of creating an evil world, are condemned at once as anti-Christian. Since it is the work of a good God, the world is not to be explained as the result of any original error, any kind of fall, lapse, ignorance, or revolt.[1] Irenaeus, moreover, quite understands, and very clearly says, that this Christian optimism is a necessary sequel to the Christian doctrine of creation. A good God, Who makes all things out of nothing, not only gratuitously bringing them into existence, but also establishing their order, allows of no intermediate and hence inferior cause between Himself and His work.[2] As sole Author He takes full responsibility ; and He is very well able to do so, for His work is good. What now remains for the philosopher is to show that it is so.

We know how cruelly this problem tormented the thought of the young Augustine. He, too, first fell in with gnosticism under the form of Manichean dualism, and shook if off on the day he left the sect that had held his early allegiance. But although delivered from the gnosticism of Mani, he was not yet clear from all his difficulties, for he still had to explain the presence of evil in a universe created by God. If there is no God, whence comes the good ? But if there is a God, whence comes the evil ? To this question Plotinus suggested an answer which had deep roots in the Greek tradition, but which doubtless took on consistence in his mind under the influence of that very gnosticism that he so often combated. Why not admit that matter is the principle of evil? Since being is the good, it follows that what is contrary to being is evil. Matter, then, in a sense, is a non-being; in a Platonic sense, however, that is to say not exactly as a non-existence but a non-good. Hence Plotinus is able to maintain simultaneously that matter pertains to non-being, and yet is the real principle of evil.[3] For the young Augustine, so full of admiration for Plotinus, how tempting this doctrine must have been! To reduce evil to matter, adding that matter is next to nothing—in what simpler way could we account for all the necessary imperfections of the world? Why, then, not adopt this very neat solution?

Well, simply because it is no solution. Plotinus' answer to the question is perfectly consistent with the rest of Plotinus' system, for his God is not a Creator in the Biblical and Christian sense of the term. He is not responsible for the existence of matter, therefore neither is he responsible for its nature, and even if it be evil it does not follow that he is evil as well. But how could Augustine clear a Creator-God from the reproach of having made matter and made it evil, or even of having merely left it evil had He found it so? Reflecting, then, on the philosophical principles of Plotinus in the light of the Biblical revelation, Augustine soon felt their insufficiency.[4] To admit that matter is at once created and evil would be an impossible pessimism, and literally contradictory in any Christian scheme. But what is of chief importance to us here is to see just how this religious optimism became a metaphysical optimism, and all we need do is to inquire the secret of St. Augustine. Now Augustine sends us once more to the text of Exodus.

What he most admirably saw and expressed is this: that matter cannot be considered evil even if we see in it

no more than a mere principle of possibility or indetermination. Suppose it reduced to a minimum, entirely unin-
formed, without any quality, it is still a certain capacity
for form, an aptitude to receive it ; and if that does not
amount to much it is certainly not nothing. But let us go
further. To be susceptible of becoming good is doubtless
not yet to be very good ; nevertheless, it is already to be
good, and in any event it is not to be bad. It is better to be
wise than merely capable of wisdom, but the mere power of
becoming wise is already no inconsiderable quality. Such
dialectical considerations have a certain value of their own,
no doubt, but not yet their full weight until they are
referred to the basic principle that underlies them, and sets
them, moreover, in their true place within the general
framework of Christian philosophy. If matter is good we
can rest assured that it is the work of God ; and here the
Manicheans were deceived : but also conversely—if it is
the work of God we can be quite sure that it is good ; and
here Plotinus was deceived. " How gloriously then and
divinely did our God say to His servant : *Ego sum qui sum,*
and then : *Dices filiis Israel, Qui est misit me ad vos.* For He
Himself most truly is because He is altogether immutable.
For this in fact is always the result of change : that what
once was, is now no more ; that, therefore, can truly be
said to be which is immutable : but as to other things
which have been made by Him, it is from Him that each
in its own way has received its being. Since therefore He is
being *par excellence* He has no contrary save that which is
not, and, consequently, as all that is good exists by Him so
everything in nature exists by Him, for everything in nature
is good. In one word, every nature is good, now all that is
good comes from God, therefore every nature comes from
God." [5] Here, then, is the principle on which rests the
Christian affirmation of the intrinsic goodness of all that is ;

and the very same principle will account for the evil that occurs in nature ; for Christianity does not deny evil, but it shows its negative and accidental character and so justifies the hope of overcoming it.

It is very certain that all things God has made are good ; and no less certain that they are not all equally good. There is the good, and the better ; and, if the better, then also the less good ; now in a certain sense the less good pertains to evil. The universe, moreover, is the scene of constant generations and corruptions in animate and inanimate nature alike. Now all these relative inferiorities and destructions make up what we may call the mass of physical evil. How are we to explain its presence in the world ?

As far as the inequalities observed in creatures are concerned, to call them evil would seem an abuse of terms. If matter itself is good, everything may be properly qualified as good, for it is not only good in itself, but even the fact that it is a lesser good may be necessary perhaps for the greater perfection of the whole. But what we must especially note is that these very limitations and mutabilities for which nature is arraigned, are metaphysically inherent in the very status of a created thing as such. For even supposing that all creatures were equal, and immutable in mode of being, they would, nevertheless, remain limited and radically contingent in their very being itself. Things, in short, are created *ex nihilo*, and because created they are, and are good ; but because they are *ex nihilo* they are essentially mutable. If therefore we insist on calling evil the ineluctable law of change in nature, we must recognize that the possibility of change is a necessity from which God Himself could not absolve His creation ; for the mere fact of being created is the ultimate root of that possibility. Doubtless the divine omnipotence could annul its effects—God can

and does maintain in being all that He has brought to being ;
and if He wished, He might cause all things to continue
indefinitely in one and the same state ; but such permanence and immutability would be merely adventitious
after all ; everything that exists in virtue of the creative
action and endures in virtue of continued creation, remains
radically contingent in itself and in constant peril of lapsing
back into nothingness. Because creatures are apt not to be
they tend, so to speak, towards non-being [6] ; all that God
makes, taken apart from the act that makes it, tends to
unmake itself ; and, in a word, the contingence of created
things in the order of existence must be regarded as the
true root of their mutability.

To accept this consequence, and Christian philosophy
cannot refuse it, is by no means to return to the position
of Plotinus for whom matter is evil, nor even to that of
Aristotle for whom matter brings disorder into the world
simultaneously with contingency. From the very outset of
Christianity the metaphysic of Exodus was carried down
beyond the plane of quality and touched the plane of
existence. If there is change it is not on account of any
particular class of beings, that is to say material beings, it
is simply because there are *beings*. In this sense, the form
and act of all that is remains open to mutability in exactly
the same way as matter, and in point of fact, evil, properly
so-called, enters the world only at the unhappy instance and
initiative of spirit. It was not the body that made the spirit
sin, it was the spirit that brought death on the body. The
whole problem now stands on a new footing : all that needs
to be made in order that it may be, is always tending to
unmake itself, so much so that what now permanently
threatens the work of creation is literally, and in the full
rigour of the term, the possibility of its *defection*. But only a
possibility, be it noted, nothing more ; a possibility without

real danger as far as concerns the physical order which has no control over itself, but a very real and practical danger indeed in the moral order, that is to say when men and angels are concerned ; for in associating them with His own divine government, their Creator requires them also to keep watch with Him against their own possible defection.

Thus so clearly to link up the possibility of physical evil with the contingency of created things, constitutes in the first place a remote preparation for one of the most important metaphysical achievements of the Middle Ages. In showing that the composition which most radically differentiates the creature from the Creator is not that of matter and form, but that of essence and existence, St. Thomas merely gives definitive expression to the thought of St. Augustine. But Augustine had already done much, for his metaphysic of evil passed wholly and almost as it stood into Thomism and Scotism. The result, in fact, of his principle is, that if you take physical evil as a positive quality inherent in any being, it is rigorously and by definition excluded from nature. The concept of physical evil is henceforth reduced to the concept of a lesser good, that is to say to that of a good. That a good should be lesser it must still be a good and, consequently, a being, for if the good entirely disappeared the being itself would vanish with it.[7] Let us go further. Even if we define evil as the privation of a due good, a good which should be there, still this privation would be meaningless save in relation to the positive good which thus lacks its proper perfection. It is as if evil were a mere *ens rationis*, a negation without meaning save in relation to positive terms, a fundamental unreality, determined, and so to speak, hemmed in on all sides by the good that limits it. It is therefore true to say that good is the subject of evil, so that one might almost be tempted to reduce one to the other,

as if the non-being of evil had no subsistence and intelligibility save with respect to the being of the good itself.[8]
Occasionally Augustine went as far as that, nor would it
be possible to go any further in the direction of optimism
without emptying the idea of evil of all meaning whatsoever
and thus suppressing the problem instead of resolving it.
Now the problem is undoubtedly there and has to be
faced—especially, we may add, when it presents itself in
the moral order.

As applied to beings not endowed with knowledge, the
ideas of happiness and unhappiness are obviously void of
meaning. It matters very little to them, or rather it matters
not at all, that some are more or less perfect than others,
or even that the greater part are condemned to corruption
to make room for others. It matters even less to the universe
as a whole, for as soon as one good is lost another replaces
it ; indeed, it is highly desirable that this should be so, for
the beauty and perfection of the universe are thereby
increased rather than diminished. A succession of beings in
which the weaker always yields to the stronger issues in a
harmony which is not to be disturbed by the death of the
individuals, to which indeed this death contributes a good
deal. A universe of this type might be likened to an eloquent
discourse, where all the beauty arises from the quick succession of sounds and syllables as if it depended altogether
on their very birth and death. Thus, as St. Augustine said,
all that we call physical evil is reduced to the harmony of a
sum of positive goods, or, as St. Thomas Aquinas would
say, the presence of corruptible things in the universe,
added to the incorruptible, only increases its beauty and
perfection.[9] On the other hand, we have to recognize
that the problem becomes more complex when we pass
from brute matter to reasonable beings ; for the latter are
aware of their destiny and suffer accordingly :   other

things merely undergo privation and corruption, but these have to face misery. And then arises the question of moral evil, that is to say of human suffering including the physical conditions on which it depends : pain, sickness, death. What shall we say of these ? The principles we have already will be sufficient to prepare the way for a solution ; but we shall have to make them more precise.

Let us remark in the first place that man, as a reasonable being, is a great good, and that not only in himself but also on account of the whole destiny that awaits him and, above all, of the beatitude of which he is capable. Created in the image of God, he is, in the words of St. Bernard, *celsa creatura in capacitate majestatis.*[10] Now to be capable of entering into society with God requires an intelligence, but to be capable of rejoicing in this society requires a will. To possess a good is to adhere to it, to absorb it by an act of will. Thus, to create a being capable of the highest of goods, that is to say of participation in the divine beatitude, is, *ipso facto*, to create a being endowed with will ; and since to will the thing known by intelligence is to be free, we may say that it would be impossible to call man to beatitude without endowing him with liberty at the same time. A magnificent gift assuredly—but a very formidable one too, for to be capable of the greatest of goods is also to be capable of losing it. St. Augustine often dwells on this aspect of human freedom with all the indefinite possibilities of greatness and misery it involves. In a world in which all that is, in so far as it is, is a good, liberty is a great good : there are lesser goods, but still greater are conceivable. The virtues, for instance, are superior to the freedom of the will, for it is quite impossible to misuse temperance and justice, whereas we can very easily misuse our liberty. The truth is, then, that free will is a good, and the necessary condition of the greatest of goods, but not the one all-sufficient condition ;

everything depends on the use—which in itself is also free—
that we make of it.

Now it happens that man, in the first transgression, made
an ill use of it. Mutable like all creatures, endowed with a
free will, capable, consequently, of rebellion, he did in fact
rebel. The fault did not consist in desiring any object evil
in itself, for the very notion of such an object is contradictory, but for the sake of a good he turned away from the
better : *Iniquitas est desertio meliorum*.[1] Made for God, he
nevertheless preferred himself to God, and in so doing
brought moral evil into the world—or rather he would have
brought it into the world had not the Angel forestalled him.
Now this evil has a quite special nature, profoundly different
even from the moral evil of Greek philosophy. When man
subverted order he did a great deal more than merely fall
away from the rationality of his nature, diminish his own
humanity, which is all that he does in Aristotle's ethics, nor
did he merely compromise his destiny by an error, as
happens in the Platonic myths ; he brought disorder into
the divine order, and presents the unhappy spectacle of a
being in revolt against Being. That is why the first moral
evil has a special name in the Christian system, which
extends to all the faults that spring from the first ; that is to
say the name of sin. By the use of this word the Christian
always intends to convey that moral evil, as he understands it,
entering a created universe by the act of a free will, directly
bears upon the fundamental relation of dependence which
unites the creature to God. The prohibition, so light, and,
so to speak, so gratuitous, which God had put upon the
use, so perfectly valueless to man, of a single one of the
good things placed within his reach, was but the sensible
sign of this radical dependence of the creature. To respect
the prohibition would be to recognize the dependence ; to
violate it would be to deny the dependence, and to proclaim

that what is good for the creature is better than the divine good itself. Every time man sins he renews this act of revolt and prefers himself to God ; in thus preferring himself, he separates himself from God ; and in separating himself, he deprives himself of the sole end in which he can find beatitude and by that very fact condemns himself to misery. That is why, when there is question of moral evil, we can justly say that all evil is either sin or the consequence of sin.[12] Perverted in soul, subjecting his reason to concupiscence, subordinating the superior to the inferior in the order of spirit, man by the same act, brings disorder on the body animated by the soul. The equilibrium of the constitutive elements of his physical being is upset, just as disorder falls upon a house when it enters the heart of its master. Concupiscence, or rebellion of the flesh against the spirit, infirmity, sickness, death, are all so many ills that have fallen on man as the natural consequence of his sin : *omne quod dicitur malum, aut peccatum est, aut poena peccati ;* a phrase of St. Augustine's which is nothing but an extension of St. Paul's : *per unum hominem peccatum in hunc mundum intravit et, per peccatum, mors ;* and again, through St. Paul, an echo of the narrative of Genesis. Once more, by revealing to man a fact that he could not naturally know, revelation opens up the way for the work of reason.

And here at last we are at the heart of the question, and if Christian philosophy can justify itself on this point, it will certainly have succeeded in interpreting, in the most optimistic manner possible, a universe in which the reality of evil is in any case an undeniable fact. Attacks on this solution of the problem were not long in coming, and naturally enough they came from the side of Pelagianism ; and since they respond to a permanent philosophical difficulty it will be simplest to consider them in their original form. The question in short is whether the Christian position is not itself tainted with Manicheanism. Admit that man is not the work of a Demiurge of more or less limited power, admit that in the beginning he was untainted by any evil principle, but that his sin sprang solely from his free will ; then, if he sinned, it was because he willed it, but if he willed it, his will itself must have been bad. And it is useless to object that his will became bad because he willed it so—to do that at all it must have been already bad. We shall not escape by going round and round ; a world reduced to universal disorder by the mere presence of a free will can never justly be called good. It would seem that Christian thought, presenting a face of optimism as against Mani, presents a face of pessimism as against Pelagius ; or even that some unexorcized shadow of Manicheanism had all unwittingly been left behind by St. Augustine.

To clear up the point, let us for the last time recall the principle that governs the whole problem. The question is not whether God could have made creatures who should not be mutable, the thing would have been just as impossible as making square circles. As we have seen, mutability is as co-essential with the nature of a contingent being, as immutability is in the case of the necessary Being. But now, when the question of moral evil arises, this principle is to be applied to the case of a free being created by God out of nothing. Suppose then that neither angel nor man had ever realized the possibility of defection inherent in their nature, none the less they would have remained radically mutable beings ; such virtuality may be unactualized, in a given case it may even be morally unactualizable owing to the effects of divine grace, but nevertheless it is always there, an indelible mark of contingency. Unless, then, we are simply going to deny the possibility of justifying creation altogether, we must accept the possibility of moral evil as its necessary correlative, as soon as we admit the presence, in the bosom of this creation, of a class of free beings. But then, it will be said, why create free beings ?

Because they are not only the noblest ornament of creation, but also, after God, its final cause. What God creates, as we have already said, are beings who may be witnesses of His glory, and by that very fact participators in His beatitude. That this beatitude should be really theirs they must will it ; but that they may be able to will it, they must also be able to refuse it. The whole physical world is there only to serve as the habitation of spirits created by God in order to participate in His own divine life, and enter into a real society with Him. Subjected to the necessity of sinning, they would be altogether absurd creatures, mere monsters, since their nature would contradict their end ; but unable to sin they would be rigorously impossible, since then they would be immutable creatures, that is to say realized metaphysical contradictions. No doubt it is a much greater good to be capable of beatitude without the power of sinning—it is the good proper to God and to His elect whose wills are confirmed in grace—but it is already no small good to be so created that in order to escape misery and achieve beatitude we have nothing to do but to will it. I make no attempt here to compel acceptance of this Christian solution of the problem of evil, for it depends on the acceptance of a certain metaphysic of being, and thereby it stands or falls ; but I wish to bring out its fundamentally optimistic character. Now it seems difficult to go further in this direction than St. Augustine went and with him all the other philosophers he inspired. For all evil comes of the will ; this will was not created evil, nor even indifferent to good or evil ; it was created good, and such that it needed only an effortless continuance in good to attain to perfect beatitude. The only danger threatening such a nature lies therefore in that metaphysical contingence inseparable from the state of a created being, a pure *possibility*, without the least trace of actual existence, a possibility that not only could have remained unactualized but ought to have done so. Thus, without taking into account the divine art which knows so well how to bring good out of evil and to remedy the results of sin by grace, considering this evil strictly in its root, it seems that we may justly claim for Christian thought that it has done everything necessary to reduce it to the status of an avoidable accident, and to banish it to the confines of this fundamentally good universe.

What is true of the problem of the origin of evil, is true also when we come to consider the worth of the world after the introduction of evil by the original transgression. The popular idea of a Christian universe corrupted in its very nature by sin owes much of its favour to the influence of Luther, Calvin and Jansenius ; but to look at Christianity through their eyes would be to regard it in a very different light from that of Thomism, or even the authentic Augustinianism. No one, in fact, could be further than St. Augustine from considering the world in the state of fallen nature as worthless. His own metaphysical principles would forbid it, to start with. Since evil is but the corruption of a good and cannot possibly subsist at all save in this good, it follows that inasmuch as there is evil, there is also good. Certainly, we have travelled very far from that degree of order, beauty and measure which God bestowed on the world in creating it, but if sin had abolished all good it would have abolished all being along with the good and the world would no longer exist. In this sense we may say that evil could not eliminate nature without eliminating itself, since it would have no subject left to inhere in, there would be none of which it could be affirmed. It is not in the least surprising therefore to find St. Augustine indulging in genuine eulogies of fallen nature. If he deplores all that we have lost he never dreams of despising what remains ; even our present miserable state has not lost all its glory in his eyes. We behold a human race that is still of such fecundity as to spread over the entire earth ; man himself, *opus ejus tam magnum et admirabile,* whose intelligence, dormant in the infant, progressively awakens and develops until it produces all these arts lit up with the prodigal splendours of intelligence and invention. How much good must there not remain in such a nature, to enable it to invent so many techniques, of dress, agriculture, industry and navigation ; to achieve these noble arts of language, poetry and music, and lastly this very moral science itself which puts it on the road to an eternal destiny ! There is nothing even in the very body but Augustine will admiringly detail its beauties, for even these remain radiant, in spite of the Fall. If, then, we misconceive him, it is because we no longer dare to rise to the height of his splendid vision of the world as it was before the Fall, as it will be again in the state of glory. If he qualifies the splendours of this world as " consolations for condemned wretches," it is not because he would belittle them, they are dearer indeed to him than they ever can be now to us.[13] But he believed that the world had known better things, and moreover awaits better things, so that, accepting all that we accept, rejoicing in all that we rejoice in, he hopes for far more. If this hope now fails us, it is not he that should be taxed with pessimism, it is ourselves.

For the technical justification of this Augustinian and Christian feeling we must turn once more to the philosophers of the Middle Ages, and particularly to St. Thomas and Duns Scotus. What was wanting to St. Augustine for the purpose of finding the definitive formula for it, was an exact idea of a nature considered as a stable essence with defined contours. He is quite sure that evil is powerless to destroy nature ; what he never succeeded in saying clearly is, that it cannot even alter nature. Nothing on the other hand is clearer in St. Thomas ; only one who has never read any one of the articles which he devoted to the question in the *Summa* could possibly understand the expression "corrupted nature" in that *simpliste* sense in which it seems to be so often misunderstood. If, in fact, you take it literally the expression is a contradiction in terms ; and it is enough to follow St. Thomas' analyses to see in what an altogether relative sense we should understand it.

When it is asked what effects were produced by original sin on the good of human nature, we must first of all define what we mean by this "good." Three different interpretations are possible. In the first place it may signify human nature itself, as determined by its constitutive principles and defined as "rational living thing." Secondly, it may mean man's natural inclination to good, that natural inclination without which he could not even continue to live since the good in general includes his own proper good. Thirdly, it may mean the gift of original justice bestowed on him by God at the time of his creation and received therefore as a grace. Understood in this last sense, the good of human nature is not a part of that nature, it is something added, and that is why it was totally destroyed by original sin. Understood in the second sense, the good of nature is a real part of nature, and is not therefore to be suppressed, but simply diminished by sin. Every act initiates a habit, that is to say the first bad act results in a disposition to commit others and thus enfeebles the natural human inclination towards good. But this inclination, nevertheless, remains, and so the way to the acquisition of all the virtues is still open.[14] As for the first and proper sense of the word "nature," that is to say the very essence of man, it can neither be suppressed nor diminished by sin ; *primum igitur bonum naturae, nec tollitur, nec diminuitur per peccatum.*[15] To deny this would be to suppose that at one and the same time man could both remain man and cease to be man. Thus sin neither adds to nor takes away from human nature : *ea enim quae sunt naturalia homini, neque subtrahuntur neque dantur homini per peccatum.*[16] Man's metaphysical status is essentially unchangeable and independent of all the accidents that may befall him.

When therefore the Renaissance is held up to our admiration for its discovery of nature and its worth, and opposed to the Middle Ages as the day of its unjust depreciation, we must carefully scrutinize the meaning of this assertion. In so far as it has any it can only be this : that the Renaissance marks the opening of an era in which man will profess to be satisfied with the state of fallen nature. And that, no doubt, happened, although to a much lesser extent than alleged ; but it would be altogether unjust to conclude against the Middle Ages that having unfavourably compared the state of fallen nature with another and a better, it had no feeling left for it at all. If anyone showed such lack of feeling, or denied its reality or value, it was certainly neither St. Thomas nor St. Augustine, it was much rather Luther and Calvin. In this sense it is true to say that if the spirit of mediæval philosophy was profoundly accordant with certain positive aspirations of the Renaissance it was precisely because that spirit was Christian.

The tradition goes back to the remotest antiquity. No one did more than Tertullian to defend the unity of the true Church, and yet Tertullian left the Church as soon as he came to the conclusion that the human body is bad in itself· St. Augustine was never guilty of this error. He knows very well that since the body was created by God it must be good ; he refuses to follow Plato in holding that the soul is imprisoned in the body as the sequel of some metaphysical fall ; he will not allow that the duty of the soul is to flee the body, but would rather counsel it to accept the body as a precious charge placed in its care to be brought up in due order, unity and beauty.[17] But just as it is not Christian to run away from the body, so neither is it Christian to despise nature. How can we possibly belittle these heavens and this earth that so wonderfully proclaim the glory of their Creator, so evidently bear on them the marks of His infinite wisdom and goodness? The true Christian feeling for nature is that which finds expression throughout the Psalms, and, above all, in the Canticle of the Three Children in the fiery furnace : *Benedicite opera Domini Domino ; laudate et super-exultate eum in saecula.* And after many centuries St. Francis of Assisi will echo that song in his Laudes and the Canticle of Brother Sun, wherein not only water, earth, and air, and stars, but the very death of the body itself, will receive their meed of praise and benediction. If anywhere the heart of man entered into fraternal communion with all that lives and breathes and has being, most assuredly it did so there ; for this purely Christian soul it was altogether one and the same thing to love the works of God and to love God.

Here perhaps we have arrived at the point where the mistake that obscured the significance of Christian optimism begins to become clear. Not even the Middle Ages knew any ruder asceticism than that of St. Francis—or any more absolute confidence in the goodness of nature. Far from excluding optimism, Christian asceticism is merely the reverse side of its optimism. Certainly there is no true Christianity without the *contemptus saeculi*, but contempt for the world is not the same thing as hatred of being—quite the contrary, it is hatred of non-being. By wrestling with the *flesh*, the mediæval ascetic sought to restore the *body* to its pristine perfection ; if he did not rejoice in the world for the world's sake, it was because he knew that the true way to use the world is to restore it to its own integrity by referring it to God ; the world that the Christian detests consists of all that mass of disorder, deformity and evil introduced into creation by man's own voluntary defection. He turns away from these, no doubt, but precisely to adhere with all his heart to the order, beauty and good which was willed from the beginning ; he works to restore these in himself and others ; with an heroic effort he would clear the face of the universe and render it resplendent once more as the face of God. Nothing could be more positive than such an asceticism, nothing could be better grounded in hope and resolute optimism. The disaccord that persists between Christian and non-Christian on this point is of another order therefore than is usually supposed. The question is not whether the world is good or evil, but whether the world is sufficient to itself, and whether it suffices. The testimony, and, we may add, the secular experience of Christendom is, that nature itself is powerless to realize itself, or even fully to survive as nature, when it attempts to do this without the help of grace. If optimism thus consists, not in denying the existence of evil, nor in accepting evil, but in looking it in the face and fighting it, then we may legitimately speak of Christian optimism. The work of creation is shattered, but the fragments remain good, and, with the grace of God, they may be reconstituted and restored. And now, if we would be fully convinced of this, we must take up another inquiry and ask in what sense it is that even in the state of fallen nature the heavens and the earth declare the glory of God. After that we shall be in a position to indicate man's true place in the universal order, and to point out the path in which he has to travel in order to achieve his end.

CHAPTER VII

THE GLORY OF GOD

WHEN we read those coherent expositions of Christian thought contained in the *Summae* of the thirteenth century we are apt to forget the long years of preparation that made them possible. Systems of ideas so carefully and minutely adjusted to each other are not to be made in a day. Centuries of resolute effort and a vast outpouring of genius were required before the Christian philosophers were able to realize all the implications of their own principles and to succeed in formulating them with precision. It was for this reason, moreover, that Christian philosophy rather profited than otherwise from the markedly academic conditions that surrounded it in mediæval times ; the constant personal contact, the close collaboration and mutual criticism that went on within the walls of the school did much to ripen the solution of current problems. And nowhere was progress more evidently realized than in the treatment of the difficult question concerning the relation of the world to God. The universe is good because it is of being, but God is Goodness because He is Being ; how, then, shall we apportion their relative perfections ? How much should be granted to things so that they may truly be said to be, and how much must we deny them lest we attribute a sufficiency that would be incompatible with a just regard for the glory of God ?

Let us at the outset lay down a principle undisputed by any of the Christian philosophers, a principle that always virtually contained the solution that finally emerged from it after thirteen centuries of speculation. Since all things are good, and all things have being in virtue of a continuous outpouring of the divine bounty, nothing that exists is, independent of God. Malebranche was quite in the right when he said, in this sense, that of all the temptations that beset the creature the most dangerous, the most firmly to be resisted, is the temptation to set up a claim to *independence*. I doubt whether Malebranche always recognized it where it exists, and I am quite sure that he often saw it where it does not ; but having a lively sense of the peril he found at least a name for it, and that name I propose to retain, taking it in its purely metaphysical sense where it attains its full truth. In a created universe such as the Christian universe, the existence of every being stands in a radical ontological relation of dependence with respect to God. It is by Him alone that they are, and by Him alone they continue to exist ; also they owe it to Him that they are what they are, since not only their existence but also their substantiality is a good that God creates. But since their causal power flows from their being, their causality also must be referred to God ; and not only causality but also its exercise, for an action is of being ; and lastly the efficiency of the causal act along with the effect produced, since all that we make, God creates.[1] Thus the radical contingence of the finite being brings it into absolute dependence on the necessary Being, to Whom all must be principally referred as to its source, and this not only in the order of existence, but in the orders of substantiality and causality. If we forget this the original transgression is re-enacted in ourselves, or rather it is just because its effects continue that we forget it so easily.

There is a reverse side, however, to this fundamental truth which certain Christian thinkers, in their zeal to refer everything to God, have shown a tendency to let slip, although it is none the less necessary on that account. If it is true, metaphysically speaking, that all that exists exists by God and for God, it is equally true, physically speaking, that all that exists is something " in itself " and " for itself." What God creates depends integrally on the creative efficacy, but if this efficacy is not to be in vain it must produce something, that is to say it must produce being. It remains true to say that created being is contingent through and through, but since it is not nothing, and since on the other hand it is not God, it must have an ontological status of its own. The being it has received is certainly its own being ; its constitutive substance is its own substance ; the causal action it exerts, the efficacy it puts forth, are certainly its own efficacy, its own causality.[2] No middle position is possible : either the creative act produces nothing, which is absurd, or we must say that creation is the act whereby Being confers being—and that supposes that, along with its radical ontological dependence, the creature enjoys its own proper existence and all consequent attributions. But the very same secret that brings us face to face with this difficulty serves also to dissipate it. It is not in spite of its ontological dependence that the creature is really something, since if it is something it is so precisely in virtue of this very dependence. *In eo vivimus, et movemur, et sumus* [3] ; that is to say it is only in God that we have life, movement and being, but, then, in Him we really have them.

A very lively realization of this double aspect of the matter appeared early in the history of Christian thought. St. Augustine, of whom perhaps we could hardly expect so much, expressed himself on the point with all desirable clearness. He does not allow, any more than will St. Thomas, that the divine government is substituted for things, so as to act or produce in their stead. On the contrary, since the being of things is not God's being, God always governs things in such a way that it is they that really perform their own operations. It is as if He were in the act of con-tinuously creating original centres of activity and efficacy, and these were then pouring forth operations which really flow from their own natural perfection and really depend on it, and so must be referred to it as to their cause. In St. Thomas this aspect of Christian thought is unfolded in splendid amplitude because the metaphysical centre where its evidence is concentrated is kept steadily beneath our eyes. How, it is asked, could a created universe be filled with efficacious causes? How, replies St. Thomas, could it possibly be filled with anything else? Since it is created, the universe is born under the sign of fecundity itself, and here once more the effect must certainly be an analogue of the cause. We feel at once the contradiction implied in the idea of a fecundity generating a sterility; but this contradiction must not only be felt, it must be clearly perceived. To create beings is to create acts of existence *actus essendi*, and since we know already that causal efficacies are rooted as second acts in the first act of existence, it follows that creatures, from the very fact that they are, must be endowed with efficiency. In a word, just as the created being is an analogue of the divine being, so created causality is an analogue of creative causality; to be a cause, is to exert a finite participation in the infinite fecundity of the creative act.[4]

So far, all the mediæval Christian philosophers are in accord with each other and in accord also with the Fathers of the Church. It is perhaps superfluous to insist on the extent to which this conception of the world differs from the Greek. Between the universe of Plato and Aristotle on the one hand, and that of Christianity on the other, a funda-mental difference is introduced by the Judeo-Christian idea of Being, and the idea of creation it carries with it. Integrally dependent on God in its very existence, the Christian world is thereby no doubt limited but also consolidated at the same time. What it loses by losing its independence in the order of existence, it regains by participating in the likeness of Being, by leaning on Being ; it is dependent, of course, but both as regards itself and its causality it now stands on a firm foundation. The question, then, is not as to what it is that God makes in all that the creature is and makes, for He makes all that the creature makes and all that the creature is, except, indeed, evil,— which, precisely, is not. The true question in debate among Christian philosophers is rather this : how far extends the causality and efficacy that God concedes to creatures in general, and, in particular, to man. It would seem that in the discussion of this problem the various different shades of religious feeling came to be reflected, in a way, in Christian metaphysics ; so that the idea of the glory of God has a history of which the two critical points coincide once more with the doctrines of St. Augustine and St. Thomas Aquinas. The principles involved, being fixed from the outset, have no history ; but the feeling for the glory of God had a history, and inasmuch as it had one so also has the idea. Any detailed review of it would be co-extensive with the history of Christian philosophy itself, but at least we may try to fix the two poles between which it never ceased, and in all probability will never cease, to oscillate.

From the deepest roots of its inspiration down to the details of its technical structure the whole doctrine of St. Augustine is dominated by one fact : the religious experience of his own conversion. I have elsewhere expressed the view, and I think it remains true to say, that his philosophy is essentially a " metaphysic of conversion." [5] The difficulty, for Augustine, was to find the point of coinci-

dence between this metaphysic of conversion and the
metaphysic of Exodus. For he appealed to both ; he knew,
indeed he even felt, that they were but one, but it is much
easier to start from Exodus and find a place for the ideas
of the conversion, than to set out from the conversion and
return to Exodus. It is precisely Augustine's effort to accomplish this difficult operation that gives their meaning and
importance to the last three books of the *Confessions*. Long
neglected for those that precede, regarded sometimes as
an almost superfluous appendix, they gain in value and
beauty, nevertheless, with every century that passes. The
recital of Augustine's youthful errors will always be the
popular part of the book—like Dante's *Hell* ; but just as
the *Paradiso* is the pearl of the *Divine Comedy*, so this conclusion is the most splendid part of the *Confessions*. " But
thou, O Lord, to whom belongs eternity, dost thou know
nought of all I say, or dost thou see in time what passes in
time ? Why, then, should I pour all these things out before
Thee ? Not indeed that thou shouldst know them, but
that my heart may be lifted up to thee, and the hearts of all
those that read, so that I and all they may exclaim : *magnus
Dominus et laudabilis valde*." [6] There then, clearly formulated,
we have the fundamental theme of God's glory, which St.
Augustine at once goes on to justify by an appeal to the
idea of creation and to the metaphysic of being. It is
because God is beautiful that things are beautiful ; because
He is good that they are good ; because He IS that they
are. [7] But then, on the other hand, his conversion had taught
him how radically impotent, how insufficient to himself
the creature is ; and this truth, an absolute, an unqualified
one in the field in which the discovery was made, that is
to say in the supernatural order where the will is powerless
without grace, always tended in his hands to overflow from
the theological into the metaphysical order, to encroach

on nature, as nature, for the benefit of supernature. To confess God is certainly, for him, to confess God's greatness and the marvel of the works that manifest it, and indeed a whole book might be compiled out of the pages where he hymns the praises of creation, and even, as we have seen, of fallen creation ; but we detect at the same time a kind of metaphysical reticence, a hesitation to grant nature a perfection that might seem to make it self-sufficing. The supernatural dependence of creatures in the order of grace and their natural dependence in the order of existence tends, in his hands, to run to a strict limitation of their efficiency. On that account he is certainly the legitimate ancestor of all those Christian thinkers who try to discern a certain emptiness in nature that only God can fill and which, when it occurs in us, attests the great need we have of Him. The less we suffice to ourselves the more we need God, so that our very misery attests His glory as eloquently and, indeed, even more persuasively, than our greatness, for the latter suggests sufficiency whereas the former urges us to seek Him out. This we shall see very clearly if we take up the examination of three questions which St. Thomas, with infallible judgment, has selected as the critical points of the debate : that is to say the question of causality in the physical order, or the doctrine of the seminal virtues (*rationes seminales*), that of causality in the cognitive order, or doctrine of truth, and that of causality in the moral order, or doctrine of virtue.[8]

When St. Augustine would represent the universe in its dependence on the creative action of God, the formula that comes most naturally to his mind is that of *Eccle-siasticus* (xviii. 1) : *qui manet in aeternum creavit omnia simul.* He regarded the work of creation as an instantaneous *fiat ;* and that does not mean only that the six days of Genesis are purely allegorical and amounted in fact to an instant, but that in that instant the work of creation was really and wholly achieved. The seventh day's rest still endures. All that seems to us to be a new production, to be born, to grow, to develop, whether in the organic or the inorganic world, was already there from the first moment of creation. In order to maintain the contrary we should have to admit either that God constantly creates an infinity of new effects, and in that case it would no longer be true to say that He created all things simultaneously, or else we should have to say that they owe their being to second causes, and then we should have the absurdity of causes at one and the same time created and creative. To extricate himself from this embarrassment Augustine has recourse to the old Stoic doctrine of seminal virtues, which for the rest he links up closely with the Christian concept of creation. Over and above those beings which God created in their perfect form, He also created the seeds of all things to come along with the numerical laws that will rule their development in time. " As a mother great with child so is the world itself pregnant with the causes of things to come to birth hereafter ; so that all these are created only by the Supreme Essence, where nothing is born or dies or begins or ceases to be." [9] These last words, moreover, reveal St. Augustine's secret preoccupation. What he especially wishes to make clear is that God alone creates, and that to admit the production of a really new effect by a second cause, would be to transform this second cause into a creative cause. What is it that second causes really do ? They awaken or excite the latent virtualities that God deposited in matter when He created it—that, and nothing more. Whenever a new being comes to birth under our eyes the very fact of creation is manifested : *creationem rerum visibilium Deus interius operatur.* The parents who engender are nothing, but God is, and it is He that creates the infant ; the nursing mother is nothing, but God is and gives the infant growth.[10] Thus even in the purely physical order there is a lack of efficiency in nature which marks a kind of void to be filled up by the divine efficiency : to the word of *Ecclesiasticus* there responds that of St. John (v. 17) : *Pater meus usque adeo operatur, et ego operor.* Every apparently new production attests it ; we have but to open our eyes to see it.

When we turn from nature in general to man, and to what is properly human in man, the reason, the same conclusion is forced upon us. The proper function of reason is the formation of true judgments. Our thought is true when, in place of merely accepting empirically the thing that is, we judge it by referring it to what it ought to be. In a sense it is correct to say that truth is *what is*, but this " what is " is not the changing appearance of things, but rather their norm, their rule, that is to say the divine idea in which they participate and strive to imitate. Either then our judgments will lean in an immediate manner on the divine idea, and we shall be in possession of the rule of truth ; or they will depend on nothing but our intellect itself, and then all true judgments will be for ever impossible. Now we do in fact make true judgments. The definition of the simplest geometrical figure, even that of arithmetical unity, is rich with elements which can come neither from experience nor from our mind. The mathematician does not simply state what circles are, or what sensible unities are, for in nature he finds neither true circles nor true unities ; but he decides what a circle or a unity ought to be so as to satisfy their definitions. Thus these definitions transcend all possible human experience. The thing given, and our thought itself to which it is given, lie in the sphere of the contingent, the mutable, and the temporary, whereas truth has its natural home on the plane of the necessary, the immutable and the eternal. How, moreover, could it be otherwise since truth implies a reference to the divine ideas ? That is why, in Augustine's doctrine, every true judgment supposes a natural illumination of the mind by God. An intellect created out of nothing could no more give birth to the necessary, that is to say to true being, than a mother could generate the mortal body of her infant. It is God who fecundates our thought by His Word ; nor is He only the interior master as a voice that whispers in the ear of the mind, He is the light whereby it sees, and more still, He is its food, as bread in the mouth ; and more, the living seed that enters the womb of thought, espouses and fecundates it that it may conceive the truth : *Deus lumen cordis mei, et panis oris intus animae meae, et virtus maritans mentem meam et sinum cogitationis meae.*[11] But God's fecundating embrace of the soul does something more than engender truth, it engenders also virtue : *cujus unius anima intellectualis incorporeo, si dici potest, amplexu, veris impletur fecundaturque virtutibus.*[12] Let us briefly consider this last point.

It offers no great difficulty since, in effect, it is a mere repetition of what has preceded. Virtues are stable habits of well doing, and, as such, they are based on true judgments of the reason. Now these judgments are altogether of the same nature, whether they bear upon what things ought to be, or upon what men ought to do. When I say that the better is preferable to the good, that all human actions should observe due order, that we should will justice and, as far as in us lies, extend its reign over the whole earth, then I utter truths that are every whit as necessary as the definition of the circle or the definition of unity. How could I draw such universal and eternal necessities either from things or from my mind, seeing that these latter belong altogether to the order of the contingent and the mutable ? It is necessary, clearly, that I should receive the seeds of

the virtues in the divine embrace just as I receive the seeds
of science. God illumines me, not only with the light of
numbers, but also with the light of wisdom, so that my moral
life, no less than my scientific knowledge, attests His intimate presence in me, at the root of this metaphysical
memory to which we shall have to return later on.[13] God
surrounds me on all sides, He penetrates my whole being,
He is more intimate to me than what is most intimate in
my being, He is in me as the sole sufficient reason of all
that I do and of all that I am not, for every void within me
attests His fullness and even my misery bears witness to His
glory no less eloquently than my greatness. A profound
and enduring sentiment, co-eternal no doubt with Christian
thought itself, true also and necessary when expressed in
the order of grace ; but is it necessarily true in the order
of nature ? This is the question we are led to consider by
another and different expression of the sense of the glory of
God.

The chief thing to note in St. Thomas' critique of the
three theses just set forth is this : that although their consequences develop in three different fields of thought, he
considers that they come ultimately to the same thing,
because he judges them from the standpoint of a certain
conception of natural causality and because they seem to
him, moreover, inconsistent with the just exigencies of the
glory of God. Doubtless the doctrine of seminal virtues
very carefully safeguards the rights of the divine efficiency,
for then all is already realized and second causes have little
enough to do ; and a similar advantage is indubitably
offered by the doctrine of divine illumination, for then it is
God who endows reasonable beings with truth and virtue :
but both these theses suffer from the same inconvenience,
for whether all is ready made in the womb of nature, or
whether all is effected for it from outside, nature itself, in

either case, does nothing at all. Now in a created universe
such as the Christian it is inconceivable that beings should
not be genuine beings and causes therefore not genuine
causes. The generosity and goodness of God are such that
He is not content to stop at giving things existence, but gives
them also the causality that flows from it : *prima causa ex
eminentia bonitatis suae rebus aliis confert non solum quod sint, sed
etiam quod causae sint.*[14] Starting from this principle St.
Thomas successively rectifies the three Augustinian theses
of seminal virtues and illumination of the soul by truth and
the virtues.

To admit the seminal virtues is to maintain that the
forms of beings to come are already latent in matter. But
what is the truth ? They do not exist already realized in
matter, nor does matter receive them ready-made from
without ; they are there, as Aristotle said, in potency—
that is, matter can receive them. That it may receive them,
there is required a second cause which, itself a being in
act, causes something of its actuality to pass into the potency
of the matter. Thus we return to the idea of causality as
above defined, and we see that the efficacy of second causes
lies in the second causes themselves, as a participation of
the divine causality. Certainly they do not create, but they
cause ; as substances themselves they generate, not indeed
being, but at least substantiality.

By a natural application of the same principle, St. Thomas
modifies the economy of the Augustinian illumination, and
invests it with a new significance. The fundamental thesis
of illumination remains intact. In Thomism, as in Augustinianism, we know the truth only in the divine ideas and
by the light with which the Word enlightens us ; but now
it enlightens us in another manner. According to St.
Thomas illumination consists precisely in the gift, made by
God to man in his creation, of that which it is of the very

essence of the Augustinian noetic to deny—that is to say, an intellect sufficing to produce truth. [From the time of St. Thomas we are henceforth in possession of a natural light, that of the active intellect, which is neither Augustine's mind nor Aristotle's active intellect. Like the latter, it is capable, on contact with sensible experience, of generating first principles, and, with the aid of these, it will gradually build up the system of the sciences ; but, like the Augustinian mind, it is capable of generating these truths only because it is itself a participation in the Truth. But instead of an intellect naturally lacking the light of truth, into which therefore this light must fall from on high, we have an intellect with which this truth is, so to speak, incorporated, or rather an intellect which has itself become this light of truth, in an analogical mode of course, and by way of participation.[15] And what holds of truths holds equally of the virtues. They are innate in us in this sense, that we are apt to acquire them ; they come from God in this sense, that we acquire them by dint of bringing into operation the principles of the practical reason, which is itself only a participation of the divine light.[16] In both cases we attain to the divine ideas through the agency of an intellect which is itself a participated likeness of the uncreated light in which dwell the Ideas ; and it is in this sense that, to the question put by man] *Quis ostendit nobis bona ?* the Psalmist replies : *Signatum est super nos lumen vultus tui Domine ;* as who should say, it is in virtue of the very imprint of the divine light in us that we know all things.[17]

How far indeed in Thomist doctrine does God seem to recede from man and the world ! The thirteenth century Augustinians felt it keenly, and that explains their often rather lively reactions against St. Thomas. However, St. Augustine's God does nothing that St. Thomas' God does not, and the Thomist creature can do no more without

God's aid than the Augustinian creature can. In both doctrines God produces all things, and creatures produce what they produce; the difference is that the Thomist God shows Himself more generous than St. Augustine's. Let us say, rather, since it comes to the same thing, that however great is God's generosity towards the world in Augustinianism, it is greater still in the philosophy of St. Thomas Aquinas. He has created an intellect which lacks nothing that it needs, and in particular lacks nothing needed for the exercise of its proper function : namely, to know the truth.

Reduced to essentials, and setting aside psychological considerations which would throw a measure of light on it, the difference between the two doctrines may be stated simply. They are two different expressions of the same sense of the glory of God. It really is the same. For St. Augustine, as for St. Thomas, *coeli enarrant gloriam Dei*, and if the heavens declare His glory it is because they bear His likeness ; only, with St. Thomas, the divine likeness sinks for the first time into the heart of nature, goes down beyond order, number and beauty, reaches and saturates the very physical structure, and touches the very efficacy of causality. The work of the Almighty can by no means be an inert world, for then the work would not give testimony to the workman. Later on, perverting the principles of Augustinianism, Malebranche will have it that the glory of God is chanted by a world without nature and without efficacy ; a radical impotency attesting the omnipotence of its Author. More faithful to Augustine's true spirit, St. Thomas would rectify his philosophy in the sense of its own principles, and reinstate creation in the whole plenitude of its rights ; because it is by the greatness of the work that we know the greatness of the workman.

And in the first place let us revert to our supreme principle. The proper effect of creative causality is the gift of being. At the basis of all that the world receives of God, more intimate to it than anything else, lies existence itself: *ipsum enim esse est communissimus effectus primus, et intimior omnibus aliis effectibus; et ideo soli Deo competit, secundum virtutem propriam, talis effectus.*[18] On the other hand, since the effect always bears a likeness to its cause, beings are analogues of God merely by force of existing; and since it is of their very nature to resemble Him, the more they resemble Him the more they are, and the more they are the more they resemble Him.[19] But to see how far the consequences of this principle go, we must take account of the type of being characterizing created things. The greater part of those that fall under our senses are concrete substances composed of matter and form, that is to say partially in potency and partially in act. As such they are altogether good, for even matter itself, as an aptitude to receive form, must be considered as good. We have said this already, and here is the ultimate reason. If matter is being only in potency it is good only in potency. Absolutely speaking, being is attributable only to that which subsists, but goodness on the other hand extends to all that enters into relation with being; and for this reason, matter, existing only in view of a form and in relation to a form, and not as a being in act, is nevertheless a good. It is one of the results of the primacy of being that even a simple possibility of being is good.[20] However, in the case of an unrealized possibility there is something wanting, a privation, consequently also a certain evil. The concrete substances of which the universe is composed are thus incomplete in being and in becoming, and it is just on that account that they need to act. They operate to complete themselves, before they can operate to give of themselves. The more incomplete they are the more numerous and diversified are the operations to be accomplished. In any case, and no matter what their degree in the scale of being, it is not enough for them simply to be in order to achieve their perfection. A man, even if vicious, is good in so far as he is a man ; but he is not yet fully man, for he suffers that privation of necessary virtues which constitutes vice. In the case of God, being suffices ; for to be Being is to be perfect ; but to be a being, and especially a being involved in the potentiality of matter, is to rest open to further possibilities of being that have to be acquired by action. We can say then that if the very being of things consists in a likeness to God, all that helps them to realize their being more completely also helps them to realize their likeness to God more completely ; now creatures cannot attain to the perfection of their being save by exercise of their proper operations ; therefore they bear the divine likeness not only in their existence but also in their causality.[21]

This opens up a vision of the universe which externally resembles that of Aristotle, but is profoundly different in inner significance. According also to the Greek philosopher things move themselves in order to acquire their own proper substantiality and to imitate in this respect the divine perfection of the unmoved movers. According to the Christian philosopher things move themselves in order to acquire the fullness of being, for to bring their own nature to its point of perfection is, at one and the same time, to perfect their likeness to God : *unum quodque tendens in suam perfectionem tendit in divinam similitudinem.*[22] The essential difference between the Greek standpoint of substantiality and the Christian standpoint of being here yields one of its most hidden, and yet most important consequences. In virtue of the very fact that it is a sequel and an analogue of creation, Christian causality so to speak prolongs and continues it. Once more, it is not a creative causality, since its source is always in being *received*, but it is truly productive of being, since, in the measure in which it is, every being can give of the being it has received, and make it pass, in the character of effect, into another being : *causa importat influxum quemdam ad esse causati*. That is why, in the high words of Dionysius, there is nothing more divine in this world than to become a co-operator with God ; a phrase in which we hear the echo of St. Paul's : " For we are God's coadjutors." Now if we are so, if we really co-operate in the creative work, it is precisely in distributing being around us, and enriching our own by the fecundity of our causal activity. To be a cause is neither to add to the sum of created being—which God alone could do—nor yet to leave created being precisely as we found it, which would be to do nothing at all ; but to realize the possibilities of the universe, to substitute everywhere the actual for the virtual, to confer on what already is, the whole fullness of which it is capable or can receive ; in a word it is to serve as an instrument for the creative work : *Dei cooperatorem fieri*, and to assist the universe of becoming that results from it to realize itself : *Dei sumus adjutores*.

In such a doctrine it is easy to see that far from derogating from the glory of God in insisting on the perfection and efficacy of creatures, we only exalt it in exalting them. For the Christian philosophers of the classical period it was always a mistake to belittle nature under pretext of exalting God. *Vilificare naturam* is a philosophical error in itself, for a nature cannot be conceived unless we attribute to it the means of acquiring its own proper perfection. But it also does wrong to God, for God is the pure actuality of being, and since it is by creating that He communicates being to creatures, it must needs be that in communicating the likeness of His being He communicates also a likeness of His causality.[23] But from the fact that He is Being, God is also the perfect Cause ; and therefore the things He creates participate in His perfection in such a way that all detraction from their perfection is detraction from the perfection of His power : *detrahere ergo perfectioni creaturarum est detrahere perfectioni divinae virtutis.* A universe without genuine causality, or with a causality not allowed its full effect, would be a universe unworthy of God.[24] Finally, since God is the supreme good, He must have made all for the best. Now it is better that the good assigned to each creature should flow over and become the common good of all. But for the good of one to become the good of others it must be communicated from the one to the other, and it can be communicated only by action. God therefore has Himself communicated His goodness to things in such a way that each one of them can transmit to others the perfection it has itself received : all injustice towards the causality of creatures becomes an injustice towards the goodness of God.[25]

Arrived at this point, we begin to see the main lines of Christian philosophy tend to their final convergence. Being Itself created all things, and created them all for His glory, in this sense, that He created all beings for their glorification. Now in this state of glorification creatures will rejoice more in the honour and glory of God than in their own glorification itself.[26] It is really therefore His glory which is their end, just as it is their principle.[27] And if a universe is destined to the state of glory and made to the likeness of the Supreme Good, how otherwise can it appear than altogether good to the mind that contemplates it ? But to acquire this beatitude, to realize this glorification, not only being is required but action ; now all action, whether conscious or not, and even whether good or bad, contributes to the glory of God, for our acts may be deprived of their good, but nothing can deprive God of His glory.[28]

Thus the Christian universe is entirely good as regards what it is, but incomplete and tending consequently to achieve its good in realizing its being. That is why the doctrines of St. Thomas and Duns Scotus on the efficacy of second causes and the consequent redressment of the Augustinian doctrine on the seminal virtues and illumination lies precisely in the true line of Christian tradition. Let us say rather, since it is from Augustine himself that their principles are borrowed, that if we penetrate beneath the letter to the spirit of history Augustinianism is itself fulfilled by these non-Augustinian doctrines of knowledge and causality. From this standpoint everything at once becomes significant in the mediæval philosophical texts, even down to the very form in which they clothe their thought. The Scriptural quotations in the first place, scattered broadcast everywhere, are no mere ornaments, superfluous confirmations, without genuine philosophic signification. They are needed as guides going along with thought on every side, warning and protecting it, avowed and visible signs, borne on the front of Christian philosophy, of the aid that revelation brings to reason. But thereby we understand also the constant solicitude of the mediæval thinkers to appeal to St. Augustine and the other Fathers, and that not merely where they follow but even where they abandon them. For they abandon them only the better to carry on their thought. It is very true that the obstinate concordism of the Middle Ages does not facilitate the task of the historian. It is impossible not to feel a certain hesitation in the face of texts which sometimes claim an accord in the name of the very formulæ that contradict it. But if the mediæval thinkers often let fall the differences it was because they knew that the differences tend to fall out of themselves ; that the resemblances alone are really fruitful. Above all, it is because they were conscious of cooperating in one and the same work ; because they felt themselves more faithful to their forerunners in abandoning them at need instead of following them blindly ; because to express their real intentions, to say what they should have said, what they would themselves have said had they enjoyed the fruit of long centuries of reflection on their own principles—because all this was not to betray them, but rather to save in and for their work what they themselves would have done anything to save. But we are very far yet from attaining the last consequences of the metaphysics of Exodus, and, in the natural sequence of ideas, we have now to watch the birth of the Christian conception of divine providence ; a new stage on a road so long that we can do little more than lay down the landmarks.

CHAPTER VIII

CHRISTIAN PROVIDENCE

THE idea of divine providence is not an exclusively Christian idea ; but there is a specifically Christian idea of divine providence. For its historical antecedents we must undoubtedly turn to Plato. Here, as often enough in other matters, he is nearer to Christian philosophy than any of the other thinkers of antiquity, and it is easy to see why the Fathers of the Church and the philosophers of the Middle Ages appealed to him so often. Reduced to its essential elements the natural theology he sketched in the *Laws* (Book X) amounts to the three following points : there are gods ; these gods watch over human affairs ; and it is impossible to bribe them or make any kind of purchase of their goodwill. The second of these three points directly concerns the doctrine of divine providence, and we must look at it a little more closely.

What Plato tries to bring out as forcibly as possible is this : that it is contradictory to admit that the gods exist and yet have no care for human affairs. For if the gods exist they are good, and if they are good they must be admitted to be virtuous ; now negligence, idleness and luxury are vices and altogether incompatible with virtue, and therefore cannot be attributed to the gods ; wherefore we must grant that they are vigilant, attentive, careful of all human affairs, and not only in great matters but also in small. To convince oneself of this more fully it is sufficient to recollect that the gods are immortal, that nothing can escape their eye, and that consequently to say that anything whatever is deprived of their care is to accuse them of negligence. Just as no good doctor and no good general will neglect even the humblest details under pretext of giving better attention to the whole, so neither does the divinity ; the immortal workman is not less prudent than the mortal. And so we must admit that everything in the universe is ordered and directed in view of the good of the whole—everything, man also included : "And thou too, O miserable mortal, hast a part to play, however small, in the universal order. But this lies hid from you, that every particular generation is for the sake of the whole that the life of the universe may be happy ; and that nothing is made simply for your sake, but you exist for the sake of universe. For every physician and skilful artisan directs his operations to an end tending to the common good of the whole ; and fashions each part for the sake of the whole, and not the whole for any one of the parts. And you murmur, and are forgetful of what is better at once for you and for the whole, that is to say what comes to pass under the laws that rule universal existence." Now what are these laws ?

As far as man is concerned it is a law which rules the series of his states in function of his acts, throughout the course of the successive generations and reincarnations he has to go through. This single and simple law is that like naturally attracts like and that when by this attraction they come together, " like behaves to like as like would naturally expect." In other words, if we are good we shall live in the company of the good, and be treated by them in accordance with their goodness ; if we are wicked, we shall live with the wicked, and suffer from their wickedness as they will have to do from ours. Such is the law from which no man is exempt either in this life or in the next. Hence we have declarations like the following, which certainly awake familiar echoes in Christian ears : " Neither thou, nor anyone, will ever, in any circumstances, be able to boast of exemption from this order, established by the gods to be inviolably observed and infinitely respected. You will never escape it, not even were you so small as to sink into the depths of the earth nor so lofty as to reach up into heaven." That, certainly is well worthy of citation in any " *Praeparatio Evangelica* " ; but is it really in the spirit of the Gospel or even of the Old Testament ?

What strikes us at once when we pass from Platonism to Scripture is this : that instead of having to do with a plurality of artisans who once for all have determined the future of their work, we are now in the presence of a God Who, having created the universe, henceforth is Master of it. And He is a jealous Master. Jahve never ceases to remind the world of His right of authorship and it is always upon this right that the Bible founds His claim to dispose of human affairs at His good pleasure. If, above all other peoples, He makes a special choice of the people of Israel, it is because the earth and all its fullness is His, and if the earth is His it is because He made it. This unceasingly asserted right of property is not limited to the universe taken as a whole, nor to the chosen people, nor to other peoples who might have been chosen but were not so in fact—it extends to the whole totality of beings and as far as the creative action itself extends : men, animals, plants, earth—all is God's because all came from God. This fundamental relationship of the thing made to its Author goes even further than one might suppose, for in the first place it explains God's miracles, which are so many public attestations of His creative power and testify to His authority to make laws for the people of Israel. It is the omnipotence of God made manifest in His works which authorizes the promulgation of what, even in the Gospel, will remain the first and greatest commandment : Thou shalt love the Lord thy God, and serve Him with all thy heart and all thy soul. This jealous God to Whom all belongs is also a God Whom nothing escapes. He searches the reins and the hearts, and every secret thought of each man lies open to His eye like all this *plenitudo universi* of which man is a part. Without a doubt He is faithful, free from all iniquity, just and righteous ; whoever acknowledges His power and respects His law has nothing to fear from Him and everything to hope ; like His works, all His ways are perfect : *Dei perfecta sunt opera, et omnes viae ejus judicia ;* but woe be to him who sets himself up against Him were it only in the secret depths of his heart. For he sees and foresees all things, what we are, what we think, and what we do—absolutely all, whether in past, present or future. Now if He brings all that marvellous and formidable knowledge to bear upon us, it is precisely because He made us. He created us and His hand rests upon us. There we have the reason why in the Biblical outlook too, man will never hide himself from God even were he small enough to sink into the depths of the earth or great enough to reach up to the heavens. In contrast, however, to the philosophy of Plato he is subject now to no mere impersonal law but to the will of a Person on Whom his existence and destiny depend : *quia tu possidisti renes meos ; sucepisti me de utero matris meae.* It is a very long cry from Plato's rational acceptance of order to the Psalmist's dread of the divine power : *Confitebor tibi, quia terribiliter magnificatus es : mirabilia opera tua, et anima mea cognoscit nimis.* Here then we have to do with no mere different cast of feeling, but a different idea, the fundamental Judeo-Christian idea of a Providence, which, because it is creative, is also elective.

Election of a people in the Bible, election of all humanity in the Gospel ; for Jesus Christ and St. Paul the whole human race is the inheritor of the promises made to the people of Israel and may claim the benefit if they will.

The Creator-God is always there, but now He veils His
creative power beneath His fatherhood. " Our Father,
Who art in heaven "—it is still to Him that we owe our
being, but now He feels towards His creatures as a father
towards a son. Also, while it retains the basis on which it
was set already in the Bible, the Christian idea of providence
assumes a new aspect. The personal relations which unite
each creature to its Creator extend over the whole of nature,
since all the works of God are God's and He loves even the
least of them. The Heavenly Father feeds the birds of the
air ; the ravens neither sow nor reap, they gather not into
barns, and yet they live ; the lilies of the field toil not,
neither do they spin, yet Solomon in all his glory was not
arrayed like one of these. And if God so cares for every
sparrow, what shall we say of man who is of more value
than many sparrows ? Why should he fear ? The very
hairs of his head are numbered. Let him therefore take no
care for the morrow ; for He who feeds the birds of the air
and clothes the lilies of the field will also care for him :
*si autem frenum, quod hodie est in agro, et cras in cibarium mittitur,
Deus sic vestit ; quanto magis vos, pusillae fidei ?* For the
Christian, one thing only is necessary, to seek the Kingdom
of God and His justice, and all these other things will be
added unto him. Hence the *Pater noster* carrying into Christianity the fear of the Almighty, and giving it at last its
true significance, is, and will for ever remain, the universal
prayer of Christendom. Let the will of the Father be done
in earth as in heaven. Let us will for ourselves what He
wills for us, knowing that it will surely be accomplished
because He is Master, but knowing also that it is well for
us that it should be accomplished because He is our Father—
that is the deepest expression of the Christian idea of
providence : *nolite timere, pusillus grex, quia complacuit patri
vestro dare vobis regnum.*

And this explains why it was that the first Christian thinkers so readily dwelt on the idea of providence as one of the characteristic traits of the new conception of the world. For they rightly felt themselves in a new world. The old Stoic destiny was dead, and dead also the Stoic resignation—passive even in its highest flights of generosity. The mechanical universe of Lucretius and Democritus had given place to a cosmos in which every element had been chosen, created, predestined, with tender love. The pure thought of Aristotle, co-eternal with a universe which was none of its making, a universe of which it knew nothing, and which strove vainly towards it without hope of any help,[1] was now replaced by the Heavenly Father Whose creative care extends to the least blade of grass that grows in the fields. The iron law of justice, which, in Plato's world, automatically gathers up the good with the good and the wicked with the wicked through the indefinite cycles of their successive existences, now became a paternal solicitude producing creatures from nothing in order to manifest the divine glory and associate them too with itself. If we grant that the best historical witnesses to Christian thought are the first Christians themselves it is sufficient to listen to them to be convinced of it ; but the list is too long to make it possible to listen to them all. The really important thing to grasp is how clearly they saw that the idea of creation is the ultimate foundation for the Christian idea of Providence. Athenagoras asserts it with the utmost force when he says that it is contradictory to believe in creation and not to believe in providence. Because God made all things in heaven and earth it follows that all, whether great or small, are under His government, and, so to speak, saturated with it through and through. Having been created individually, things are the object of an individual providence supplying the natural needs of each,

and leading each to its particular end.[2] The same doctrine
is found in Irenaeus. Things subsist only by the will of the
Creator Who made them ; they therefore depend upon
Him for their government no less than for their being :
unum esse qui creaturam fecerit et regat.[3] As for Minucius Felix,
the whole of his apologetic may be said to consist in showing
how necessarily bound up with monotheism is the Christian
idea of providence. But why insist on these preliminary
doctrinal sketches, when a complete synthesis is offered us
by St. Augustine ?

God, because He is Being, created all things, but He
created them by His Word, and it is also by His Word that
He conserves them in being : *portansque omnia verbo virtutis
suae.*[4] If it is true to say that in the beginning God created
the heavens and the earth, that is because in the beginning
was the Word, and by Him all things were made, and without
Him was made nothing that was made. Before the world
was made, and from all eternity, God expressed Himself
in His Word ; He uttered it to Himself, and in so uttering
it He expressed at once the totality of His being and that
of all its possible participations. Subsisting eternally in the
Word, the expressions of the possible participations of God
are, like God, uncreated, immutable, and necessary with
the necessity of His being. These are the Ideas. Thus the
Platonic ideas which subsisted in themselves, as an intelligible world independent of the Demiurge, are now
gathered up in God, generated from all eternity by the
fecundity of His being, born of His intimate life, are themselves life, and creative in their turn of everything else.
The divine ideas are thus the initial forms in which things
have their principle : *formae principales ;* the laws to which
things are subject : *rationes rerum stabiles atque incommut-
abiles ;* the rules that preside at creation : *creandi rationes ;*
the causes, finally, of things to be created : *causae rei*

creandae.[5] If this is so the world is very far from being the effect of a blind fate, it is the work of a supreme wisdom, which knows all that it makes, and can make it only because it eternally knows it.

Creation, thus understood, implies not only the thorough-going optimism described above—for all that has any degree whatever of being is a likeness to God—but carries with it also the affirmation of providence—since to govern things is to create them and to create them is to govern. It is not in the least necessary to introduce any new princi-ples here in order to explain the order of the universe, it is sufficient to grasp the radical contingence and mutability of beings. What is contingent and subject to becoming cannot possibly give itself what it does not possess ; neither its form, nor the place in the universal order assigned it by that form. It must therefore be admitted that every con-tingent thing receives its form : *omnis enim res mutabilis, etiam formabilis sic necesse est.* The contingent thing receives this form from God, Who is the immutable and eternal Form of Whom we read in Scripture : *mutabis ea et muta-buntur ; tu autem idem ipse es, et anni tui non deficient* (Ps. ci. 27–28). Now, to understand this is to understand that the world is subject to the providential government of God. Since, in short, all that exists exists only in virtue of its form, and since, were this withdrawn, it would wholly cease to be, to say that God is the immutable form by which all the contingent subsists and develops according to the rhythm and law of its form, is to say that God is its providence. Things would not exist if providence did not exist.[6] Resuming in striking formulæ the whole chain of the principle and its consequences, Augustine asks himself why things are subject to evil ? Because they are changeable. But why are they changeable ? Because they are not Being. And why are they not Being ? Because they are inferior to

that which made them. And who made them? He Who
Is. And who is this *He Who Is?* God, the immutable Trinity,
Who made them by His sovereign wisdom and conserves
them in being by His supreme goodness. Why, then, did
He make them? That they might be; for merely to be,
no matter how insignificant the being, is a good, since the
supreme Being is the supreme good. Out of what did God
make them? Out of nothing, for since all that is is good in
the measure in which it is, since the very least of forms,
the slightest measure of beauty is good, it must needs be
that from God alone all being, beauty, goodness and order
come: *omne autem bonum, aut Deus aut ex Deo est.*[7] Thus it
is altogether one and the same thing for God to create, to
form, and to govern. In a universe which holds all its being
from God, all is foreseen, willed, ordered—nothing is left
to chance.[8]

Thus the personal character of the Christian idea of
providence passed from the Bible and the Gospel into the
doctrine of St. Augustine; and thence, naturally enough,
the Middle Ages took it up and supplied the technical and
systematic interpretation. St. Bonaventure and St. Thomas
Aquinas will no longer be concerned merely to find a law
which assures the good of the whole by putting each part
into its proper place; nor even merely to show that each
part falls into its place and accepts it as a good, because the
good of the whole demands it: what they now set out to
define is a providence which wills the good of the parts as
parts, and governs the whole universe in such a way that,
taken simply as a part, the lot of each part shall be good. To
reach this conclusion they had to return to the source of the
difficulty and explore the depths of the Augustinian doctrine
of ideas. Let us see, then, how the greatest of them did it.

All are at one with St. Augustine in affirming the existence
of the ideas, and in putting the doctrine of ideas at the

very heart of philosophy ; at the same time they do not all take the ideas in quite the same way, and their differences of detail are of considerable interest to the present question. According to St. Thomas Aquinas, the ideas are in God as the forms to the likeness of which all things were made. They are in God ; they do not exist outside God as Plato thought.[9] Since all that is in God is God, the ideas, in this sense, are identical with God : *idea in Deo nihil est aliud quam Dei essentia.* However, they are the essence of God in so far as this is known under a certain aspect. For God is self-existent ; He had not to be made, and therefore, as Malebranche will say later on, He has no archetype. It cannot then be said that inasmuch as He knows Himself in Himself and with respect to Himself God knows Himself as something to be made. His essence is the principle of the production of all things save Himself, and since the idea is the model of the thing to be made, God does not know Himself by way of idea. The idea appears when God knows His essence as the principle of creatures which would be His possible participations,[10] and, in this sense, although the essence of God is one and known by Him as such, there are as many ideas in Him as there are creatures. To recall a phrase of St. Augustine's, *singula propriis rationibus a Deo creata sunt.* For God knows His own essence perfectly ; and therefore He knows it in all the ways in which it is knowable. Now this essence is knowable not only as it is in itself, but also as it is capable of participation in any way whatsoever. But each creature is a certain mode of participation in, and likeness to, the divine essence. Thus, then, inasmuch as God knows His own essence as *imitable* by a creature, He knows it as the model and idea of this creature. This multiplicity of ideas within the divine unity constitutes precisely the divine art, one as God Himself is one, not caused by the creatures, but their cause.[11]

Clearly, in such a doctrine the conception of idea would be meaningless except as referred to a possible creation. So true is this that there are no ideas in God of anything that is not susceptible of existence on its own account. Thus, since genera do not exist apart from the species, they have no other idea in God save that of the species. Similarly, those accidents which are inseparable from their substances have no other idea than that of these substances. Lastly, matter, since it never exists without its form, is, in God's knowledge, included in the idea of the concrete substance.[12] God knows every one of these things, but He knows them as they are, and this precisely because they are as He knows them. That is why the Christian doctrine of ideas differs far more profoundly from that of Plato than is commonly believed. Not only must we remember that the ideas, which in Plato's doctrine subsisted independently of the Demiurge are now gathered up in his thought, but we must realize above all that it is precisely because the Demiurge has given place to the Creator that His thought has now become the *locus* of the ideas. The Platonic ideas would remain exactly what they are, whether there was a real world or whether there was not ; and this would still be the case even were no universe outside them possible at all, since, as supreme reality, these pure intelligible essences suffice to themselves, look to nothing but themselves, and are themselves their own proper end. In the thought of St. Augustine, St. Thomas Aquinas, St. Bonaventure and Duns Scotus it is quite otherwise. That which, in their philosophies, has no respect to anything but itself is the divine essence itself; as for the idea, it appears only with the possibility of a creation and as an expression of the relation of possible creatures to the creative essence. It is this which shows that all that is, no matter by what title it is, has its idea in the Being from Whom it holds existence, that in God there are ideas of individuals themselves, and, we might say, above all of individuals, because it is individuals that are truly real and it is only in them that accidents, species and genera subsist.[13] The Christian doctrine of a providence that goes to singulars rests wholly on this metaphysical basis.

These principles may be said to be common to all the classical philosophies of the Middle Ages ; but although the direct connection between the doctrine of ideas and the concept of creation is everywhere maintained, it is expressed in different formulæ in different systems. In St. Thomas, the idea is essentially God's knowledge of His own essence as capable of being participated ; this knowledge is an emanation of the divine essence which includes the relation of possible beings to God.[14] In the doctrine of St. Bonaventure the idea is conceived rather as an *expression* of the divine truth ; it is saturated with all that fecundity which the theology of the Word supposes in the act whereby God eternally utters Himself. The ideas appear, then, as included in the act by which God, in expressing Himself in the Word, expresses the totality of possibles. True, we always have to do here with a similitude identical with God Himself, and consequently with an expression *expressing* rather than expressed.[15] However, the Bonaventurian expressionism adds to the Thomist doctrine of ideas a kind of internal generation of the notion of realizable beings. Whereas, with St. Thomas, it is sufficient that God knows Himself as capable of being participated and at once we have the ideas of all things, God now has to put forth within Himself, by His eternal Word, the notions of the possible participations of His essence. This, no doubt, is a mere nuance, but this particular care to emphasize the act which eternally brings to birth the ideas in God only brings out more clearly their fundamental character : they are the expression of a possible creation. Lastly, with Duns Scotus, the

connection between the conception of idea and the conception of creation becomes still more evident, or, if one
may say so, more palpable. Instead of reducing ideas, like
St. Thomas, to God's knowledge of His own essence, and,
like St. Bonaventure, to the expression of this essence
considered in its possible participations, Duns Scotus now
takes them as the creatures themselves as creatable by God,
and existing in Him in virtue of their concepts as possibles.[16]
In this doctrine, as in St. Bonaventure and St. Thomas, the
idea has its source in the depths of the divine essence, but
now it bears no longer on this essence, were it only as capable
of participation, it bears directly on the eventual participations. It is by His essence, certainly, that God knows
possible creatures, but His ideas of these creatures are not
views of His essence nor even of its imitability, but of the
imitations. Thus in Scotism the essence of God, taken in
itself, is wholly enclosed in its own splendour, unclouded by
the shadows which the multiplicity of its finite imitations
might cast upon it, even were these considered as simply
realizables ; God conceives the ideas because He thinks
of creatures, although it is only with respect to Himself
that He thinks of them.

The fundamental unity of Christian thought is preserved
throughout all these divergencies ; its expression varies,
but only that it may be the better formulated ; what it
always sets out to explain is why the Heavenly Father, in
virtue of the very fact that He has drawn all things out of
nothing, cannot but be conceived as a providence : *Tu
autem Pater, omnia providentia gubernas* (*Sap.*, XIV, 3). We
have, in fact, placed the ideas, that is the divine knowledge,
at the root of His creative action ; so that God's knowledge
must of necessity extend as far as His causality extends.
Now His creative virtue may properly be so called precisely
because it is not limited to the transmission of forms, but gives

being to matter itself. Whether, then, we admit with Duns Scotus that the principle of individuation lies in the form, or with St. Bonaventure in the union of the matter and the form, or, with St. Thomas, in matter alone, it must equally be granted in all three cases, that God knows individuals, and that He knows them in their very individuality. Plato's gods may throw the task of regulating the lot of individuals on some general law; Aristotle's unmoved movers may be wholly uninterested in what goes on in the universe; nothing could be more natural, for neither one nor the other has created matter, and consequently need not know it. Now if they have no knowledge of matter they must inevitably be without knowledge of the beings which it individualizes. But in a universe in which all being is created, the material and the individual must of necessity fall within the grasp of the divine intelligence.

If, then, the case stands thus, providence can by no means stop short at the universal; rather we should say, as in the case of the divine ideas, that it is essentially on the particulars[17] that it bears. But the particular is not to be divorced from the order into which it enters, the order of the work is part of the work; and therefore He Who made the world must have known, foreseen and willed what the world would be down to its least details. Nothing is more remarkable than the perfect continuity of the tradition throughout the whole Judeo-Christian doctrine of creation. The God of the mediæval philosophers remains identical with the God of the Bible, that is to say Being, the Creator, the Lord, and finally, in consequence, the Free Orderer of all things. St. Thomas has synthetized the *ensemble* of these views in a page so perfect, that we cannot do better than let him present them in all the rigour of their order : " We have shown that there is a First Being, possessing the full perfection of all being, Whom we call God, and who also, of the abundance

of His perfection, bestows being on all that exists, so that
He must be recognized to be not only the first of beings,
but also the first principle of beings. Now this being bestows
being on others not by any necessity of His nature, but
according to the decree of His will, as we have shown
above. Hence it follows that He is the Lord of all things
He has made, as we too are masters of those things that are
subject to our will. And this dominion which He exercises
over all that He has made is absolute, for since He has
produced them without the help of an extrinsic agent, and
even without matter as the basis of His work, He is the
universal efficient cause of the totality of being. Now everything that is produced through the will of an agent is directed
to an end by that agent ; because the good and the end are
the proper object of the will, wherefore whatever proceeds
from a will must needs be directed to an end. And each
thing attains its end by its action, but this action needs to
be directed by Him Who endowed things with the principle
whereby they act. Consequently God, Who in Himself is
perfect in every way, and by His power endows all things
with being, must needs be the Ruler of all, Himself ruled
by none : nor is anything to be excepted from His ruling
as neither is there anything that does not owe to Him its
being. Therefore as He is perfect in being and causing, so
also is He perfect in ruling."[18] The whole Augustinian
metaphysic of creation is resumed in these lines, but now it
has achieved a perfect consciousness of itself, comes face
to face with itself in its own clear limpidity.

To arrive at an exact conception of the nature of Christian providence all we need now do is to connect the conclusions just reached with those of the preceding chapter.
God created and ordered all things in view of an end. But
what is this end ? As we have already seen, it is God Himself :
Omnia propter semetipsum operatus est Dominus. To say that He

rules the world by His providence is therefore simply to say that He orders all things in view of Himself, by His knowledge and His love.[19] Here again the principle must be applied to the totality of being and also to the singular itself. For if God does not direct each individual towards Himself as towards the universal end of creation, we must suppose either that He does not know this individual, or that He cannot do it, or that He does not will it. Now it cannot be said that God does not know singulars, since He has ideas of them ; nor can it be said that God cannot order them to Himself, since His power is infinite, like His being ; nor again can it be said that He does not will it, since His will is a will to the whole totality of good.[20] Therefore all beings, whatsoever they are, are ordered towards God by His providence, for just as He is their principle, so is He also their end. Thus does the Christian theme of the glory of God end by investing the idea of providence with all the fullness of its meaning. If God orders all things towards Himself as towards their end, it is not that He expects the least addition to His own perfection, but that He may invest them on the contrary with a share in that perfection, communicate it to them in the measure of their capacity.[21] Now human beings are eminently capable of a share in this perfection, and therefore God surrounds them with a very special care.

When we say that nothing is exempted from God's providence we must never forget that God does not substitute Himself for things so as to act in their place. As we have seen, His concourse is the foundation of their being and their causality. In the case of natures without knowledge, consequently without freewill, since all individuals of the species act of necessity and infallibly according to the nature of the species, it is sufficient to establish the law of the species, to ensure that the individuals that compose it shall attain their end. In this sense, divine providence may watch over each individual sparrow without adverting to what distinguishes it individually from other sparrows. Created, willed and directed by the will which endows it with a nature which it cannot but obey, the animal is set in action towards its own good, but it is certainly set in action. It is otherwise with man, who excels all other creatures in this world, as much in perfection of nature as in dignity of end. In nature first ; for reasonable beings are masters of their own acts and free to direct themselves in their own ways ; in end next, for while other creatures have only to realize a certain divine likeness by being what they are, the end of the reasonable creature is to attain the last end of universal nature by acts of knowledge and love. The ruler of an orderly city governs each according to his condition ; how, then, can we believe that God fails to govern each created being according to its condition ? [22] The only question, then, is in what particular way is the providence of God applied to human beings ?

It is clear to start with, that in virtue of his very rationality man is able to use other things as his instruments. They may destroy him eventually by sheer brute force, but they can never use him ; on the contrary, he uses them. Things, therefore, are ordained to man as to their end, not man to things, and that amounts to saying that the rest of the universe is directed towards its end by man and through man. Reasonable beings are there, in a sense, for their own sake, the rest are there only for the sake of reasonable beings. The case is like that of an army, of which the whole object is the achievement of victory ; those who are to achieve it are the soldiers who fight, and as for the auxiliary services, they exist only for the soldiers and participate in the victory only through the soldiers. It is the same with the universe and man ; for the end of the universe is beatitude, and since only reasonable beings can enjoy it, the rest are called to participate it in them and for their sake.[23] Providence, therefore, has specially chosen the human species and leads it towards its end in an altogether special manner, since God is the end of the universe and it is through humanity that it is to attain Him. But then we must go further still and consider how providence watches over each individual of the human species.

The world is full of various beings, but among them man alone is immortal. There needed to be different and unequal species in order that all possible degrees of being should be represented and the law of order satisfied, but in all species other than the human the individuals that represent it are destined to perish. They are born, live and die once and for all. We may say, therefore, that providence does not will these individuals for themselves, but for the species they perpetuate. It is altogether otherwise in the case of man. Here the very individuals are immortal and inde-structible, and so they do not exist simply for the sake of the species, but also for themselves, and God therefore wills and governs each of them for himself.[24] This very well appears, moreover, from the way in which men act when they act as individuals. Precisely because they are free their actions are unforeseeable, and each acts differently. Now nothing escapes the attention of divine providence, not even the very least differences and variations of individuals : *divina providentia ad omnia singularia se extendit, etiam minima.* It is therefore insufficient to say that God watches over the human species in general, or even over each man in particular—He watches over each particular free action of each particular man.

And thus we are brought to the consideration of something that confers one of his highest dignities on man, but also confronts him with the most formidable of all problems, since it is that of his own destiny. From the very fact that they are rational, human beings are called to the noblest of ends, but then they have the responsibility of attaining it. They too, in a manner analogous with that of God, are capable of conceiving an end, which is the good, and of ordering the means necessary to achieve it. As God created the world, so man builds up his life ; nor can he choose the materials, that is to say his acts, nor fittingly dispose them to their end, that is beatitude, without the use of wisdom and prudence. Human foresight is to the providence of God what human causality is to divine creation. God, therefore, not only controls man by His providence, but also associates him with His providence ; while all the rest is simply ruled by providence, man is ruled by it and rules himself ; and not only himself but also all the rest. To say everything in one word, each human being is a *person ;* his acts are *personal* acts, because they arise from the free decision of a reasonable being, and depend only on his own initiative. It is therefore precisely as such that divine providence has to bear upon them. God, Who directs all according to His will, has bestowed on each of us an unique privilege, a signal honour, that of association with his own divine government.[25] Here, more than anywhere else, it is a great thing to cooperate with God, or as St. Paul puts it, to become the coadjutor of God. But what, then, is this idea of " person " which invests us with so high a dignity ? It seems that by introducing this idea into philosophy Christian thought did a great deal more than merely transform the Greek conception of divine providence. *Quid est homo,* asks the Psalmist, *quod memor es ejus ?* (Ps. viii. 5). If it is because he is a person that each man, instead of simply submitting to the law of the world, collaborates in making that law reign in the world, the Christian con-ception of " person " must be something very different from the Aristotelian and Platonic conception of " man." Thus, in the necessary sequence of ideas, the metaphysic of being brings us to the problem of Christian anthropology. Only there, seeing exactly what man is, shall we see how God's providence governs him ; and, knowing how it governs him, we shall learn in what way he ought to govern himself.

CHAPTER IX

CHRISTIAN ANTHROPOLOGY

TAKEN in himself man is no more than one of those beings that go to make up the universe as above described ; an analogue of God, endowed with activity and causal efficacy in the measure of his being, and providentially led towards his proper end. His dependence with respect to God, much more intimate and deep rooted than that of man with respect to the Platonic idea, or than that of the movable thing with respect to Aristotle's First Mover, carries with it differences of metaphysical structure which differentiate the Greek and Christian man even more profoundly still. Here, and perhaps more than elsewhere, the differences are hidden under an identity of terminology, and we shall have to look close to see them.

One of the surprises in store for the historian of Christian thought lies in its insistence on the value, dignity and perpetuity of the human body. The Christian conception of man is almost universally taken to be a more or less thoroughgoing spiritualism. What does it profit a man to gain the whole world and lose his own soul ? To cultivate the soul, to purify, and thus to liberate it, and finally to save it—there, it would seem, is the whole end and effort of Christianity. Add to this that the Christian God is spirit, that man therefore can hold communion with God only in the spirit, that God, in short, would be adored in spirit and in truth—and how, after all that, should we not expect to see the Christian philosopher bringing all his forces to bear on the spiritual part of man, that is to say on the soul, and making little account of the perishable part, the blind and mindless body that knows nothing of God ? But to the no small scandal of a goodly number of historians and philosophers the contrary turns out to be the fact. St. Bonaventure, St. Thomas Aquinas, Duns Scotus, I will even say St. Francis of Assisi himself—one and all were men who looked benignly on matter, respected their bodies, extolled its dignity, and would never have wished a separate destiny for body and soul. How shall we explain this fact ? What does it teach us concerning the true Christian conception of human nature ?

The problem may be formulated in properly historical terms. At first sight it might well seem that Platonism would be the philosophy, and very especially the anthropology, most naturally adapted for Christian purposes. The Fathers of the Church, needing a doctrine of the spirituality of the soul, found it in the *Phaedo*, found also along with it several demonstrations of the soul's immortality and the conception of a future life, complete with rewards and punishments, heaven and hell. Certainly without the *Phaedo* the *De Immortalitate Animae* of St. Augustine would never have been written. How came it then that, having followed the Platonic tradition for so long, the Christian thinkers gradually yielded to the growing influence of Aristotle, and, after numberless hesitations no doubt, finally defined the soul as the form of the body ? It happened because here, and here more than anywhere else, an inner ferment was working at the heart of Christian philosophy, more and more it was submitting itself to the regulative influence of revelation.

We have more or less forgotten the original sense of the word " Gospel." It means, of course, the " Good News." Just as the Bible was the Book *par excellence*, so the Gospel brought man the Good News *par excellence*. Jesus Christ announced that the Messiah was come, that salvation was at hand for Israel, and that the just were called to reign with God. St. Paul soon made it clear to all that this salvation was no exclusive privilege reserved for Jews, that it was thrown open to all humanity, rich and poor, master and slave, learned and ignorant alike. What we must especially note is that the salvation announced by the Gospel was not simply the salvation of souls but the salvation of men, that is to say of each of these individual beings with flesh and members, and the whole structure of corporeal organs without which none would feel himself more than a shadow of himself, would hardly even be able to conceive himself at all. When Jesus Christ announced to the Jews that they would reign with Him, He certainly spoke of themselves, not simply of their souls ; and it is hardly necessary to mention that in the eyes of St. Paul the resurrection of Christ was the promise and proof of their own resurrection to come : " Now if Christ be preached that He rose again from the dead, how do some among you say, that there is no resurrection of the dead ? But if there be no resurrection of the dead, then Christ is not risen again. And if Christ be not risen again, then is our preaching vain, and your faith is also vain. . . . If in this life only we have hope in Christ we are of all men most miserable." [1] But since the Christian's hope is not for this life only, he is, on the contrary, of all men most happy. " The dead shall rise again incorruptible and we shall be changed : for this corruptible must put on incorruption ; and this mortal must put on immortality." [2]

All this is familiar ; but it seems that we often overlook its influence on the development of mediæval philosophy. The faith of St. Paul and the first Christians in the individual salvation of the concrete man carried with it two consequences ; and first, the permanence and eminent worth of the individual as such. Nowhere is this feeling better expressed than in the moving *Mystery of Jesus*, wherein Pascal speaks from the heart of the Christian confidence in the destiny of a being redeemed by the death of God : " In my agony I thought of thee : for thee I shed these drops of blood." And then Pascal goes on to draw the second consequence : " It is I Who heal the body, and bestow upon it immortality." Christianity affirms the worth and permanence not of the soul only, but of the concrete being called man made up of body and soul, and it does so because Christ came to save not the soul only but the man. What Pascal said in the seventeenth century had already been said by Christian writers at the end of the second or the beginning of the third century ; and they brought out with no less force the necessary connection between faith in the resurrection of the body, and the philosophical thesis of the substantial unity of the human composite : " For God calls even the flesh to the resurrection and promises it eternal life. To announce the good news of salvation to man was in effect to announce it to the flesh. For what is man if not a reasonable being composed of soul and body ? Shall we say that the soul in itself is the man ? No, it is the soul of the man. And the body alone—is that the man ? By no means ; we should rather say that it is the body of the man. Since, then, neither soul alone nor body alone are man, but the thing called man arises out of their union, when God called man to the resurrection and the life, He called no mere part of man but the whole man, body and soul together in one." [3]

The fundamental importance of this connection between soul and body was so deeply felt by the first Christian thinkers, that what we regard to-day as one of the essential theses of Christian philosophy seemed to them of comparatively little significance. Before everything else we are concerned in these days to establish the immortality of the soul, the pledge of our future beatitude. It would probably surprise a good many modern Christians to learn that in certain of the earliest Fathers the belief in the immortality of the soul is vague almost to non-existence. This, nevertheless, is a fact,[4] and a fact to be noted, because it casts so strong a light on the point on which Christian anthropology turns and on the course of its historical development. A Christianity without the immortality of the soul is not, in the long run, absolutely inconceivable, and the proof of it is that it has been conceived. What really would be absolutely inconceivable would be a Christianity without the resurrection of the Man. The man dies, his body dies, but nothing would be irretrievably lost, the Good News would not be rendered vain, if the soul died also, provided always that we were assured of a resuscitated body and soul, so that the entire man might rejoice in eternal beatitude. There is no occasion therefore for surprise if certain Fathers admitted the death of soul and body pending the resurrection and the judgment. That, however, was only a passing hesitation in the history of Christian anthropology. It was very soon understood, chiefly owing to the influence of Platonism, that compelling philosophic reasons exist for the immortality of the soul. And from that moment the question assumed a new aspect, for now there was needed an idea of man which would leave the immortality of the soul conceivable and safeguard at the same time the future destiny of the body. The Greek philosophic tradition offered a choice between two, and only two, possible solutions of the problem, namely that of Plato and that of Aristotle ; the Christian thinkers tried first one and then the other, and it was only after twelve centuries of hesitation that the question was settled, when, passing quite beyond both Plato and Aristotle, mediæval philosophy revealed all its creative originality in the system of St. Thomas Aquinas.

At first sight no philosophy could seem to hold out better promise for the future of Christian anthropology than that of Plato and his disciple Plotinus. According to Plato the soul is essentially the source of movement, or even, as we read in *Phaedrus* (246 A), automotive. One step only is needed and we shall conclude that this " movement moving itself " which we call soul, is naturally endowed with life (*Laws*, 894c–895b) and therefore, by that very fact, immortal. It is this fact that accounts for the Platonic tendency to emphasize the radical independence of the soul which gives life, with respect to the body which receives it from the soul. In the human composite the soul represents the permanent element, changeless and divine, while the body is transitory, changeable, perishable ; as a man is independent of his clothing, survives what he casts off, and casts off a good deal during a lifetime without any special inconvenience, so the soul progressively emancipates itself from the body by means of philosophy, dies voluntarily to matter, until the death of the body delivers it altogether from matter and leaves it free for the contemplation of the Ideas. There is no doctrine in which the independence of the soul with respect to the body is so strongly marked as it is in Platonism, and that explains why the Fathers, as soon as they awoke to the importance of the immortality of the soul, so readily turned to Plato as a natural ally.

It was all very well as far as concerns the soul ; but then what becomes of the man in Platonism ? Since the soul is all that counts, the man becomes what the philosophers of the Academy, faithful interpreters in this respect of the thought of their master, will define as " a soul using its body." St. Augustine himself, although on his guard against certain consequences of the basic principles of Platonism, saw that he could hardly refuse to accept that. The man, he says, such at least as he appears to himself, is

" a reasonable soul, using a terrestrial and mortal body," [5] or again, as he writes elsewhere, he is " a reasonable soul possessing a body." But we see very well that the Christian philosopher feels the difficulty hidden under these formulæ, for he soon adds that the reasonable soul and the body it owns do not make up two distinct persons, but a single man.[6] Undoubtedly he intends to safeguard the unity of man, but can he do it with these principles ?

It is easy to see that he cannot from the mere fact that the Augustinian definition of the soul is identical with the Augustinian definition of man. St. Augustine tells us that man is neither a soul taken separately nor a body taken separately, but a soul which uses a body. But when asked to define the soul in itself he answers that it is " a rational substance apt to rule the body." [7] In the end, then, the man is only his soul or, if you prefer it, it is the soul itself that is the man. No doubt St. Augustine would still have something to say to this criticism ; he might perhaps reply that a soul is a soul only because it has a body to use, and that a body is a body only because it is at the service of a soul ; in which case, in fact, the definition of the soul alone is equivalent to that of the entire man. It is no less true that this piece of Platonism, taken up undigested into Christian thought, remains somewhat of an alien body, and that its presence gives rise to difficulties unknown to the authentic doctrine of Plato. It is quite natural that Plato's man should be the soul, for Plato was not concerned to safeguard either the unity or the permanence of the human composite. For him, the union of soul and body is the acci- dental result of a fall ; the soul is shut up in a body as in a prison or a tomb by a violence done to its nature, and that is why the whole effort of philosophy is directed to effect its deliverance from the body. A Christian on the other hand must admit, in the first place, that the union of soul and body is natural, for it is willed by that God Who Himself declared that all His works are good : *et vidit quod erant valde bona ;* and no natural state can possibly be the result of a fall. And since, furthermore, it is the integral man that is to be saved, Christian philosophy cannot propose the salvation of the soul of the body but rather the salvation of the body by the soul ; and this demands that man shall be a substantial composite, and something very different from the accidental juxtaposition imagined by Plato. That, moreover, explains why St. Augustine, firm as he was in asserting the unity of man, found himself quite unable to justify it. Are man and his body like two horses harnessed to the same chariot ? Are they united like the torso and trunk of a centaur ? Is the soul the man in the sense that a horseman supposes the horse ? So many questions to which he does not risk a reply,[8] and in the face of which he can do nothing but confess his embarrassment.

Thus, then, in spite of all its indubitable advantages, Platonism carried a latent but insurmountable difficulty into the very heart of Christian philosophy. We see at once why the Aristotelian definition of the soul, so little Christian in appearance, was soon taken into consideration by certain philosophers and theologians. The thing was not done without resistance and the opposition of the Platonists was naturally tenacious. A severe criticism of Aristotle's definition left behind by Nemesius and passed on to the Middle Ages under the authority of Gregory of Nyssa, had a deep and lasting influence on the history of the controversy. But by the beginning of the thirteenth century it is quite common to find this definition given along with a number of others, and all in some sense regarded as acceptable. The soul, according to Aristotle, is the act or form of an organized body having life potentially. Thus the relation of soul to body is a particular case of the general relationship between form and matter. There is a sense in which the form and the matter may be said to be separable, since no particular form is of necessity destined by its nature to inform any particular portion of matter, but in another sense we must consider them as inseparable, at least in concrete substances, because forms of this kind could not exist apart from any matter whatsoever. Since the human soul is precisely a form of this kind, the advantages and disadvantages of the definition become at once apparent.

The advantages are quite evident. If we adopt the Aristotelian position we shall no longer be troubled with any difficulty about the substantial unity of man, for body and soul are not two substances but the two inseparable elements of one and the same substance. It is easy to see that the Christian philosophers, concerned as they were to safeguard the permanence and unity of the human composite, would feel secure on this side of the problem. On the other hand it was difficult to concede the Aristotelian principle that the soul is the form of the body, without at the same time accepting all the necessary consequences involved in it. Now it happens, just as in the case of Plato, that some of these consequences are rather troublesome for a Christian philosopher. If we follow Plato in his proof of the substantiality of the soul we put the unity of man in jeopardy ; if we follow Aristotle in his proof of the unity of man, we risk the substantiality of the soul, and its immortality along with it.

From the moment, in fact, that concrete reality is defined as a union of form and matter it becomes impossible to consider either one of these two elements as a substance properly so called. As long as the union of soul and body endures the man endures, for the man is the substance ; but as soon as the union is dissolved it is not simply the man, it is just as much his body and his soul that cease to be,

A corpse is no longer a human body, it is something organic returning to the inorganic, and as to the soul of the animal, how should it survive the animal seeing that it forms a part of it and that this animal has ceased to exist ? It is true that Aristotle himself felt the difficulty of his own doctrine when he came to consider the case of man ; and the fact is so much the more remarkable since he was evidently in no debt to any theology in the Christian sense of the term. It is not because God has said it, but because reason demands it, that he admits man to be no mere animal like other animals. The human being is a rational animal, and his soul seems to be something more than the form of the body ; it includes a principle of operation which, as regards its exercise, is independent of the body, and is consequently superior to any mere substantial form. That principle is the intellect. And when Aristotle comes to define the relation of this intellect to the body his embarrassment becomes extreme. He asks himself whether it bears any analogy to the relation that exists between a pilot and the ship, and gives no answer [9] ; he adds that the nature of this intellect is not yet clear to him, but it might perhaps be considered as another kind of soul, the only one to be separate from the body as the immortal is from the mortal ; [10] in a brief and fugitive formula he says that this intellect comes to the soul " from without," [11] and that, no doubt, suggests the possibility of its ultimate return to its source and its survival. But what, then, becomes of the substantial unity of the human being ? How can that which is the form of the individual body, at the same time be separate from this individual body ? On this capital point Aristotle vouchsafes us no answer. Everything seems to point to the conclusion that in his thought, the man is indeed no more than the union of his soul with the body of which it is the form, and that this intellect he speaks of is another intellectual substance, in contact and communication with our soul, separate from our body by the very fact that it does not enter into the composition of our concrete individuality, immortal in consequence, but with an immortality which is its own, and is not ours.

Such, in fact, was the interpretation of Aristotle proposed by Averroes, who passed in mediæval eyes for the Commentator *par excellence*. It is not without sufficiently strong texts to fall back upon,[12] and it is not to be denied that it accords with the general orientation of the system : but no Christian thinker can possibly accept it, for what Christianity promises man is his own individual immortality and not the immortality of a separate substance that would not be his at all. The whole anti-Averroist controversy headed in the thirteenth century by St. Bonaventure, St. Albert the Great, and St. Thomas Aquinas sufficed to show up the complete incompatibility of the two doctrines ; historians are sufficiently well aware of the fact, and it need not be dwelt on here. What is important to note, on the other hand, is the effort made by the Christian philosophers to break out of the double impasse in which Plato and Aristotle had involved them ; and what was its outcome.

However embarrassing the problem might be, it was never felt to be desperate—or perhaps we ought rather to say that there was always a solution that might be accepted in default of a better, while the world awaited the further progress of philosophic reflection. St. Augustine was always there, upholding at once the immortality of the soul and the unity of man, and it was possible to hold on, like him, to both ends of the chain even if the point of junction remained out of sight. On the other hand, it happened that Avicenna, whose works were available by the beginning of the thirteenth century in a Latin translation, had had to resolve an analogous problem on his own account, and it was natural to look to him for the elements of a solution. Avicenna, like his master, Al-farabi, believed that the neo-platonic compilation known as the *Theology of Aristotle* was a genuine work of the Stagirite ; his whole philosophy therefore presupposed an effort to effect a synthesis of this apocryphal work with the doctrine contained in the authentic writings of Aristotle, and this would amount to an attempt at a synthesis of Aristotle and Plato. Now on the one hand the Christian thinkers were naturally glad to recognize in Avicenna those elements of Platonism which, thanks to St. Augustine, were already incorporated in their own tradition ; and, on the other hand, what they needed here was precisely a doctrine capable of saving both the Platonic immortality of the soul and the Aristotelian unity of the human composite. Avicenna, then, came at an opportune moment, and it is just on this account that he exerted so deep an influence on the Christian philosophy of the thirteenth and fourteenth centuries. An attempt was therefore made to rid his doctrine of those elements which were not assimilable with Christianity, to reduce his principles to those of St. Augustine, and to subordinate to these principles the Aristotelian elements which it was necessary to retain. What were the results of this work as far as our present point is concerned ?

As to the definition itself of the soul, it might seem at first sight that the whole work was done. For Avicenna there were two equally possible ways of considering the soul. Taken in itself, that is to say in its essence, it is a spiritual, simple, indivisible, and therefore indestructible substance ; and under this aspect, therefore, Plato's definition is entirely satisfactory. Considered in its relation to the body it animates we may say that the first and most fundamental of the functions it there fulfils is to be its form ; and in this respect it is Aristotle who is in the right ; the soul is certainly the form of an organized body having life potentially. A simple illustration taken from Avicenna himself will serve to make the meaning of his doctrine clear. I see, for instance, a passer-by, and I ask you who he is. You say, perhaps, " a workman." Suppose the answer is correct, suppose that the man is really a workman ; nevertheless, it is not a complete answer, nor even the most fundamental that could be given, for the passer-by is a man first and a workman afterwards ; a man by essence, a workman by function. The case of the soul is altogether similar. In itself it is a substance which fulfils the function of a form, and that moreover is why we need have no fear of its being affected by the destruction of the body it animates ; when the body dies, all that happens is that the soul ceases to exercise one of its functions.

At first sight nothing could seem to be more satisfying for Christians than such a doctrine. Since the soul is a substance it is immortal ; since the soul is a form, man is one. And in fact we shall find this solution of Avicenna's in the *Summa* attributed to Alexander of Hales, in the *Commentary* of St. Bonaventure and in several theologians of lesser importance. In order the better to safeguard the substantiality of the soul St. Bonaventure reinforces the doctrine of Avicenna with that of Gebirol and teaches that the soul is composed of a form and an incorporeal matter. Although he will not admit this hylemorphic composition of the soul, St. Albert the Great maintains that the true definition is that given by Avicenna ; none understood its composite character better than he, and the fact that it represents a desperate attempt to reconcile irreconcilables. "When we define the soul in itself," he writes, " we follow Plato ; when, on the contrary, we define it as the form of the body, we follow Aristotle." [13] Evidently Christian thought is here tending to a facile eclecticism, but the facility is merely apparent, as is always the case with eclecticisms. They last their time : until we realize the difference between reconciling the principles needed to justify our conclusions and merely reconciling the justifications which necessarily flow from these principles. Here, as elsewhere, it was the Thomist doctrine that intervened to disturb the complacency of easy-going solutions and to mark fresh progress.

At bottom the whole difficulty comes to this : man is a unity given as such and the philosopher is bound to take account of the fact. When I say that *I know*, I do not mean that my body knows by means of the soul, or that the soul knows by means of the body ; but that this concrete being " I," taken in its unity, performs an act of knowing. The same thing holds when I say that *I live*, or simply that *I am ;* " I " means neither the body nor the soul, but the man. That is why the eclecticism of St. Albert the Great is no true answer to the question. Inasmuch as the animating functions of the soul are not included in the definition of its very essence, the union of soul and body remains an accident, without any kind of metaphysical necessity. As substance, the soul remains exactly what it is, whether it informs a body or whether it does not. Doubtless we might try to save the situation by saying that it is " natural " to the soul to be the form of a body : and we can go further and credit it with a kind of natural inclination towards its body, but it will always remain true, nevertheless, that union with the body is not included in the essence of the soul as soul. Now, if the case stands thus, man is not a being *per se* but a being *per accidens ;* I can no longer say that I am, or that I live, or that I think, without thereby understanding that I am a soul that is, and lives, and thinks, and would be, live, and think just as well, or better, without the body to which it is united. It is not in the least sur-

prising then that Avicenna anticipated Descartes' position :
if the soul is a substance essentially separable from the
body then, whenever I say " I think," it is of my soul
alone that I speak ; and it is the existence of my soul alone
that I affirm when I conclude " therefore I am."

It was precisely for this reason that St. Thomas had to
go to the root of the matter and effect a complete reconstruction. It was a typically Christian philosophic effort,
and the whole history of ideas presents few more beautiful
examples, for never has it better appeared how a philosophy, in becoming more truly philosophic, becomes
thereby more truly Christian. But never was philosophic
effort so ill repaid. And the reason is the same. St. Thomas
adopts Aristotle's terminology and avails himself of Aristotelian principles : for him, too, as we all know, the soul is
the form of an organized body having life potentially ; but
refusing on the other hand to accept the principles of Platonism, even eliminating from mediæval philosophy those
elements of Platonism introduced on this point by the
tradition of St. Augustine, he proposes nevertheless to
maintain the immortality of individual souls—which only
Plato's philosophy is able to justify. Are we not forced to
say, as some, indeed, have not hesitated to say, that it is
precisely Thomism that is the real eclecticism, and even
the worst of all since it is incoherent into the bargain ?

The only possible reply is to be found in a critical
examination of the formulæ St. Thomas uses, but, above
all, of the significance he gives them. In order to make the
drift of the discussion clear in advance we may say at once
that the Thomist soul is neither a substance playing the
part of a form, nor yet a form which could not possibly be
a substance, but a form which possesses and confers substantiality. Nothing could be simpler ; and yet we should
look in vain for any philosopher before St. Thomas who

conceived the idea. There are few enough even to-day who are capable of setting out its meaning correctly.

To understand it we must first of all remark that although this solution was never discovered by Aristotle it stands in no sort of contradiction with his principles. Quite the contrary ; when it is objected that the soul could not be at once a substance and the form of a substance, it is forgotten that even in the authentic Peripateticism there are substances which are pure forms. It might even be said, in a sense, that the more a form is purely form, so much the more is it also substantial. It is because the divine thought is supremely immaterial that it is supremely formal, and it is because it is supremely formal that it is pure act. The case is the same, on a lower plane, with Aristotle's separate Intelligences ; these are pure immaterial forms, wholly free from matter, and nevertheless they subsist ; and that amounts to saying that they are substances. As soon as we recollect these well-known facts the problem at once appears open to another solution than the one so injuriously attributed to St. Thomas. If there are positive reasons for considering the soul as a substance, there is nothing to prevent us admitting at the same time that it is a form, *since, on the contrary, it is precisely its formality that lies at the basis of its substantiality.* The question to be settled is therefore quite a different one : we need not ask whether the soul is a form, but why this form is the form of a body ; and the first question therefore that has to be resolved is that of its substantiality.

By a coincidence that is not in the least fortuitous, it is a text of St. Augustine's that St. Thomas invokes as the basis of his demonstration of the substantiality of the soul, and this text affirms precisely that the only thing that prevents people admitting this truth is the common incapacity for conceiving a substance that would not be

corporeal.[14] The fact is that in passing from the hands of
St. Augustine into those of St. Thomas the question has
undergone a transposition. For the young Augustine himself, and for those he criticizes in his maturity, the question
is not whether the soul is a substance, but rather whether
we can conceive a non-corporeal substance. For St. Thomas,
on the other hand, if the soul exists its incorporeal nature
admits of no doubt, but it is not altogether the same as
regards its substantiality, and it is on this precise point that
he brings all his forces to bear.

The principle that underlies his demonstration, and it
seems to have been insufficiently noticed, is this : that every
distinct operation supposes a distinct substance. It is only
in fact by their operations that substances are known, and,
conversely, operations are not to be explained except by
substances. We know what being is, and, notably, that it
is actuality as by definition ; it is precisely there that we
found the ultimate root of causality in concrete beings ;
when we said that every being is in so far as it is in act, and
operates only in so far as it is in act, we identified the principle of the activity of beings with their actuality. If, then,
there are acts of intellectual cognition, their cause could
lie in no abstract principle such as thought in general, but
of necessity in some concrete principle, real, and consequently subsisting in a determinate nature. In a word,
wherever there are acts of thought there are thinking substances. We may give them any name we please ; it is
of no importance to the present question whether we call
them minds (*mentes*) as in the Augustinian terminology, or
intellects (*intellectus*) as in the Thomist terminology—in
any event they are things that think, and all that remains
is to get a precise idea of their nature.

The operations of which intellects are the principle are
cognitive operations. By the intellect we are capable of

knowing the nature of all corporeal things. Now the power of knowing all things demands as a first condition that we ourselves shall be no one of those things in particular, for if the intellect had a determinate corporeal nature, it would be but a body among bodies, limited to its own proper mode of being and incapable of apprehending natures different from its own. In other words, a thinking substance that knows bodies could not itself be a body. From this it results that if there are so many beings incapable of knowledge, it is precisely because they are nothing other than bodies, and that if there are corporeal beings that think, the principle of their cognitive activity is not to be found in their corporeity. That is why the human intellect, on account of the very fact that it is an intellect, must be considered as an incorporeal substance, and this, too, in its being as well as in its operations.[15]

If then this is so, what is there to prevent us from going further and identifying the intellect with the man? Precisely the fact that men have other operations besides the intellectual. Even if we wanted to identify the man with his soul we should be faced with a difficulty of the same kind. Such an identification would be possible in the doctrine of Plato or in that of St. Augustine, because, for these philosophers, sensation is an operation proper to the soul and one in which the body plays no part. And that, moreover, is why we have seen them defining man as a soul using a body.[16] But since we refuse to disassociate the unity of man into two accidentally united halves, we are bound to admit that the substantial form we have described is only a part of man. Everything therefore goes to show that man is a being composed of a corporeal matter organized by a form, and of an intellectual substance which informs and organizes this matter. So much we must admit if we would merely remain faithful to the data of the problem :

it is the intellect itself, an incorporeal substance, that is the form of the human body.[17]

It is just here perhaps that the internal difficulties of Thomism begin to look most formidable. Grant that the intellect is an incorporeal substance ; how, at the same time, could it form part of another substance, and form with it nevertheless a whole that would be no mere accidental composite ? It is difficult to see, in short, how a doctrine which begins with Plato can possibly end with Aristotle, and it seems that Christian philosophy is in greater danger than ever of falling into an incoherent eclecticism. However, St. Thomas' position is very different from that of his predecessors, and a great deal more favourable. He never loses sight of the fact that just as it is the man, and no mere sensibility, that feels, so it is the man, and no mere intellect, that thinks. Like all facts, this does not need to be deduced, but simply apprehended. Beings of another kind exist and are consequently conceivable. The pure Intelligences are subsisting forms altogether free from corporeity ; animals and plants are non-subsistent corporeal forms ; men are subsisting corporeal forms. Why ? Precisely because these substances could not subsist save as forms of certain bodies. They are intellects, and, consequently, substances capable of apprehending the intelligible ; in this sense there is nothing lacking to their substantiality. But an intellect without a body is like a hand cut off from its body,[18] a mere part of a whole, impotent and inert when separated from the whole. Let us then simply suppose that there are certain spiritual substances too feeble for the direct apprehension of any other intelligibles save those enclosed in bodies ; intellects like ours, that is to say dazzled and blinded by the pure intelligible, but open, nevertheless, to the intelligibility involved in matter—and then it is at once evident that such substances would be unable to get into touch with the world of bodies save by the intermediation of a body. In order to apprehend sensible forms and thence elaborate intelligibles, they must themselves become the forms of sensible bodies ; descend, so to speak, into the material plane in order to communicate with matter. They would have to do this— and here is the essential point—precisely because they are the kind of substances they are.[19]

For a correct interpretation of the meaning of this Thomist reply, it will not be without advantage to remark that its apparent difficulty is due to an illusion of the imagination. To say that man is a concrete substance and complete in himself is in no way to contradict the thesis of the substantiality of the soul. The error of interpretation into which we are liable to fall on this point is due to imagining body and soul as two substances and attempting out of these two to construct a third which would be the man. Then, indeed, the Thomist man really would be a mosaic of fragments, some selected from Plato and some from Aristotle. In reality, however, the Christian man is another thing altogether, since, as we shall soon have occasion to see, although the man alone fully deserves the name of substance, he nevertheless owes all his substantiality to that of the soul. For the human soul is act, and is therefore a thing for itself and a substance ; the body, on the contrary, although without it the soul cannot develop the fullness of its actuality, has neither actuality nor subsistence, save those received from its form, that is to say from the soul. That, moreover, is why the corruption of the body cannot involve that of the soul, for if the principle which gives the body its actual being is withdrawn the body dissolves : but the dissolution of that which owes its being to the soul cannot affect the being of the soul itself. Thus the substance " man " is not a combination of two substances but a complex substance which owes its substantiality to one only of its two constitutive principles. Here, then, we see the added significance with which Christian thought invested the Aristotelian formulæ, even when it took them in their literal tenor. Souls have now become immortal substances, which cannot develop their activity without the co-operation of sensorial organs ; in order to obtain this co-operation they actualize a matter ; it is due wholly to them that this matter is a body, and yet they are not them-selves save in a body ; the man, therefore, is neither his body, since the body subsists only by the soul, nor his soul, since this would remain destitute without the body : he is the unity of a soul which substantializes his body and of the body in which this soul subsists. And this brings us to the threshold of a new problem, the discussion of which can hardly fail to shed some light on the preceding con-clusions ; that is to say, the problem of individuality and of personality.

CHAPTER X

CHRISTIAN PERSONALISM

THE modern mind is quite sufficiently familiarized with the ideas of individuality and personality. We might well ask ourselves whether their importance has not been somewhat exaggerated—no doubt as a kind of reaction against the evils of mass production. Whenever the collective comes to be regarded with a kind of religious awe, as if a suppression of the individual was all that was needed to attain the divine, the individual and personal, in their turn, begin to lay claim to a kind of sanctity and are even put forward as the sole possible bases of any genuine religion. Men seem to be incapable of facing an antinomy without worshipping the terms.

There is no absolute necessity for this. We might be permitted to dream of a philosophy which would recognize the collective where it exists without holding itself bound to sacrifice all the individuals to the collective, or the collective, for that matter, to the individuals. It would still be possible to put the emphasis on one of the two terms and we might form a somewhat different philosophical estimate of each ; but an area of accommodation would remain open and perhaps some genuine philosophical progress might be realized. That is just what took place in the interchange between Greek and mediæval speculation : because they never denied the reality of the individual the Greeks opened the way for the Christian recognition of the eminent worth of the person ; not only did they not prevent it or even simply retard it ; they did a great deal to forward its success.

It is not to be denied, without disregarding obvious historical facts, that the Cynics, the Stoics, and even, as V.
Brochard has eloquently reminded us, the very Epicureans
themselves, knew how to bring the conduct of the interior
life to a high pitch of perfection. The names of Epictetus
and Seneca should absolve us from all commentary ; the
Middle Age, as we know, was quite ready to regard them as
precursors of Christianity and, occasionally, even as saints.
It is not to be doubted therefore that a keen sense of this
philosophic truth existed long before its open recognition
and technical justification ; the only question that remains
is whether Christianity hastened the process by which it
came to maturity and, fully realizing the need, called forth
the speculative effort that was still required to justify it.
Now on this precise point a great deal remained to be done.
Neither the Stoics nor the Epicureans ever succeeded in
rising out of the plane of ethics to a metaphysic of personality, although nothing stood in the way of the attempt.
Neither Plato nor Aristotle, although they held all the neces-
ary metaphysical principles in their hands, ever had a
sufficiently high idea of the worth of the individual as such,
to dream of any such justification.

In a doctrine like Plato's it is not at all this Socrates,
however highly extolled he may be, that matters : it is
Man. If Socrates has any importance at all it is only because
he is an exceptionally happy, but at the same time quite
accidental, participation in the being of an Idea. The idea
of Man is eternal, immutable, necessary ; Socrates, like
all other individuals, is only a temporary and accidental
being ; he partakes of the unreality of his matter, in which
the permanence of the idea is reflected, and his merely
momentary being flows away on the stream of becoming.
Certain individuals, no doubt, are better than others, but
that is not in virtue of any unique character, bound up with

and altogether inseparable from their personality, it is simply because they participate more or less fully in a common reality, that is to say this ideal type of humanity which, being one and the same for all men, is alone truly real.

In the system of Aristotle the unreality and accidental character of the individual physical being as compared with the necessity of the pure acts and the eternity of species are no less evident. Aristotle's world is certainly a very different one from Plato's, since the Ideas, far from constituting the typical reality are now refused all proper subsistence. In Aristotle's philosophy the universal is far from nothing, but it never enjoys the privilege of subsistence, only particulars can properly be said to exist ; and it is therefore only just to say that the reality of the individual is much more strongly marked in his doctrine than it is in Plato's. Nevertheless, in both, the universal is the important thing. Although the only real substances he will recognize are the men, that is to say the specific form of humanity as individualized by matter, Aristotle considers the multiplicity of the individuals as a mere substitute for the unity of the species. In default of an Humanity which cannot exist apart, nature contents herself with the small change, that is to say with men. Each one is born, lives a brief span, and disappears for ever without leaving a trace behind ; but what does that matter if new men are born, live and die and are, in their turn, replaced by others ? Individuals pass away, but the species endures ; so that when all is said and done the individual that subsists and passes away is there only to assure the permanence of that which does not subsist, but also does not pass.[1]

How deeply this depreciation of the frail and fleeting individual offended Christian feeling for the permanent worth of human persons, may be easily seen in the earliest witnesses to the tradition. There is a wide field of choice, and so I will turn for preference to one of the most unjustly neglected, to Athenagoras, whose treatise on the *Resurrection of the Dead* is the first known attempt at a justification of the great Christian hope. Having shown that the resurrection of human bodies, scattered to the four winds by death, is a work which is neither impossible to nor unworthy of God, the apologist undertakes to prove that positive reasons exist for expecting the event.

The first is taken from the final cause of man's creation. God made us only in order that we might partake of a life of wisdom consisting in the contemplation of His perfection and of the beauty of His works. Since this contemplation cannot possibly be perfect here below, the very purpose of man's birth is a guarantee of his perpetuity, and his perpetuity in its turn is a guarantee of his resurrection, in default of which he could not subsist. That is the fundamental principle on which the whole doctrine rests, and it leads, as we shall see, to an important epistemological consequence, that is to say, the primacy of contemplation. For the moment we will be content to analyse its implications in so far as they bear on the nature of man.

If God had simply created souls, their end would be as just defined. In fact, however, He has created men, so that in reality we cannot properly speak of the end of the soul, but only of an end of the man. That the end of the man may be identical with that of his soul, the body must necessarily be called to participate it : " If men have been gifted with mind and reason, that they may know things apprehensible by reason, and not only their substance but also the goodness, wisdom and justice of Him who gave them substance, it must needs be that since those things on account of which this rationality has been granted remain the same, the power of judging, which is inseparable from it, must subsist. Now it could not subsist if the nature which has received it and in which it resides, did not subsist. *But that which has received mind and reason is the man, and not the soul by itself.* Therefore the man, a composite of body and soul, must always subsist, and this could not be if he does not rise again." [2]

Athenagoras' expressions, well weighed, show how deeply the Good News influenced philosophy. Created by God as a distinct individuality, conserved in being by an act of continuous creation, the man is henceforth the protagonist of a drama, which is none other than that of his own destiny. As it does not depend on us to exist, neither does it depend on us to cease to exist. The divine decree has condemned us to be ; made by creation, re-made by redemption—and at what a price !—we have now but one choice, between a misery and a beatitude, both equally eternal. Nothing could be more resistant than an individuality of this kind, foreseen, willed, elected by God, indestructible as the divine decree itself that gave it birth : but nothing also could be more alien either to the philosophy of Plato, or to that of Aristotle. Here too, as soon as it aspired to a full rational justification of its hope, Christian thought found itself constrained to originality.

Two different ways of assuring the subsistence of the individual suggested themselves to the thinkers of the Middle Ages. Neglecting the chronological order of set purpose, I will consider that which seems the easier first, and to show once more how Christian philosophy carries on the work of Greek philosophy even when it finds it insufficient, I propose to borrow from one of the best modern interpreters of Aristotle certain considerations which go to show that the problem of individuality was already a problem of pure philosophy several centuries before it became a Christian one.

According to Aristotle's principles, an individual is a concrete being, made up of a form analogous in all the individuals of the same species, and of a matter individualizing this form. If, for example, we consider the case of men, none of them could be considered as different from the others as man ; all are man in the same degree and in the same manner. In one word, precisely because the form is *specific* it is of the same nature in all the individuals of the same species. This is not the case with the matter that individualizes them. The same quantity of matter can only exist once over, it cannot repeat itself, for it is of the very nature of the parts of extension to exclude each other, to exist outside each other, to have *partes extra partes*, so that every form which enters into union with one part of matter becomes distinct from every form united to another portion of matter, in virtue of the very division of the matter to which it is united.

And here, as has been justly pointed out, the difficulties begin. There can be no question of regarding prime matter as the principle of individuation, for matter which is only matter, matter as matter, is pure potentiality, absolute indetermination ; so that even the specific form, uniting itself with something so wholly indeterminate, would then remain just as indistinct after the union with this matter as it was before. In order to overcome the difficulty we fall back on another position and say that the form individualizes itself by entering into union with a matter already qualified, for example with such and such a matter determined by extension. Well and good !—but then, in this case, the only reason why matter becomes the principle of individuation is because it is itself made individuating by a form. In other words, if, in the last analysis, all difference is a formal difference, it is not easy to see how matter can play the part of individuating principle. And let us add that the difficulty becomes graver as our Aristotelianism becomes the more rigorously consequent. If we admit the unity of form in the composite the matter must necessarily have received from the form itself that very quantity and inpenetrability by which this same form is supposed to be individualized. In such a case, and we know that this will be the case in Thomism, are we not forced to say that it is the form that individualizes itself, in virtue of the quantity with which it invests the matter ? A troublesome and ultimately useless *détour* which might, it seems, have been avoided by frankly admitting at the start that the principle of individuation must be placed in the form itself.

But there are serious objections to that in Aristotle's doctrine, and especially if we consider it in its spirit. His description of an humanity in which the individuals are distinguished from each other only by accidental differences is in full accord with his firm conviction that the individuals exist only for the sake of the species, and that in the long run the individuals, as individuals, do not count. But further : these individuals cannot be made to count as individuals without ruining the unity of the species itself. If an individual difference is introduced into the form of each individual, that individual at once becomes a species, and henceforth irreducible to any other species. Socrates will be as different from Callias as both Socrates and Callias actually are from an animal or a tree. In short, in order the better to safeguard the originality of the individual we should have destroyed the unity of the species, without taking account of the fact that in order to have men we must first of all have humanity.

So strong is the Christian desire to save the subsistence and originality of the individual that Duns Scotus did not hesitate to incur the risk. He took the view that each human form, precisely as form, is marked with an individual character that distinguishes it from all the others. In an isolated text Aristotle himself seemed to say something of the kind. " The causes and the elements of different individuals," he writes, " are different : your matter, form, and moving cause and mine." [3] But it seems clear that in so expressing himself Aristotle did not intend any modification of his own principles, for if my form is different from yours, it is probably because it is individualized by a different matter. It was enough, however, to encourage Duns Scotus to follow his own bent and to seek the ultimate principle of individuation in the form as form.

If the form of a man, he says, is not individual in its own right, then, since matter as such possesses none of the characters of individuality, there will be no human individuals at all. Now human individuals exist. Therefore the very form of the individual must of necessity be the principle of his individuation. Of course, there can be no question of uselessly multiplying the number of forms in a given individual, for this cannot be done without destroying his unity ; but there is no necessity to add a new form to each form in order to individuate it. No one doubts that from a certain standpoint, God is an individual ; and certainly His individuality cannot be attributed either to matter or to any supplementary individuating form. The Angels are individuals, and nevertheless they are immaterial, and simple in their form. Why, then, should it be otherwise in the case of man ? It is enough to admit that taken in itself, in its root reality, the soul is individual and the cause of individuality. Of itself, and in virtue of its own proper essence, it is not simply *a* soul but *this* soul, and it is this essential individuality which individualizes the matter of the body and, along with the body, the whole man.

The Scotist solution of the problem was not without its obvious advantages. It was economical, in the sense that it was easily grasped, and transformed the Greek man into the Christian man with a strict minimum of supplementary hypotheses. But however skilful the mediæval philosophers were at finding confirmation for their views in texts of Aristotle, it was difficult to find many that would justify them in invoking his authority for a doctrine of that sort. Duns Scotus, moreover, was quite alive to the difficulty, in his doctrine, of maintaining the unity of the human species, and in order to reconcile that of the species with that of the individual he had to modify the notion of unity itself. In this sense, Scotism finds itself eventually faced with as many complications as it avoids, and perhaps it was some presentiment of these latent difficulties that kept St. Thomas from entering on this path, however tempting it might have seemed at the outset.

At first sight it would seem that the Thomist solution of the problem is altogether indistinguishable from the Aristotelian. Their principles are the same and their conclusions formulated in identical terms. The formal distinction is that whereby one species is distinguished from another species, the material distinction is that whereby one individual is distinguished from another individual. Now since matter is inferior to form as potency is to act, the material distinction must of necessity be there for the sake of the formal distinction, and that amounts to saying that the individuals are there for the sake of the species. When the species is realizable in a single individual, as in the case of the pure Intelligences, there is no need to distribute it into a plurality of numerically distinct individuals ; and that is why, in the Thomist system, every angel constitutes a distinct species in himself. Where the specific form cannot subsist by itself in its fullness, as in the case of man, it endures and is perpetuated by means of the generation and corruption of a series of numerically distinct individuals individuated by matter.[4] It would seem impossible to imagine two philosophies in more complete accord, or perhaps we ought rather to say that they are but one and that between the philosophy of Aristotle and that of St. Thomas Aquinas there is only a numerical distinction. But when on the other hand we remember the original character of the Thomist anthropology it seems difficult to believe that the substantiality of the soul, which is there so strongly accentuated, had no repercussion on the problem of individuality. What, then, is an individual according to St. Thomas Aquinas ?

An individual is a being divided off from all other beings, and not itself divisible into other beings. In that respect therefore it is very different from a species, for although a species is formally distinct from every other species, it can be divided into a plurality of distinct individuals without thereby losing its nature. Humanity exists in each man, and, indeed, it is because there are men that the human species exists. A man, on the contrary, is distinct from every other man and we could not divide him into several men without destroying him ; it is just for this reason that we call him " individual."

When we compare this definition with our metaphysical analysis of the human being we shall soon see that, in spite of an apparent contradiction, they mutually confirm each other. For it is at one and the same time true to say that man is not a simple substance, and that nevertheless he is indivisible. Neither the soul nor the body is the man, but the composite of both. Now this composite, precisely on account of the incommunicability of the extended matter which enters into its substance, is, by definition, a unique exemplar in itself, therefore original and irreducible to any other. It goes perhaps rather against the grain to have to admit, as it seems we must according to this conception of the individual, that what makes each one of us precisely himself, and endows our personality—this unique and proper character that we look on so complacently and protect with such jealousy—that this should be, not the spiritual element in our nature, but the accidental fact that the portion of matter that goes to make up my body is not the portion allotted to yours. That seems neither human nor Christian. And nothing could be truer ; but then we must observe that no such thing is in question. The mediæval thinkers pushed the analyses of some problems a good deal further than we usually do to-day. What the Thomist doctrine of individuation sets out to explain is individuation itself and nothing other than individuation. From the fact that there would be no human individuals if there were no human bodies, it does not at all follow that it is from the body that the dignity of the individual derives, or even that it is the body that defines his originality. Let us recollect that while there is no concrete substance without matter, the substantiality of the human composite is not due to the matter, but is that substantiality which is communicated to the matter by the form. The question will then begin to appear in an entirely new light.

At this point in fact it becomes clear that even while we retain all the Aristotelian data of the problem we are not bound to resolve it precisely in Aristotle's way. What, in point of fact, is our principle ? This : that the form of man cannot subsist as an individual subject ; but none the less it is in virtue of its form that the quality of substance belongs to the individual subject, for it is form that gives the matter its actual being and thus permits the individual to subsist.[5] From this it follows that the being of the matter and the being of the composite are none other than the very being of their form, but also, conversely, that the being of the form is the being of the whole composite. Nothing could be more natural, for it is the form which gives being, and it is just for that reason that I said, in the last lecture, that it is absurd to imagine that the destruction of the body involves that of the soul, since it is the soul that bestows being on the body, not the body on the soul. We must therefore strictly maintain the principle of the non-individuality of the soul as soul, since two forms of this kind which would be numeric- ally distinct as forms, would be an absurdity ; and even if we attributed a numerical distinction in virtue of a capacity for union with different bodies the metaphysical difficulty would remain the same. But we must grasp at the same time the whole truth involved in Aristotle's principle, the truth which he failed to extract from it for himself because he did not share the Christian solicitude to base the unity of the individual on the spirit.

This truth appears in all its clearness only when we dis- tinguish, in the light of the preceding analyses, the idea of individuality from that of individuation. The principle of individuation is matter, and it is therefore certainly matter that causes individuality ; but it is not in his matter that the individuality of the individual consists ; on the contrary he is only individual, that is to say undivided in himself and divided from all else, because he is a concrete substance taken as a whole. In this sense, the individuating matter is such only in virtue of its integration with the being of the total substance, and, since the being of the substance is that of its form, individuality must of necessity be a property of the form as much as of the matter. Indeed, it belongs to the form even much more than to the matter, since, like matter, the form partakes of the individuality of the sub- stance, and since further, in this substance, it is the form and not the matter which is the source of the substantiality. To express the same idea in another way we might say that it is indeed matter that individualizes the form, but that, once individualized, it is the form which is individual. In short, the soul is an individual form, although not precisely *as* form,[6] and it is the subsistence of this individual form which, investing matter with its own proper existence, permits the individual to subsist.[7] In what, for the rest, this individuality of the soul consists we shall understand fully only when we rise from the plane of individuality to that of personality.

Every human person is, in the first place, an individual ; but also much more than an individual. We speak of a person, as of a personage, only in the case of an individual substance considered as possessing a certain native dignity of its own. Animals are individuals, but certainly not persons. We must go further : it is always on account of his highest dignity that any being whatever is a person, and we may find a sign of this in the fact that when we would render due honour we salute a dignitary with the highest of his titles as if the rest were of no account at all. Now, if we seek the highest dignity of man we find it in the reason, so that we must define the human person as " an individual substance of a rational nature." The definition is Boetius' : few mediæval philosophers found it otherwise than satisfactory, for it fits the thing defined, and fits nothing else ; but it is of some importance to determine its meaning with precision, and to follow up its consequences.

The distinction between the universal and the particular is applicable to beings of all kinds. In a sense it even applies as much to accidents as to substances, but since accidents are particulars only in virtue of appertaining to particular substances, it is these latter that we ought to regard as the true particular beings in the full sense of the term. Taken, then, in their concrete subsistence, substances are individuals ; but there are certain individuals distinguished from others by the remarkable character of being autonomous sources of spontaneous activity. Natural inorganic bodies submit passively to the laws of nature, they are channels for the forces of nature which pass through them and move them without calling forth either co-operation or reaction. Plants and animals are already on a higher plane, especially the latter, because they react to external stimuli ; but the nature of the reaction is still determined by the action undergone, so that even of these we may still say that they are acted on rather than that they act. The case of man is altogether different. Gifted with reason, capable of apprehending a varied multiplicity of objects, he has possibilities of choice before him that are not open to others ; as we shall see more fully in the sequel, his rationality is the root principle of his liberty. A man, then, is distinguished from individuals of any other species by the fact that he is master of his acts ; others are acted on by natural forces, he alone acts in the fullest sense of the term. To designate the individuality proper to a free being we call him a person.[8] Thus the essence of personality is one with that of liberty ; on the other hand liberty has its root in rationality, and since it is this very rationality that lies at the basis of the subsistence of the soul, and the subsistence, therefore, of the man, it follows that, in us, the principle of individuality and the principle of personality come back in the end to the same thing. The actuality of the reasonable soul, in communicating itself to the body, determines the existence of an individual who is a person, so that the individual soul possesses personality as by definition.

Thus we are carried far beyond Greek thought, whether it be Plato's or Aristotle's. For if the human soul is a substance and principle of substantiality, it is because it is an intellect, that is to say an immaterial being by definition and consequently incorruptible. After that, St. Thomas can turn to his own account, and does so unweariedly, the famous Aristotelian principle that the individual exists for the sake of the species ; only, by a now inevitable reversal, the consequences that favoured the species in the

Aristotelian system work out in favour of the individual in the Christian system. That to which the intention of nature now tends is much less the species than the incorruptible. If, sometimes, it looks to the good of the species rather than that of the individual it is only in those cases where the individuals are corruptible and the species alone endures ; but in the case of incorruptible substances, it is not only the species that permanently endures, but also the individuals. And that is why the individuals themselves fall within the principal intention of nature : *etiam individua sunt de principali intentione naturae*. Now it is the soul that is the incorruptible part of man ; and consequently we must admit that the multiplication of human individuals is a primary intention of nature, or rather of the Author of nature, Who is the only Creator of human souls : God.[9]

Thus firmly based henceforth on the substantiality of the intellect and the immortality it carries with it, the Christian individual is invested with àll the dignity of a permanent being, indestructible, distinct from every other in his very permanence, and an original source of rational activity responsibly deciding his own future destiny. We must not conceal from ourselves that here we stand at the source of the whole life of the spirit in its two-fold activity, theoretical and practical, since it is precisely as rational, and therefore also as a person, that the individual is able to discern the true from the false, that is to say to acquire science, and to distinguish good from evil, and achieve morality.[10] The whole interior life of the Christian man thus consists in gradually building-up, constantly rectifying, unweariedly perfecting a personality which will only attain its full stature in the future life. For it is indeed true to say that the person is constituted in being and exists in virtue of the sole fact that an intellect, a principle of free determinations, is united to a matter so as to constitute a rational substance. Kant himself, with all the personalism that pervades his doctrine, was but an heir of the Christian tradition when he recognized in the person the identity of a thinking substance which remains one and the same beneath all its multiple acts, and is predestined to immortality by its very unity. But it must now be added, that although the principles have been laid down, our picture of this Christian person is still very far from complete. It looks, at first sight, as though the mediævals made few efforts to connect the whole development of the interior life with the idea of personality, and thus we may partly surmise the inexhaustible fecundity that still lies hid in their principles. St. Thomas exalts personality above any other reality in all nature ; nothing can possibly surpass, in his eyes, those beings whom he defines as reasonable individuals : *singularia rationalis naturae.* How could it be otherwise, seeing that in common with all the Latin theologians when they treat of the Three Divine Persons, he extends the idea of personality to God Himself? Indeed almost all that we know of the philosophy of personality is found in the mediæval thinkers in the questions they devote to the theology of the Trinity. It is in Boetius' *De duabus naturis*, that is to say in a treatise on the two natures in Christ, that there occurs that definition of the person that inspired the whole Middle Ages, and weighs so heavily on the development of modern ethics.[11] It was in order to know whether they had the right to apply it to God that St. Bonaventure and St. Thomas examined and explored the definition of Boetius. And finally, to return once more to our starting-point and basic principle, how could personality be anything but the mark of being at the very summit of its perfection, in a philosophy like the Christian philosophy where everything is suspended from the creative act of a personal God ? For all things were made by the

Word, and the Word is with God, and the Word is God ; that is to say precisely this being Who presents Himself as personal in virtue of the sole fact that He presents Himself as Being : *Ecce personalis distinctio ; Exodi tertio, ego sum qui sum.*[12] Christian personalism also, like the rest, has its roots in the metaphysic of Exodus ; we are persons because we are the work of a Person ; we participate in His personality even as, being good, we participate in His perfection ; being causes, in His creative power ; being prudent, in His providence ; and, in a word, as beings in His being. To be a person is to participate in one of the highest excellences of the divine being. But then it seems, when that is said, all is said. Not a word throughout the whole of mediæval moral philosophy on what the mediævals themselves held to be highest in man and therefore in all nature. How shall we account for the fact that in the very moment of a discovery of such immense importance, Christian thought seems to stop suddenly short and renounce all effort to exploit its success ?

All that is merely apparent. It is true that the idea of person seems to play no part in mediæval moral philosophy, but we can regard it as absent only if we forget the very definition that the Christian thinkers gave it : an individual of a rational nature. If, as we shall show later on, Christian morals require man to live in perfect accord with reason, there can be no word said on morals that does not directly concern the history of personality. It is the person, as practical reason, whose activity weaves the web of human life ; it is the person which, destitute in its essence, ceaselessly enriches itself with new knowledge, with new moral habits, that is to say virtues, with practical habits, that is arts, and thus gradually building itself up issues at last in those human masterpieces whom we call sage, hero, artist, saint. And these masterpieces endure, imperishable as the

person they constitute, deeply sculptured in the very substance of an immortal soul destined one day to regain its
body in immortality. Of all admirable things in nature,
says the Greek p et, I know none so admirable as man.
But from the opening of the Christian era it is no more of
man that we speak but of the human person : *persona
significat id quod est perfectissimum in tota natura.*[13] To follow
up the development of this eminent being we have next to
study the conditions of his theoretical and practical activity,
the acquisition that is to say of knowledge and the achievement of a moral life.

In concluding this first series of lectures I feel more keenly
than ever how inadequate they are to the grandeur of the
theme, and, to say all in one word, how merely schematic.
But perhaps this very schematism which I would be the
first to deplore, may be suggested as a kind of excuse. What
I have tried to do, as far as my power allows, is to set forth
a small number of philosophical ideas which, as far as our
present knowledge goes, would seem to have had a religious,
and, in particular, a Judeo-Christian origin. If what I
have maintained is true as regards essentials, we may for
the present defer the study of what Christian philosophy
borrowed from Greek philosophy, being well assured that
every element, whether of Hellenic or other origin, which
has received a welcome into the Christian synthesis, has
entered only by way of assimilation and has undergone also,
in consequence, a transformation.

To put the problem in these terms is to make it clear at
once that we have no thought of making Christian philosophy a kind of creation *ex nihilo*, without deep roots in the
past. It would be an endless task to enumerate and classify
its Greek sources, that is to say all that Christianity had to
borrow before it could assimilate. Indeed we might fairly

ask whether there would have been any Christian philosophy at all if Greek philosophy had not existed. Nothing stands in the way of imagining a Christianity reduced to the contents of the two Testaments, as little speculative as Judaism had been, and yet strictly identical in religious content with the actual Christianity of to-day. Philosophy, even Christian philosophy, never was, is, or will be any necessary element in a doctrine of salvation. Far from denying therefore the large debt that Christian thought owes to the Greek, it is but just to say that Greek philosophy, that is philosophy pure and simple, still lives on in it. The sole question was whether this philosophy did not take on a new aspect, whether its development did not receive a new impulsion, from the fact that it was carried forward under Christian conditions. It is precisely to this question that the concept of Christian philosophy seemed to me to contain an answer ; subject of course to the observation that if it is due to Scripture that there is a philosophy which is Christian, it is due to the Greek tradition that Christianity possesses a philosophy. Without the Biblical revelation there would have been no metaphysic of pure Being, but then, also, without Greek philosophy, no metaphysic would have issued from the revelation.

If, then, the case stands thus, to say that the spirit of mediæval philosophy, in so far as it was constructive and creative, was none other than the very spirit of Christian philosophy, is to affirm simultaneously that the Middle Ages were a period of philosophic progress and that this progress rested on the continuity of a tradition. To put the emphasis, as I have tried to do throughout, on the Christian elements and the originality that was due to them, is to mark the distinguishing factors ; it is in no wise to forget, but rather constantly to presuppose, all the older elements there preserved and the characteristics that make it a genuine philosophy. In short, just as Christian philosophy is not the whole of mediæval philosophy, so neither are the elements due to Scriptural influence the whole of Christian philosophy ; they do not even exhaust its essence, but they mark its specific difference and reveal its spirit. In this sense an historical discussion of the concept of Christian philosophy is indispensable for the interpretation of mediæval philosophy and for any sane historical appreciation of the Middle Ages as a whole. For the rest, at this mid-point of our inquiry, we must regard all our conclusions as provisional. Was the Middle Ages able to complete its metaphysic and anthropology with a Christian noetic and a Christian ethic ? That is a question we have still to discuss before our task can be completed, and its examination will be the object of the rest of these lectures.

CHAPTER XI

SELF-KNOWLEDGE AND CHRISTIAN SOCRATISM

" How foolish of Montaigne to set out to depict himself." The truth is that Montaigne's doubts and Montaigne's miseries had but little interest for Pascal considered precisely as Montaigne's ; but they interested him a great deal considered as those of mankind in general, as the *Entretien avec M. de Sacy* abundantly shows. For, indeed, " one must know oneself : even when it avails nothing for the discovery of truth it serves at least as a means by which one can rule one's life ; that is entirely true." In thus basing morals on self-knowledge Pascal remained true to the oldest of philosophic traditions, but his mode of interpreting this self-knowledge was new, and history could never follow the transition from Socrates to Pascal without devoting a chapter of some importance to the consideration of the Christian *nosce teipsum.*

In no other case that the historian of Christian philosophy has to consider is he so likely to expose himself to the accusation of confusing the order of philosophy with the order of religion. It was from the Delphic oracle, however, that Socrates himself took the famous precept, " Know thyself," [1] and certainly the Oracle made no pretence of teaching philosophy. Interpreting the utterance as a programme and a method, Socrates recommended his successors to seek after self-knowledge in order to become better men. He had no illusions at all as to the difficulties of the task ; he foresaw them, indeed he experienced them for himself—it was precisely in this cause that he met his death. But if in the adoption of such a method there are philosophical difficulties to be surmounted and even social risks to be run, it presented no particular problem in itself—there was nothing mysterious about the decision to study man. Perhaps we might say, more precisely, that the precept ceased to be mysterious the moment it was transferred from the lips of the Oracle to the lips of Socrates ; but would certainly become mysterious once more as soon as Christian thinkers took it up and interpreted it in their turn.

Whatever aspect of human nature the Christian philosopher considers, he invariably ends by referring man to God, by bringing him into subordination to God. Now even on the very nature of man the Bible has something to teach philosophers. On the sixth day of creation God said : " Let us make man to our image and likeness, and let him have dominion over the fishes of the sea, and the fowls of the air, and the beasts, and the whole earth, and every creeping creature that moveth upon the earth. And God created man to His own image ; to the image of God He created him " (Gen. i. 26–27). Thrice affirmed in so many lines this divine likeness, stamped by the creative act in man's very nature, rules his being to its most intimate depths. It is quite out of the question here to attempt to follow up all its consequences ; their ramifications are endless ; but at least we may try to indicate how it transforms the problem of self-knowledge.

The first characteristic of the divine image is its universality, and this universality is explained in its turn by the fact that we have to do here with no mere accidental character superadded to human nature, but with human nature itself taken in its constitutive essence. That is the reason why St. Gregory of Nyssa declares that all men without exception are made to the image of God.[2] The same doctrine is to be found in the *Summa* of St. Thomas, in St. Bonaventure and in Duns Scotus ; and, in short, suggested by the text of Genesis it naturally forms part of the common heritage of Christian philosophy. But as soon as they set out to explain in what this image consists, and to try and arrive at a definition, then at once the various schools begin to diverge.

If we are content to take Genesis simply as it stands, the answer would seem easy enough. According to the Bible man is made to the image of God as being the vicar of his Creator on earth. Because God made the world, He owns it as His property, and also governs it at His good pleasure ; but makes over a share of the government to man, who thus has a dominion over things analogous to God's. These fundamental conceptions have already been sufficiently discussed, and we have only to recall them to see in what sense man is an image of God on earth ; he represents Him as a lieutenant represents his Sovereign. The real philosophic problem lies here : why is man capable of exercising this dominion, this quasi-divine sovereignty over the earth ? Evidently, in the first place, it is because he is free and the other things he rules are not. But what, then, are the ultimate roots of this freedom ? They lie in his intellect and understanding, in all that makes him able to shape his own course and exercise the power of choice, makes him, moreover, a possible subject of divine virtues and graces. This brief enumeration, proposed by St. Ephrem of Nisibis, suffices to show how far philosophic reflection had gone by the middle of the fourth century.[3] Theologians will always henceforth situate the divine image in what is highest in man, that is to say, either in his intelligence or in his freedom.

St. Bernard, foreshadowing Descartes, finds the image of

God *par excellence* in human freewill, for since it would be impossible to diminish this without destroying it altogether it may be regarded as a kind of inamissible perfection, eternal in a way, and thus like that of God Himself.[4] St. Augustine, on the other hand, insists rather on the eminent dignity of mind, open as it is to the illumination of the divine ideas, and the mediæval Augustinian school will preferably attach the idea of image to this immediate contact of the soul with God.[5] Whatever be the tradition followed—and often enough they interpenetrate—it remains true in any case to say that the Biblical idea inspiring it undergoes a philosophical development of great importance, and that the philosophical conception of man, on the other hand, is profoundly transformed.

To say, in fact, that thought and freedom constitute the image of God in us, does not merely signify that the soul is a kind of representation or picture which only needs to be looked at to reveal the nature of its Author. On a certain lower level of imagination and analysis this may be well enough, but speculations that run on these lines, however legitimate in their kind, avail less in their detail than in their spirit. They are so many different answers to one and the same question, but what is really important is the fruitful presence of the question. Mediæval authors, as we know, revel in a luxuriant growth of Trinitarian symbols, a sort of symbolic psychology in which faculties, their functions, their objects, and the various ways of attaining their objects, all go in threes. No one is likely to exaggerate the scientific importance of this kind of thing, and St. Thomas has taken care to limit its theological scope. The historical facts, moreover, are too complex to warrant any estimate that could be applied indiscriminately to all. The profound psychological speculations of St. Augustine himself in his *De Trinitate* are to be regarded now as an elucidation of theology in the light of psychology and now as a deeper understanding of psychology effected in the effort of interpretation of the dogma to which it is applied. The concept of mental word and the concept of the Word, the psychology of the will and the theology of the Holy Spirit, here influence each other reciprocally. Important questions are involved in this, and when some agreement is reached on the spirit and principles of Christian philosophy, they will call for detailed study. For the moment we will be content to note the direction in which all these speculations move ; not, that is, from God to man, but from man to God. To put this more precisely, it is not solely or chiefly in virtue of the divine image that man effectively resembles God, but in virtue of his consciousness of being an image, and the movement whereby the soul, passing in a way through itself, avails itself of the factual resemblance in order to attain to God. " The image of God," says St. Thomas, " is found in the soul according as the soul turns to God, or possesses a nature that enables it to turn to God." [6] Duns Scotus, although forcibly insisting on the reality of the image in itself as stamped upon mind, agrees that it remains imperfect so long as it is merely shut down on itself, and only becomes fully itself when it explicitly refers itself to its model.[7] Here, then, theology has influenced our reading of human nature in the most extensive and decisive manner, in giving birth to what may, not too inappropriately, be called Christian Socratism.

In the Socratism of Socrates, and in the Socratism drawn out of it by the Fathers of the Church or the mediæval philosophers, there remains in fact a common element of anti-physicism. Neither the one nor the other condemns the study of nature as such, but everyone agrees that it is much more important for man to know himself than to know exterior things. Socrates' own personal experience, as recorded in Plato's *Phaedo* (98b *et seq.*), was decisive in the matter and, in a sense, he stands behind all those who, from the Stoics to Montaigne and Pascal, have regarded man as the proper study for mankind. We must add, moreover, that if they considered the science of man as of supreme importance to man, the reason that weighed with them was that it is the only science that provides a basis for the precepts that rule the conduct of life. Their anti-physicism was not intended to clear the way for any psychologism, it was rather the reverse of their moralism. Now from either standpoint it proved attractive for Christians—what does it profit a man to gain the whole world and lose his own soul?—but it could not possibly attract them without having to undergo a profound transformation itself. When Socrates advised them to concentrate on self-knowledge, Christians would at once take this to mean that they must learn to know the nature God gave them, and the place He marked out for them in the order of the universe, so that they in their turn might order themselves towards God. All these elements, then, were at work within Christian thought : let us watch them, and seek, if possible, the law of their composition.

As we might naturally expect, it is the moralists rather than the metaphysicians who interest themselves in the matter. Man finds himself set down in the midst of things, some being at his own level, some above him, and some below. Multifarious sciences claim his attention ; only one is immediately necessary, that is to say the science of himself. As soon as it is subordinated to the interests of salvation the necessity for self-knowledge becomes absolute ; it might almost be regarded as the beginning of knowledge, the sole object of knowledge, and the last end of all knowledge. Not that the rest is useless—far from that ; it is useless, however, unless based upon true knowledge of

man. " He is not wise," says St. Bernard, " who is not wise
for himself ; let everyone be the first to drink at his own
well. Begin by considering thyself and, better still, end
with that. When thy consideration wanders elsewhere
recall it to thyself : and this will not be without fruit of
salvation. Thou for thyself art the first, and also the last." [8]
So the *nosce teipsum* here becomes an imperative of vital
import and tragic sanctions. But although all our knowledge is ultimately to be referred to self-knowledge, it is
soon clear that we cannot know ourselves truly if we propose
to know ourselves alone.

Order, says St. Augustine, is that disposition whereby
things similar or different are assigned to their proper
places.[9] Would we know ourselves we have to discover
our proper place, that is to say above lower things, and below
the higher ; and this, at bottom, is the true meaning of the
Socratic precept. Were the soul's own self-awareness the
only thing in question there would be little difficulty in the
matter and virtually no need for a precept ; the soul is
always present to itself and aware of itself, or, at any rate,
feels its own presence behind the sense illusions which cast
a veil over its own nature. But then, precisely, it has to
deliver itself from these illusions in order to recognize itself
for neither more nor less than it really is. St. Bernard,
therefore, taking his stand on this Augustinian doctrine of
order, interprets the Delphic Oracle with a naïve assur-
ance that might very well occasion a smile did we not
remember those great *Pensées* he would eventually inspire.
What, he asks himself, did Apollo mean ? This : that lack
of self-knowledge springs from a twofold source : undue
timidity on the one hand leading to an unwarrantably low
view of human nature, and something considerably more
dangerous on the other, that is to say temerity, leading to
presumption. Pride and presumption, pusillanimity and

despair of self, these are the twin dangers that beset us. Experience makes an end of pride, and realization of our true nature delivers us from despair,[10] and the true goal of self-knowledge is reached when we learn to reconcile both truths, and to keep either error at arm's length.

Here we catch a glimpse, or even already the clear outlines, of a theme familiar to all literary historians in the magnificent language of Pascal : " *Grandeur et misère de l'homme.*" Man, God's creature, was not ignorant of his true nature because he exactly knew his proper place in the universal order. " To know his own condition and place, what he owes to things above him and beneath and to himself, to understand what he has been made, how he should conduct himself, what he should do and not do— in this for man consists self-knowledge," [11] a science he has unfortunately lost and ought above all things to try and recover. Let him mark then what he knows, but also what he does not know. In virtue of his intelligence he is superior to the brutes, but in virtue of his ignorance he remains lower than the angels. Man is neither brute nor angel, but stands midway between the two.

In the hands of the moralist this theme lends itself to the most varied developments. The grandeur of man lies in the fact that he is created to the image and likeness of God. Being free, he holds dominion over nature and forces it to supply his necessities ; being intelligent, he understands and therefore commands it, but with all this he very well knows that it is not to himself that he owes his greatness, and as soon as he realizes that he catches the first glimpse of his misery. If he fails to recognize his dignity he does not know himself ; if he insists on it without referring it to a greater than himself, he founders on the rock of vainglory. *Utrumque ergo scias necesse est, et quid sis, et quod a teipso non sis, ne aut omnino non glorieris, aut inaniter glorieris :* there we have the first form of the two-fold danger to be avoided. And the second is this : that being created in all the glory of the divine likeness, man loses this glory as soon as he forgets it. Doubtless the fact of his dignity remains, but in the measure in which he fails to recognize it he lowers himself to the level of the brutes, becomes in effect as one of them. But the guilt of the opposite error is infinitely blacker, when forgetting his humble estate he aspires to set himself among the angels, and even to usurp the place of God.[12] Thus it becomes a matter of course, from the twelfth century onwards, to analyse in detail all that goes to make man's greatness, a soul stamped with the divine image, and all that goes to make his misery, the soul's wounds and the accompanying bodily suffering. And to what, then, is examination of conscience to be directed ? To ascertain the point reached on the road to the realization of our true nature, the progress made, and setbacks encountered, in the work of restoring the divine image in ourselves ; and it is precisely in the pursuance of this work that each man will learn best to realize the deep bonds of fraternity that unite him to his companions in greatness and misery. Few pages afford us a better insight into the intimate depths of the mediæval soul than those in which an unknown author expands the sense of the Delphic Oracle into Christian charity :

" Return therefore to thyself, if not always or even often, at least sometimes. . . . Set thyself face to face with thyself as with another, and weep over thyself. Weep over thy sins wherein thou hast offended God ; open to Him all thy weakness, all the malice of thy adversaries, and then, when thou art dissolved in tears, I pray thee, remember me. For I, since I have known thee, have loved thee in Christ. I make mention of thee there where every unlawful thought calls for chastisement, and every good thought a reward.

Standing before the altar of God, a sinner yet a priest, I do not fail to remember thee. And thou also wilt remember me, wilt love me, and grant me a share in thy prayers. When thou offerest prayer for thyself and thy neighbours I too would be present with thee in memory. And be not surprised at this word ' present,' for if indeed thou lovest me, and if it be because I am an image of God that thou lovest me, then am I present to thee even as thou art to thyself. *Quicquid enim tu est substantialiter, hoc ego sum* : all that thou art substantially, I am. Every reasonable soul, indeed, is an image of God. He then who seeks God's image in himself, seeks there his neighbour no less than himself ; and whoso seeking finds it, finds it as it is in all mankind. For man's vision is his intelligence. If then thou seest thyself, thou seest also me, who am not other than thyself ; and if thou lovest the image of God, then, as image of God, thou lovest me ; and I too, loving God, love also thee. Thus seeking the selfsame, tending towards the selfsame, we are always present to each other, but in God in whom we love." [13]

I would not dream of concealing the fact that Cistercian mysticism has thus led us, as by insensible steps, far beyond the limits of philosophy ; if, on the contrary, I have gone along voluntarily with the tendency it is because, whatever be the problem studied, it becomes henceforth characteristic of Christian thought. Philosophy looks up to a higher order of its own spontaneity, lends itself to it, opens up the path. What moralists and mystics develop in the sphere of grace, philosophers know how to express in concise formulæ addressed simply to the reason ; and if they speak in other tones than St. Bernard, they entirely agree with him nevertheless in reserving a mid-position for man. In conformity with Dionysius' principle that what is highest in the inferior order touches the lower confines of the superior order, St.

Thomas puts man between angel and animal, reaching up to the former at the summit of his intellectual powers, and down to the latter in respect of his perishable body. It is in this very precise sense that man is a " microcosm," a whole universe in little, " for in him everything else in the world is, in some way, represented." That, moreover, is why St. Thomas adds that the soul is a kind of frontier or horizon between two worlds, the meeting place of pure spirits and irrational animals.[14] Thus Christian philosophy extends in a two-fold direction the field to be explored by man in the study of himself : the moralists oblige him to examine conscience for the sake of progress in the interior life, the philosophers reintroduce a certain measure of that physicism which the moralist tends to ignore. The miniature world of man will never be known unless we know something of the greater. Hence the repeated efforts made by the men of the Middle Ages, and even by those of the Renaissance, to evolve complete anthropologies, in which a detailed description of the body leads to that of the soul, and this in its turn to the knowledge of God.

However, this two-fold enlargement of the field is but a prelude to a third of more considerable import still. Hard as it is to know one's own state of conscience, and to grasp man's relation to his surroundings, it is nothing like so hard as what still remains to discover. What a man finds *circa se* or *sub se* is overwhelming in amount, what he finds *in se* is embarrassing in its obscurity, but when from his own being he would obtain light as to what is *supra se*, then indeed he finds himself face to face with a dark and somewhat terrifying mystery. The trouble is that he is himself involved in the mystery. If, in any true sense, man is an image of God, how should he know himself without knowing God ? But if it is really *of God* that he is an image, how should he know himself? There are depths in human nature, unsuspected by the ancients, that make man an unfathomable mystery to himself. Who knows the spirit of God? asks the Apostle. " But for my part," answers Gregory in the Sermon on the Image, " for my part I would ask further, who knows his own spirit ? " Certainly, if our own spirit escapes our ken much more does God's ; but, then, if this is so, if we are inscrutable to ourselves, it is only because we participate in the incomprehensibility of God. That is why Scotus Erigena, commenting in his turn on Gregory's text, adds : " The spirit in which the whole virtue of the soul consists is made to God's image and mirrors the supreme good, because the incomprehensible form of the divine essence is there reflected in an ineffable and incomprehensible manner." But why should we fall back upon such texts when St. Augustine has put all that needs to be said in a clear-cut formula ? " It is in the mind that God has made man to His image and likeness, there it is that His image is stamped. If the mind is not to be fathomed even by itself that is because it is the image of God." [15]

Mens ipsa non potest comprehendi, nec a seipsa, ubi est imago Dei : although, like many another phrase, this too might seem to have fallen by chance from Augustine's pen, we feel at once that it arises out of the whole trend of his philosophy. Recollect merely his doctrine of memory, as set out in those unforgettable chapters of the Tenth Book of the *Confessions*. In quest of its own essence, the soul plunges deeper and deeper through successive planes of being. First the memory of sense perceptions, an ample mansion, wherein in some uncomprehended manner the whole immense universe finds place ; a secret retreat, mysterious, vast, nay even infinite—who has ever sounded it ? It is merely one of the spirit's faculties, says Augustine, yet it escapes me, I myself cannot wholly grasp the thing I am : *nec ego ipse capio totum, quod sum.* There, for the first time in the history of Western thought, man becomes to himself a wonder, an amazement : *stupor apprehendit me.* What shall we think when to the memory of sensible things is added the memory of the sciences, when pure ideas come before the eye of the mind ? *Quid ego sum, Deus ? Quae natura sum ?* What am I, then, O God, what manner of man am I ? But even that is not all. Beyond the ideas themselves there is the truth that rules them, and since this truth is stamped with the divine characters of necessity and eternity, God Himself must be present to the soul, whenever by His help it sees the truth. And now it no longer suffices to speak of the depths of the spirit ; for here it opens on the infinite, and reaches back into God ; full of fear and a kind of sacred horror—*nescio quid horrendum*—at the sight of this divine presence, man stands awestruck at himself, perceiving the mystery that underlies the appearances of his human nature [16] ; if, as divine image, he eludes his own grasp, it is because the last word of self-knowledge is the first word on God.

The mediæval moralists were thus led to seek in mysticism the final conclusions of their study of man. Richard of St. Victor has all the less hesitation in doing this since the divine character of the Oracle seems to justify the interpretation he proposes for it. The command to know ourselves came down from heaven ; well, then, if we obey it, would not this be to ascend in spirit towards heaven ? [17] Christian thought was so fully aware of what it was adding to Greek philosophy in this matter that it may well be left to speak for itself: " No matter how great our knowledge of creatures, how otherwise does it compare with knowledge of the Creator, than as earth with heaven, as the centre with the entire circumference of the circle ? Like this earth itself, this lower knowledge of lower things has its mountains and hills, its plains and valleys. Sciences, then, will be diversi-

fied according to the diversities of creatures. Let us begin
with the lowest : the difference is doubtless great between
one body and another, for there are celestial bodies as well
as terrestrial ; but the distance between body, and soul is
far wider than that between body and body, however dissimilar these may be. But even amongst souls some are
reasonable and some are not. He therefore who considers
only bodies seems to keep his eye fixed on the lowest things,
but when he turns to the study of spiritual things he ascends
at once, so to speak, on high. The mind that would attain
to this high knowledge must first and foremost seek to know
itself. To know oneself perfectly is most high knowledge ;
full comprehension of a rational spirit is a mighty mountain.
It overtops all human sciences, it looks down on all philosophy, on all knowledge of the world. Did Aristotle ever
discover it ?   Did Plato ? Did the whole brood of philosophers ever invent the like ? Nay, but in truth, if they had
ever ascended to this summit of their own intelligence, if
their studies had ever sufficed them for true self-knowledge,
they would never have given themselves up to the worship
of idols, never bent the knee before a creature, never lifted
up the head against the Creator." [18] Clearly, then, Richard
of St. Victor discerned the links that still subsisted between
even the highest Greek view of human nature and the pagan
cults ; he judged it not without significance that Aristotle
put up statues to Zeus and Demeter. The Christian man
alone knows what it is that raises him above all creatures,
for he alone knows of what Creator he is the image ; but
he knows also by that very fact that the depths of his own
being surpass his own powers of vision, that the full grasp
of the soul by the soul is an ambition to be realized only in
mystic union or in the future life.[19]

Once more, what the mystics suggest the philosophers
distinctly say, and we shall find, moreover, that what the

mediæval thinkers here develop has had a marked influence on modern metaphysical speculation. No one can study the mediæval texts from this standpoint without being struck by the extreme importance attached to the question of the soul's self-knowledge. We should have to look a long way for any such documentation among the pre-Christian Greeks. In the thirteenth century there is only too much to choose from. Philosophers were not then preoccupied with the question of the existence of a spiritual soul ; no one doubted it, especially after Augustine had put it on the footing of an indubitable fact of experience. The question then agitated was rather how, and how far, the soul can penetrate its own essence. " How the intellectual soul knows itself and all within itself "—that is Question 87 of the first part of St. Thomas' *Summa Theologica*. " Whether the soul knows itself and its dispositions by its essence or simply by its acts "—that is the question as put by Matthew of Aquasparta. About the same time Roger Marston puts it more precisely thus : " How the soul knows itself and its dispositions ; whether by its own essence or that of its dispositions, or through some distinct medium." A little later Olivi takes up the matter in Book II, Question 86 of his Commentary on Peter Lombard—and the list could very easily be lengthened. What chiefly concerns us is to know why the list is so long ; and what it is that makes the question so vitally important.

As the Christian philosophers conceived it, the soul is something other than the image of an idea in matter, or the perishable form of a composite ; it is a spiritual substance, immortal and endowed with an indestructible personality. Whatever solution of the problem of individuation we adopt, the soul is certainly all that, and is so in itself. In the quest of self-knowledge man therefore comes face to face with his own soul's substantiality, as something that calls for investigation, but at the same time resists it. The nature of the difficulty will become clearer, perhaps, if we say that the soul is never for itself an object of knowledge that can be apprehended like a thing, but an active subject whose own spontaneity is always presupposed by any knowledge of itself that it can have. The intuition of the soul, in short, is never the full equivalent of the soul that intuits. It is easy to recognize this trait in all the great mediæval doctrines.

We need say nothing of Scotus Erigena for whom all essences are unfathomable and incomprehensible, and spiritual ones *a fortiori*. Nor need we insist on the point in the case of St. Thomas. In a philosophy like his where all knowledge presupposes a sense intuition, the soul can know itself only indirectly. No doubt it is immediately present to itself, but it cannot apprehend itself immediately, since between the spirituality of its essence and the knowledge it has of it, a veil of sensible images is always interposed. It knows that it is immaterial, but does not see itself as such. What is really much more remarkable is that even the Augustinians, although obstinately maintaining against St. Thomas that the soul knows itself by its essence, and not through sense images, nevertheless expressly held that this spiritual essence does not apprehend itself without intermediaries. And the reason? It lies precisely in the fact that this essence is an image of God. The creative Being is an infinite reality eternally knowing itself in an adequate act : the Word. Very different in this respect from Aristotle's pure Thought, He is not the thought of thought, but the thought of Being. In Him, therefore, we should always recognize the presence of this infinite reality, revealing itself, so to speak, to itself, by the integral knowledge it has of itself. Man, made in God's image, is also an intellectual substance which not only has to express other things in order to know them, but also expresses itself to itself when it would know itself. That is why even the Augustinian soul, which knows itself immediately, does not grasp itself as an object. It is formally the sufficient cause of its own self-knowledge, but nevertheless, and precisely as cause of this knowledge, it remains beyond the reach of its own most immediate apprehension. Thus every human soul reproduces on the plane of the finite the fecundity of the divine knowledge : it " expresses " from itself the internal presentation of its own essence, and refers it to itself by an act of will, just as in God the Father generates the Word, and links it to Himself by the Holy Spirit.[20]

Consider now what it is in the soul that thus generates its own knowledge of itself. Always it is what is highest in it, the *apex mentis* or summit of the soul as the Augustinians say, its " fine point," as St. Francis de Sales will call it later on. Now all Christian philosophers agree that it is this summit of the soul that bears the image of God. St. Thomas himself puts thought, or *mens*, at the point where the intellect, the form of the composite, lies open to the regulative influence of the divine Ideas.[21] The whole fecundity of thought, the whole constructive power that enables it to build up the fabric of knowledge in the light of principles, comes from what it is, by what is highest and deepest in it, that is to say a created participation in the divine light. Doubtless St. Thomas' conception of illumination differs from St. Augustine's ; he expressly says so himself, and very particularly where it concerns the science of the soul. All the more must this be the case with St. Bonaventure, for whom man, as image, is a kind of intermediary between God and creation. And these mysterious prolongations of the soul that the mediævals invite us to sound, have not been altogether forgotten by modern philosophy. Cartesian as he is, Malebranche is far too

much of a Christian not to be aware that the very perfection of beauty in which the soul is arrayed renders its
essence for ever inaccessible to ourselves. Even in Descartes,
the *Cogito*, which seems at first sight so wholly transparent
to itself, is soon charged with mysterious undercurrents of
meaning that affect the character of the whole system.

One cannot help asking at times why it is that ideas which
historians refuse to take seriously when they find them in
some thirteenth-century theologian, seem suddenly to
acquire an incomparable value when expressed by Descartes. The father of modern rationalism is so full of the
dogma of the independence of reason that one almost
comes to believe in the end that his own, at least, was independent. Nevertheless, his whole physics is bound up with
a certain metaphysic ; this metaphysic hangs altogether
from the idea of God ; and his idea of God is virtually
one with his idea of the infinite. But what then is the
origin of this idea of the infinite ? Now that is a question
that never preoccupied either Plato or Aristotle, nor indeed
any one of those philosophers who preceded the Christian
era ; it was only when the Greek idea of the infinite had
undergone a transformation at the hands of Christianity
that the question of its origin became actual. Now the
answer of Cartesian rationalism is precisely that of St.
Augustine, St. Bonaventure and Duns Scotus. Descartes,
too, like the mediævals, had his own interpretation of the
Psalmist's words : *signatum est super nos lumen vultus tui
Domine*, and uses it precisely to situate in the divine light
the source of our idea of the infinite, this idea, " of which we
meet with no example either within ourselves or elsewhere
. . . it is, as I have already said, *like the mark of the workman
stamped upon his work*." [22] Thus the philosophy of clear and
distinct ideas throws back its principle of principles on to
the divine Being, and behind this pure thought, which

seems at first sight to be wholly expressed in the *Cogito*, it retains its Christian depths. Thereby also the Man of René Descartes does not refuse to be signed upon by the light of the Divine Countenance ; the way is open for man according to Malebranche, in whom the signature is constituted by the idea of Being, and above all for man according to Pascal ; in whose thought the two theses which hitherto we have treated separately are gathered up in one.

Pascal, writes one of his most thoughtful historians, " distinguishes two uses for self-knowledge ; the one speculative where it serves as a bàsis for knowledge of truth, which, for Pascal, is God Himself ; the other practical, to which sphere Socrates confined the efficacy of his method of interior reflection." [23] If Pascal went further than Socrates and added the speculative application, it was precisely because between Socrates and Pascal there had intervened the doctrine of creation, and its corollary, the divine image. But even Pascal's practical application of self-knowledge is very different from Socrates' : " Know, proud man, what a paradox thou art to thyself. Down then feeble reason ; and let this foolish nature keep silence ! Know how much more than merely man is man, and learn from your Master your true condition of which you are wholly ignorant. Listen to God." [24] Such words are quite enough to prove that something had happened in philosophy between Plotinus and Pascal. And this " something " was the Cistercian mysticism which inspired that of the Victorines, and, as history will doubtless one day reveal, came down in uninterrupted succession to Pascal : " To know God, and yet nothing of our wretched state, breeds pride : to realize our misery and know nothing of God is mere despair : but if we come to the knowledge of Jesus Christ we find our true equilibrium, for there we find both human misery and God." [25] In these few lines are condensed the Christian statement and solution of the problem ; and such also was the mediæval response to the precept of the Delphic Apollo. The seventeenth century had no occasion for surprise at Balzac's *Socrate Chrétien ;* but Balzac only hit upon the title of this book : it was Pascal who wrote it.

Thus we have arrived at the heart of the problem of Christian philosophy. Quite early in the course of these lectures, attempting to define it provisionally as its representatives themselves conceived it, I had occasion to cite Bossuet's phrase : " Wisdom lies in knowing God and knowing oneself. From knowledge of self we rise to knowledge of God." The foundations of this two-fold science have now been laid, but the hidden springs of human action are still to be laid bare. Would we know man through and through we must study both body and soul. The mediæval theologians, as I have said, by no means neglected to study the body, but what they had to say about it belongs rather to the history of the sciences than to the history of philosophy. They gave all their most careful attention to the soul, and this precisely because it is the soul that bears the divine image, and it was especially in this field, of course, that the influence of Christianity showed itself most fruitful. It remains for us, then, to follow them when they set out to determine the last end to which all human activity is to be directed and how these activities are to lead up to it ; and this will bring us to the consideration of some basic ideas on the subject of knowledge first of all, and then on love, on liberty, and on morality in relation to the last end : which is nothing other than to live in the society of God.

CHAPTER XII

KNOWLEDGE OF THINGS

IT is worthy of remark, it seems to me, that all the great mediæval epistemologies were, as we should say to-day, realisms. After more than three centuries of idealist speculation we have in neo-scholasticism, a neo-realism once more, a doctrine that refuses to fall in with the method foreshadowed by Descartes, or at least if it does so tries hard to avoid its conclusions. How did it come about that what seems so obvious to so many of our contemporaries, was never so much as suspected by St. Augustine, St. Thomas Aquinas, or Duns Scotus ? How, especially, shall we explain that once the necessity of starting from thought itself had been proclaimed, it should now once more be denied by so many modern thinkers, and that all those who attach themselves to the mediæval tradition should be found among the number ? Why, in a word, is every Christian thinker a realist, if not by definition at least by a sort of vocation ? This is the problem which, in one of its most important aspects, I would take up to-day ; by showing that in a Christian universe, the object of knowledge is of such a nature as to be capable of supporting a realist epistemology.

There was nothing to show, *a priori*, that this must inevitably be so ; and St. Augustine's hesitations are there to prove it. Plato, who inspired him, did not think that the nature of material things is sufficiently consistent to allow them to become the objects of any certain knowledge. For pure thought there is the world of Ideas, and these are the objects of science ; but the world of sense, hovering on the borderland between being and non-being, is no basis for anything better than mere opinion. Feeling the need of escape from the uncertainties of the sceptics, St. Augustine was led to embrace this doctrine. Through Plotinus he early understood that pure sensism leads inevitably to universal doubt ; if reality is in the end reducible to sensible appearance, then, since this is in a state of perpetual flux and self-contradiction, no kind of certitude will any longer be possible. Hence those conclusions of uncompromising severity which all the great thinkers of the thirteenth century will be called on to consider : " Whatever is attained by sense, what we call the sensible, never for an instant ceases to change. It matters not what age the body has attained ; whether the hair of the head be ungrown, whether it be in the bloom of youth or verging to old age, always it is in a state of uninterrupted becoming. Now what does not remain steady cannot be perceived ; for to perceive is to comprehend by science, and the perpetually changing cannot be comprehended. From our corporeal senses, then, no genuine truth is to be looked for." [1] The *non est igitur expectanda sinceritas veritatis a sensibus corporis*, stands there as a solemn warning, and we pause on it the more readily on account of the numerous mediæval philosophers who were quite willing to come to terms with it.

Truth is necessary and immutable ; but in the sensible order nothing necessary or immutable is to be found ; therefore sensible things will never yield us any truth. That may be said to be almost a commonplace in the Augustinian schools of the thirteenth century. St. Bonaventure, Matthew of Aquasparta, Roger Marston and many others taught it—not without some perception of the grave difficulties inherent in the position. It was, in fact, an ever-growing consciousness of these difficulties that led at last to its abandonment, after numerous attempts to save the situation by means of a copious supply of complementary hypotheses. In a doctrine of this kind the knot of the problem lies in the part still left to the object to play ; and since the examination of this point led mediæval thought as near as possible to what we now call idealism, it is instructive to watch it at grips with the temptation and to see just why it refused to succumb.

To simplify the study of the matter let us take it up at the point where it entered on its decidedly critical phase, that is, in St. Bonaventure's most original disciple, Matthew of Aquasparta. Persuaded that the instability of sensible things is such as to provide no basis for any certain knowledge, he was led, naturally enough, to ask whether our knowledge depends on the existence of its object, and he answered this question in the negative. His whole doctrine drove him to this conclusion. Some certainties we undoubtedly possess ; science, in short, is possible ; but there is nothing in material things that can serve as a basis for this science, and therefore it does not depend on their existence. " Thanks to its active power, the *intellectus agens* abstracts the universal from the particular, intelligible species from sensible species, essences from actually existing things. Now it is quite certain that universals, intelligible species, and the essences of things are not bound up with any actually existing thing. On the contrary they are quite indifferent to the existence or non-existence of things, they take no account of place or time, so that the existence or non-existence of the thing has nothing to do with our knowledge of it. Thus, just as the intellect may know the essence of a thing by its intelligible species when the thing exists, so also it can know it in the same way even when it does not." Doubtless it might be objected that in that case the object of the intellect would be a mere nought ; which would seem to be contradictory. But this gives little trouble to Matthew of Aquasparta. He has, not one only, but two replies.

The first is, that if this *nought* be taken absolutely, then, no doubt, it is contradictory to suppose that what is nothing can be an object for intellect : it is not contradictory, however, to say that the object of the intellect is the essence of a non-existing thing. Quite the contrary ; we might say that the object of the intellect is never the being taken in the sense of existence. In possession of the intelligible species of an essence, that of man, for example, the intellect draws out of it the corresponding concept, but without representing the man either as existing or as not existing. The object of the intellect is thus the essence of the thing independently of its existence : *nam, nec re existente, quidditas ut est in rebus, est intellectus objectum.* That, then, is one possible solution of the problem. It goes to the point and in a philosophical manner ; but Matthew of Aquasparta himself has certain doubts about it, and asks himself whether the question can be wholly set at rest on purely philosophical lines, and without having recourse to theology : *iste modus est philosophicus et congruus ; non tamen puto, quod sufficiat, et fortassis hic deficiunt principia philosophiae, et recurrendum ad principia theologica.* What are his reasons ?

The chief one is that if we hold strictly to the philosophic standpoint the sole object that can be guaranteed to the intellect is the concept. Suppose then, for a moment, that our cognition attains nothing but the intelligible species and the concepts drawn from them, then it would have to be admitted that it attains to no reality. It would be a science without object, therefore empty. The most we could say would be that the intellect possesses the science of its own concepts, but since it is indifferent to the question whether any real things correspond to the concepts, its knowledge remains without real content. It is a most remarkable thing that our philosopher is here directing against his own conception of a science of pure essences the very same objections that Aristotle had already directed against the Platonic theory of Ideas. If it is these that constitute the proper object of our science, then, since things are other than the Ideas, our science is no longer a science of things.[2] The only difference, an important one to be sure, is that Matthew of Aquasparta does not count on things to furnish the object of any certain science, so that, in order to reply to the Aristotelian objection that he puts against himself, he has to come back to Plato, or at any rate to Plato as completed by Anselm and Augustine.

What, in fact, is the content of our concept? Not an existence, as we have just said, but then neither is it a pure possible, or pure knowable, but rather a necessary, immutable and eternal truth. As St. Augustine says in his *De libero arbitrio* (II, 8, 21) : " As for all that my bodily senses reach, whether earth, sky or anything else, I know nothing at all about how long they will last ; but seven and three make ten, now and for ever, they never have made, and never will make, anything else but ten." Hence it is apparent at once that a necessary science exists, and that it cannot have its source in the contingence of the sensible. What remains then is to recollect that the truth of created things is only a kind of expression of the uncreated truth. St. Augustine says, and St. Anselm puts it on a firm basis in his dialogue *De Veritate*, that each thing is true only in the measure in which it conforms to its divine exemplar. Let us leave on one side the metaphysical implications of this doctrine, to which indeed we shall soon have to return, and retain only what throws light on our immediate problem. Well, it does nothing less than give us that object of human cognition we need ; not by any means the concept, which is an empty form, nor the sensible thing, too unstable to be grasped ; but the conceived essence referred, however, to its divine model : *quidditas ipsa concepta ab intellectu nostro, relata tamen ad artem sive exemplar aeternum.*[3]

Well and good !—and the problem, it must be admitted, is handled in a masterly fashion. But then it must be added that this very mastery only brings out more clearly the dangers of the position. If you begin with Plato, then, to remain coherent, you must end with Plato. Matthew of Aquasparta is perfectly coherent, and his science remains always a science of ideas. The ultimate upshot of his analysis is that the object which sensible reality fails to supply to the human intellect is supplied by the divine illumination. Not, certainly, by the divine illumination alone, but the nature of its collaboration with the sensible is such that it is indeed this illumination, and not the sensible, that furnishes that element of stability and necessity of which our science stands in need. This becomes clear when Matthew of Aquasparta asks whether the divine ideas would suffice as basis for this science in the absence of the objects. He replies, as he is bound to do, in the affirmative. That in a certain sense things are the cause of our knowledge is doubtless a fact ; but no more than a fact, since the cause wherefore our knowledge becomes science does not lie in things, but in the Ideas. Now these Ideas do not depend on things, the fact is the other way round. Therefore the intellect might very well know by means of the Ideas, even if things did not exist.[4] Things, in short, are not the necessary cause of our science ; if God should imprint the species directly upon the intellect, as He does in the case of the angels, we should still know them and know them for such as we do know them.[5]

To reach conclusions of this kind is to admit that pure philosophy is unable by its own resources to secure the foundations of science, and since Matthew of Aquasparta regards the doctrine of illumination as essentially theological, we may say that his epistemology is a philosophical scepticism saved by fideism. That amounts to saying that there is no guarantee of certitude save in faith, and that mediæval thought is here tending towards the theologism of Occam. The importance of the problem was so much the greater inasmuch as the case of Matthew of Aquasparta did not, and could not, stand alone. His conclusions are bound up with a way of stating the question which is not peculiar to himself, and they follow by way of necessary consequence. Nothing is more interesting than to watch the Augustinians themselves taking note of the danger that threatens them. Olivi would wish to follow the Augustinian tradition, but to follow it, if possible, *sine errore*, and here is the first error into which he fears to fall : " As regards the intellect we must be careful lest we deny its power of forming true and certain judgments as the Academicians did, careful, too, not to credit it with an original science of all things possessed naturally, as Plato did, who said on that account, that to learn is nothing else than to remember." [6] Faced with the alternative of an innatism that no one desired, and the scepticism that manifestly threatened, what path was left open to mediæval thought ? None, save to rehabilitate the sensible order ; and this was what Duns Scotus set out to do, following St. Thomas Aquinas.

For all his desire to stabilize the sensible order and to give it some intelligibility worthy of the name, St. Thomas was never led on that account to derogate from the rights of thought. Nor was he even tempted to do it. Truth, in the full and proper sense of the term, is found in thought alone ; for truth lies in the adequation of thing and intellect. Now in this connection it is the intellect that becomes adequate to the thing, it is in the intellect that the adequation is set up, and therefore it is certainly in the intellect that truth resides : *ergo nec veritas nisi in intellectu.* That said, two other things have to be added if we would understand the Thomist attitude to the problem : first, that the adequation of intellect to object is a *real* adequation ; and secondly, that the things to which the intellect conforms are themselves conformed to another intellect. Let us consider each of these points in turn ; for they are essential to any understanding of the Christian position in this matter.

Truth, to put the case precisely, is in the intellect affirming that things are or are not, and judging them to be this and not that. It is commonly objected to-day that this conception of truth as a " copy " will not stand examination, since the intellect is wholly unable to compare the thing as it really is, which it never reaches, with the thing as represented to itself, which is all it can ever know. If modern idealism has nothing more than that to oppose to mediæval realism, then I venture to say that it no longer has any conception of what a genuine realism is. Undoubtedly, the unfortunate habit we have acquired since Descartes of proceeding always from thought to things, leads us to interpret the *adequatio rei et intellectus* as if it involved a comparison between the representation of a thing and this mere phantom, which is all that the thing outside all representation can be for us. It is easy to amuse oneself by denouncing the innumerable contradictions in which epistemology involves itself when it enters on this path, but we should have the justice to add, since it is a fact, that the classical mediæval philosophy never entered on it at all. When it speaks of truth it does indeed refer to the truth of the judgment, but if the judgment is conformed to the thing it is only because the intellect putting forth the judgment has first itself become conformed to the being of the thing. It is of its very essence to be able *to become all things* in an intelligible manner. Its affirmation, therefore, that such and such a thing is, or is this and not that, is due to the fact that the intelligible being of the thing has become its own. It is true, of course, that we do not find our judgments in things themselves, and it is just on that account that our judgments are not infallible ; but the content of our concepts at least we find in things, and if, in normal conditions, the concept always represents the real thing apprehended as it really is, it is because the intelligence would be unable to produce the concept at all had it not itself become the thing expressed by the concept, the thing to the essence of which the judgment is to conform itself. In short, the adequation between thing and intellect set up by the judgment, always presupposes a prior adequation between concept and thing, and this, in its turn, is based upon a *real* adequation of the intellect and the object informing it. Thus it is in the primitive ontological relation between intellect and object and in their real adequation, that there resides, if not truth in its perfect form, which appears only in the judgment, yet at least the root of this equality of which the judgment takes cognisance and which it expresses in an explicit formula.

Thus in the philosophy of St. Thomas the word " truth " has three different but closely allied meanings ; one proper and absolute and the other two relative. In a first relative sense, the word " true " designates the basic condition without which no truth would be possible, that is to say being. There could in fact be no truth at all without a reality which can be said to be true when brought into relation with an intellect. In this sense it is therefore quite correct to say with St. Augustine that truth is what is : *verum est id quod est.* In its proper and absolute sense, truth consists formally in an ontological accord between being and intellect, that is to say in a conformity of fact set up between these two, as it is set up between the eye and a

colour perceived ; and this is expressed in the classic definition of Isaac Israëli : *veritas est adaequatio rei et intellectus*, or
again in that given by St. Anselm and also adopted by St.
Thomas : *veritas est rectitudo sola mente perceptibilis;* for the
rectitude of a thought which conceives that what really is,
is, and that what really is not, is not, is indeed an adequation of fact. Finally there is the logical truth of the judgment
which is but a consequence of this ontological truth : *et
tertio modo definitur verum secundum effectum consequentem*, so that
here knowledge is the manifestation and declaration of the
already realized accord between intellect and being ;
knowledge results, and literally flows, from truth as an effect
from its cause. And that is why, being founded on a real
relation, it has no need to ask how it shall rejoin reality.[7]

Hence in the first place we see that the modern critics
of scholasticism do not so much as suspect either the nature
or the depth of the gulf that divides them from it ; and we
also perceive how vital it was for mediæval thought to maintain the intelligibility of the sensible order intact. Truth,
properly speaking, is in the human intellect, but must in a
way reside also in things, albeit in things only as related to
an intellect. And primarily it must be in things as related
to the divine intellect. This is what is definitive in the conclusions of the *De Veritate* of St. Anselm. When man is in
question truth is above all in the human intellect, but,
absolutely speaking, it resides in the divine intellect. There
is therefore one sole truth for all things, in this sense : that
the truth of the divine intellect is one and that it is from this
one truth that the multiple truths of particular things derive ;
but there is, nevertheless, a truth proper to each thing
appertaining to it in the same right as its very entity. It
is communicated to each being by God and is therefore
inseparable from it, since things subsist only because the
divine intellect brings them to being. The creative action,

in producing beings, bestows therefore an inherent truth
upon each in the very act : namely the entitative truth of
things to which our intellect achieves an adequation, or of
the intellect itself which achieves it.[8] In any event there
could be no intellection unless the sensible object known
were endowed with its own proper intelligibility.

If therefore we put Matthew of Aquasparta's question to
St. Thomas there can be no doubt at all about the answer.
When things themselves are eliminated can any truth
remain ? Certainly : the truth of the divine intellect subsists, but the relative truth of things would disappear with
things, or with the human intellect which alone can perceive it. The Christian doctrine of creation carries, of
necessity, this consequence : for either the term of the
creative act is null, in which case there would be no being
and no formal truth of things to know ; or else this term is
positive, and then things are certainly real along with their
inherent truth. In any Christian philosophy fully conscious
of the meaning of its own principles, there will be an order
of truth contingent as to its existence, but as stable, as to its
essence, as that of the being to which it belongs : *nulla res
est suum esse et tamen esse rei quaedam res creata est, et eodem
modo veritas rei aliquid creatum est.*[9] This created truth, the
concept of which is characteristic of Christian philosophy,
belongs therefore of full right to sensible things, as well as
to the perception apprehending them. What remains of
the Augustinian critique of sensation is this : that sense,
because bound up with a corporeal organ, cannot achieve
that complete return upon its own act which would be
necessary for knowledge of the act. The return indeed
begins, for the animal feels that it feels ; but is not completed, for it does not know what feeling is. Nevertheless, if
truth in its explicit form is not to be found in the sensible,
it has a foundation there : *veritas est in sensu sicut consequens*

actum ejus, dum scilicet judicium sensus est de re, secundum quod est.[10] From the fact that sense does not know truth it does not follow that what it knows is not true ; quite the contrary, the intellect has only to turn to the data of sense and it disengages truth.

For my part I think that the greatness and vitality of the Franciscan school is nowhere more apparent than in the ready way in which its representatives grasped the necessity of this conclusion, and, having grasped it, set out to demonstrate it in accordance with their own principles. Many years ago a friend of mine, who happened to be a sociologist, congratulated me for treating the philosophies of the Middle Ages as those of groups, of collectivities. But the fact is that the groups were all formed around individuals, and nowhere does this more clearly appear than in the present case. By all Franciscan tradition Duns Scotus should have taken up the cudgels against St. Thomas and voiced the suspicion thrown on the sensible by the Augustinians of his Order ; but judging the position to be philosophically untenable, he purely and simply rejected it, reserving the right to find some other means of saving whatever truth there is in St. Augustine's doctrine.

He, too, had narrowly scanned the problem raised by the ninth of Augustine's *Eighty-three Various Questions.* He knew all the other texts, become by his time classical, even hackneyed, alleged in support of the one in question : but the whole mass of Augustinian citations was outweighed by the single sentence of St. Paul : *Invisibilia Dei a creatura mundi per ea quae facta sunt intellecta conspiciuntur* (Rom. i. 20). For, on St. Augustine's own showing, these *invisibilia Dei* are the divine Ideas : " Therefore the Ideas are known from creatures ; therefore also prior to any vision of these ideas, there must be some certain knowledge of creatures." [11] In thus reversing the Augustinian presentation of the

problem, Duns Scotus commits himself to a justification
of the validity of sense cognition, and we may watch him
actively employed in the task.

This is so much the more remarkable inasmuch as Duns
Scotus' whole noetic tends to strengthen as much as possible the independence of the intellect in respect of the
sensible order. It is just this that distinguishes it from the
noetic of St. Thomas. Throughout all its emendations of
Augustine's doctrine a vein of Augustinianism persists ;
sense knowledge is never anything more for him than the
" occasion " of intellectual knowledge. But even while he
thus limits its *rôle*, Duns Scotus considers it absolutely
necessary to establish its validity : it must be beyond
attack in the exercise of its proper function. Why this
solicitude ? Because Duns Scotus very clearly saw that the
New Academy is the legitimate daughter of the Old ;
Plato's transcendent idealism is quite compatible with a
complete scepticism as regards the world of bodies ; and
on account of his very Christianity Duns Scotus could
never reconcile himself to that. He finds himself set down
in a world of created substances ; and he must first acquire
a knowledge of these if he is to rise to that of the Ideas ;
he cannot possibly install himself in the world of Ideas and
look down complacently from that height on the world of
bodies. Aristotle warns him that behind the Augustinian
criticism of sense he will find not only Plato but Heraclitus.
For indeed it is Heraclitus who furnishes the major proposi-
tion of the syllogism, and this major is false : *antecedens
hujus rationis, scilicet quod sensibilia continue mutantur, falsum
est : haec enim est opinio quae imponitur Heraclito.*

In thus taking up an attitude of opposition to the Platonic
depreciation of the sensible, Duns Scotus is naturally led to
criticize the Augustinian conception of illumination. His
chief objection is that it tries to build on a sinking founda-

tion. If the sensible element in knowledge is such that it eludes the grasp of science, if in short it is essentially fugitive, how can even the divine illumination give it the stability it needs ? Either the divine light will be drawn along with the sensible and become no less fugitive itself, and in that case nothing will be gained, or else it will so transfigure the sensible that our knowledge will no longer have any relation to the true nature of its object ; and we shall fall back into the error with which Aristotle reproaches Plato. If the situation is to be saved there is but one resource left, and that is to admit an empirical certitude based upon experimental reasoning. It is true, of course, that an induction of laws from a starting point in experience will lead to no absolutely necessary conclusions ; there is no contradiction in supposing that things might be produced otherwise than they are in fact produced, but our knowledge of the laws of reality is none the less certain and exempt from error for all that, since it rests precisely on the stability and necessity of natures. The great principle that guarantees the validity of experimental science is that all that happens regularly in virtue of a non-free cause is the natural effect of that cause. *Natural*, that is to say non-accidental, necessary ; so that even the natural sciences themselves, though only to be acquired through experience, will nevertheless contain an element of necessity.[12]

The whole of Duns Scotus' doctrine on this important point may be summed up by saying that he is opposed to the view of Heraclitus on the ground that if natural things are fugitive their natures are not. It was therefore not by chance, nor in virtue of any mere freak of history, that the thirteenth century gave birth to the work of Roger Bacon, and the fourteenth to the first tentative developments of positive science. We should deceive ourselves, however, if we credited the mediævals with any love of science for

science's sake, with any " disinterested " science as we like
to call it to-day. Of course, as far as practical ends are
concerned, their love of science was as disinterested as ours,
often indeed a good deal more so, but they never regarded
the knowledge of things as an end in itself. They turned
towards nature because they were Christians, and as Christians they studied it ; for what they saw and loved in it
was nothing other than the work of God. Hence comes that
vein of religious tenderness with which they numbered all
its marvels, and hence too their ever watchful care to safeguard its intelligibility.

Mediæval realism thus became the heir to Greek realism
for quite another motive than that which inspired the philosophy of Aristotle ; and it is this that gives it its own peculiar
character. Aristotle turned away from Platonic idealism
because man's kingdom is a kingdom of this world, and
because above all else we need to know something of the
world in which our lot is cast. Christians turned away more
and more resolutely from Platonic idealism because the
kingdom of God is not of this world, but because the world,
on the other hand, is necessary as a starting-point from which
to rise to the kingdom of God. To dissolve it into a flux of
inconsistent appearances is to snatch from us our best means
of rising to the knowledge of God. If the work of creation
were not intelligible what could we ever know of its Author ?
Were we presented with nought but an Heraclitean flux,
would a work of creation be even imaginable ? It is just
because all is number, weight and measure that nature
proclaims the wisdom of God. It is precisely in its fecundity
that it attests His creative power. Because things are of being,
and no mere quasi-nought, we know that He is Being. Thus
what we learn concerning God from revelation the face
of the universe confirms : " The creatures of this visible
world signify the invisible attributes of God, because God

is the source, model and last end of every creature, and because every effect points to its cause, every image to its model, every road to its goal." [13] Suppress all knowledge of the effect, the image, and the road, and we shall know nothing of the cause, the model, and the goal. The philosophical realism of the Middle Ages was nourished on Christian motives, and a realism there will always be as long as the influence of Christianity continues to make itself felt.

We often speak, nevertheless, of a Christian idealism, and no doubt the expression has an acceptable meaning ; but the type of Christianity then in question is something very different from mediæval Catholicism. For Lutheranism, characterized by a theology that does not lead to any great interest in a nature irremediably corrupted by sin, idealism is a philosophical issue ready made to its hand ; perhaps we had better say, to be more exact, that it was natural that it should be Lutheranism that issued in idealism. Reducing the history of the cosmos to the drama of the individual soul's salvation, the true Lutheran is not drawn to seek God in nature, he feels God at work within his soul, and that is enough for him. Of course mediæval Catholicism never overlooked the fact that nature is at grips with sin, and stands in urgent need of grace ; it insists again and again that the material world is made for man and man is made for God. This anthropocentrism, and the geocentrism that seems to go along with it, have been too often cast up against mediæval thought to allow any room to-day for the reproach that it underestimated the importance of the human standpoint. It remains true to say, nevertheless, that in a deep sense the Middle Ages always kept itself at arm's length from that type of anthropocentrism so dear to the heart of our contemporary idealists.

We have here, indeed, one of the most extraordinary

spectacles in the history of thought ; the first act opened
with Descartes and the play still proceeds. The Christian
philosophers were certainly persuaded that nature was made
for man, and in this sense it is true to say that man stands
at the centre of the mediæval world ; but they recognized,
none the less, that since the universe was created by God
it is endowed with an existence proper to itself, it is something that man can know but could never pretend to have
created.[14] Guaranteed by the efficacy of the divine action,
the nature of things always remains a distinct reality for
thought, something to be received from without, accepted,
mastered finally, and assimilated. When Kant proclaimed
that he was about to effect a Copernican revolution by
substituting his critical idealism for the dogmatic realism
of the Middle Age, the thing he really brought about was
the very opposite.[15] The sun that Kant set at the centre of
the world was man himself, so that his revolution was the
reverse of Copernican, and led to an anthropocentrism a
good deal more radical, though radical in another fashion,
than any of which the Middle Age is accused. It was only
in a local sense that mediæval man thought himself to be at
the centre of things ; the whole creation of which he was
the destined crown and end, which he recapitulated in
himself, was none the less something outside himself,
something to which he had to submit and conform himself
if he would know anything of its nature. But modern man,
brought up on Kantian idealism, regards nature as being no
more than an outcome of the laws of the mind. Losing all
their independence as divine works, things gravitate henceforth round human thought, whence their laws are derived.
What wonder, after that, if criticism has resulted in the
virtual disappearance of all metaphysics ? If we would pass
beyond the physical order there must first of all be a physical
order. If we would rise above nature there must first of

all be a nature. As soon as the universe is reduced to the laws of the mind, man, now become creator, has no longer any means of rising above himself. Legislator of a world to which his own mind has given birth, he is henceforth the prisoner of his own work, and he will never escape from it any more.

If we consider the spirit of the Kantian reform we go far to grasp the spirit it proposed to make an end of. For through the dogmatism that Kant attacked it was mediæval realism that he aimed at. Criticism shall make all things new, *cum revera sit infiniti erroris finis et terminus legitimus*.[16] But what he never seems to have suspected is, that mediæval realism can never be uprooted from the mind save along with the Christian spirit that ruled its evolution and assured its growth. The mediæval thinkers learned from Genesis that the world is God's work and not man's ; and from the Gospel that man's end does not lie in this world but in God. Turning the whole problem the other way about, critical idealism made it for ever insoluble. If my thought is the condition of being, never by thought shall I be able to transcend the limits of my being, and my capacity for the infinite will never be satisfied. Even if my thought does no more than provide the *a priori* conditions of experience, there will always remain interposed between myself and God the veil of the categories of the understanding, shutting off all knowledge of His existence here on earth and all beatifying vision of His perfection hereafter. Doubtless it is possible to imagine a complete transmutation of man which would qualify him for a mode of knowledge foreign to his nature, but this is just the sort of thing which the Christian thinkers regarded as very dubiously philosophical and considered it necessary to avoid. Taken simply as he is, the wayfaring man should be found to be *en route* for an end which, however much it may transcend the forces of his nature, is not essentially beyond its capacity ; his intellectual activities, remaining precisely what they are, should be capable of leading him to this ; docile in face of reality, enriching without ever being able to satiate itself with the essences of beings, human thought should be gently led towards Being and should thus prepare itself for the sole Object which will finally saturate and fulfil its nature.— But how our thought stands to this Object, and in what sense it is capable of it, are questions that remain to be examined in the following chapter.

CHAPTER XIII

THE INTELLECT AND ITS OBJECT

THE question of the object of human knowledge covers in reality two questions—distinct, and yet at the same time closely connected. The first relates to the *natural* object of knowledge and may be stated thus : what class of beings fall directly and of full right within the grasp of an intellect like ours ? The second relates to the *adequate* object and asks whether what is naturally knowable to us suffices in itself to fill the whole capacity of the intellect ? Two powerful metaphysical syntheses, those namely of St. Thomas Aquinas and Duns Scotus, resolved the question by two different methods, but quite in the same spirit ; and our purpose cannot be better served than by considering these two, with an eye on their differences as well as on their agreement.

When philosophers are concerned to safeguard man's power to know God, the simplest way of setting about it might well, at first sight, seem to be to make God the natural object of the intellect. The mediæval systems are often in fact interpreted in this sense ; that is to say when the critic has grown tired of setting them aside as mere theologies, and prefers to condemn them instead for mysticism. Nothing, indeed, could be more wide of the mark. That God is made the natural object of our cognitive powers was, on the contrary, an accusation freely levelled by the mediævals against each other ; and that precisely because the danger was felt to be so serious and because everybody flattered himself that he had avoided it more completely than his neighbour. St. Bonaventure makes the accusation against Grosseteste, Duns Scotus makes it against Henry of Ghent, and more than one modern Thomist makes it to-day against Duns Scotus. Let us make no mistake about it : the point is one that decides the fate of any Christian epistemology and is well worth the trouble of a little careful attention.

As in other matters, so also in this ; there were Christians who were strongly tempted to follow the line of least resistance and turn to Platonism for the principles of a solution. Not so St. Thomas Aquinas. Far from agreeing with Plato that the proper and natural object of our intellect is the intelligible Idea, to which we should endeavour painfully to rise by a violent effort of detachment from sense, he declares himself at one with Aristotle—and with experience —in affirming that, in this life, we can form no concept unless first we have received a sense impression, nor even return later on to this concept without turning to the images that sense has left behind in the imagination. There is therefore a natural relation, an essential proportion, between the human intellect and the nature of material things ; whence it results that even granting the existence of purely intelligible Ideas, such as those of Plato, the very fact that their nature puts them out of the reach of our senses makes it impossible to regard them as the natural object of the intellect. Now to eliminate the Platonic ideas from the normal field of human cognition is to eliminate all cognate objects whatsoever, that is to say all objects that transcend sense experience.

In the first place we shall have to deny, against certain philosophers, that by means of concepts abstracted from the sensible we could ever arrive at any proper knowledge of purely intelligible substances. Pure intelligibles are in fact of an entirely different nature from sensible essences ; wherefore we may abstract, refine, and purify as much as we will, but we shall never succeed in making what is given us as sensible into a representation of the intelligible. An abstract knowledge of the pure intelligible is doubtless better than none at all, but never for a moment could we take it as a proper knowledge of the intelligible as such. Now what is true of non-sensible objects, such as the pure Intelligences, is obviously so much the more true in the case of God. For the pure Intelligences are distinguished from ourselves only by immateriality ; they are substances of another kind than man because not composed like us of matter and form, but they remain substances nevertheless, because composed of essence and existence. Although they and we, in short, do not belong to the same natural genus, both fall, nevertheless, within the same logical category, namely that of substance, or, more simply, of creatures, that is to say of all that which, because it is not Being, does not contain in its essence the sufficient reason of its existence. But God falls neither within the same natural genus as ourselves, since He is not composed of matter and form, nor into the same logical category of substance in general, for He is not even composed of essence and existence. How then can He Who at all points transcends both the human soul that knows and the sensible object that it knows, fall within the natural grasp of our intellect ? Thus, then, the Thomist answer to the whole question will be both simple and clear : an intellect which bears naturally on sensible things cannot naturally have God for object.[1]

The Scotist position is more complex and less easy to seize, but as far as conclusions go, it is in entire accord with those of Thomism. In either doctrine it is equally true to say that the human intellect cannot apprehend the pure intelligible, *pro statu isto ;* [2] but they diverge in their conception of man's actual state and of the reason why he is in this state. When St. Thomas declares that the intellect must necessarily turn to the sensible *secundum statum praesentis vitae* he presupposes that man's state in this life is also his natural state, that in which he is placed in virtue of the mere fact that his nature is a human nature. No doubt that nature is wounded by original sin, but it could not have been changed, since to change a nature amounts to destroying it. How, moreover, could it be otherwise ? Since the union of soul and body is a natural union, the state that results from the union is a natural state, and knowledge by way of abstraction from sense, which is bound up with this state, is natural too : *anima ex sua natura habet quod intelligat convertendo se ad phantasmata*. So far does St. Thomas go in this direction—and he always goes as far as his reason demands—that when he recognizes that the soul, when separated from the body, must be capable of direct knowledge of the intelligible, he adds that the state in which it then finds itself is no longer its natural state. The thing is obvious. Soul is united to body precisely in order that it may operate in accordance with its nature—*unitur corpori ut sic operetur secundum naturam suam*—and the mode of cognition open to it when separated from the body may perhaps in itself be nobler, but certainly cannot be the soul's natural mode of knowing.[3]

But what Duns Scotus calls a *state*, on the other hand, is not quite the same thing as that to which St. Thomas applies the word. For him, the state of a being is not defined in function of a nature that suffices to assure the state ; it is not even something demanded as of right by the internal necessity of a nature, it is simply a stable mode of being, the permanence of which is assured by the laws of the divine wisdom : *status non videtur esse nisi stabilis permanentia legibus divinae sapientiae firmata*. Hence, in the philosophy of Duns Scotus the state of the intellect may be pre-

cisely what it is in the philosophy of St. Thomas, without
on that account authorizing us to draw the same deductions ;
and that is precisely how it turns out in connection with the
present problem. It is very true, according to Duns Scotus,
that the human intellect, taken in its actual state, is unable
to form any concept without first turning to sense ; but
that is simply due to the state in which it finds itself in fact,
not at all to a state in which it should be as of right. Without
an initial movement of the intellect by sensation there is
no intellectual knowledge. But why ? Perhaps—and this
is an hypothesis to which Duns Scotus returns with a certain
complacency—as a divine punishment for original sin ;
perhaps again, simply because God desires this strict collaboration between our various cognitive faculties. Whatever the reason may be supposed to be, this at least is
certain ; that nothing in the nature of the intellect as such,
nor even in the nature of the intellect as united to the body,
makes it a matter of necessity for it to turn to sense in order
to exert its activities.[4] But if the case stands thus, we can
no longer say that the proper object of our intellect is the
essence of the sensible thing ; still less can we maintain
that the pure intelligible is out of its reach on account of its
very intelligibility ; nor, finally, can we pretend that the
reason why God is not the natural object of knowledge is
because sensible reality alone is naturally proportioned to
our intellect. And that is also why Duns Scotus, thus
brought up short in the epistemological field, turns instead
to metaphysics for a solution. The path on which he enters
here is not without its thorny places, but I venture to ask
you to follow him down it for a while, being confident that
you will have no regrets.

Let us agree to take the word " motion " as meaning the
action exercised by one real being on another ; and let us
then ask whether God can be considered as an object

capable of moving our intellect naturally ? To this question two answers only are possible, for there are but two conceivable kinds of motion, natural motion and voluntary motion. A natural motion is that which one being exerts on another in virtue of an internal necessity of its very nature ; it has only to be and at once the motion is produced, and once it is, it cannot but produce it. A voluntary motion flows from a free decision, and consequently it depends wholly on this decision whether it is put forth or not. Thus the question is open to two answers only : either God moves our intellect as one natural being acts on another, in which case we shall have to say that He is its natural object : or else God moves our intellect to know Himself by a decree of His free will, and then we shall have to say that between Him and us no natural relation is conceivable, not even that which obtains between a knowing subject and the object known. Of these two answers which must we adopt ?

One of the points most frequently misapprehended in the philosophy of Duns Scotus concerns the *rôle* of the will. " Scotist voluntarism " is a very well-worn expression. And, indeed, it is true that the will plays a considerable part in his doctrine, but there are limits to this, and the most important of all, the thing that altogether divides Duns Scotus from anyone who puts the will at the root of being, lies in his refusal to admit that any voluntarism is possible in God with respect to Himself. To put this more precisely : since we class all conceivable motion as either natural or voluntary it is quite impossible that the absolutely first motion, that from which all the rest is suspended, should be voluntary. What is at the bottom of things can be none other than a nature ; it can by no means be a will. The reason for this is easy to grasp. In order to will, it is first of all necessary to know ; the act by which the will wills presupposes, therefore, the act by which the object willed is made known to the intellect. Now the movement of the intellect by its object is a natural one. Therefore there must be a natural motion prior to any act of the will. If now we apply the preceding distinctions, however improper they may be when God is in question, we shall see the solution of our problem emerge from them by way of necessary consequence.

The first of all natures capable of putting forth any motion, that prior to which, absolutely speaking, none other is possible, is the divine essence itself, since this precedes all the rest as the infinite precedes the finite. Now this first of all natural motions can be received by nothing save the divine intellect, so that the first of motions is the natural motion of God's intellect by His essence : *omnino prima motio est naturalis motio divini intellectus a suo objecto.* Then, and only then, can the divine will intervene, to seize in an act of love the infinite essence thus known and expressed : the voluntary act follows the natural act, and just as God's intellection of Himself generates the Word, so, from the love of God for Himself, proceeds the Holy Spirit. The circle of the immanent operations of the first essence is thus closed : but, closed as it is, we already catch a glimpse of what will make possible the existence of all the rest.

From the standpoint of metaphysical analysis in fact, the first of all conceivable motions does not terminate in the intellection of the divine essence taken solely in its infinity and necessity, but in the same act it attains to the finite and creatable participations of this essence, to those beings, namely, whose possible existence, if actualized, would be analogies of the Being Who alone can bring them to existence. But by a deep intuition, which takes us into the very heart of his thought, Duns Scotus calls attention to the fact that the knowledge of the possibles in God cannot be of the same nature as God's knowledge of His own essence. For His essence is necessary, while, by definition, the possibles are not. Taken in themselves, in their pure possibility, they lack the requisite determination to fall under a distinct knowledge. This, moreover, will be easily seen if we consider the case of future contingents. Suppose, in fact, that the divine intellect, taken in itself, and wholly apart from any voluntary determination, knows that of two possible events one must happen ; then, either that event remains contingent in itself, in which case the divine intellect might possibly be deceived ; or else the divine knowledge of it is infallible, and in that case the event would cease to be contingent. Thus, then, as long as we remain strictly within the order of intellect, a distinct knowledge of the contingent remains impossible. What, then, is required to make it possible ?

First of all, from the standpoint of our present metaphysical analysis, the divine will must decide to will a certain determinate being or certain determinate event. Secondly, the divine intellect, seeing this infallible determination of the divine will, must know, in consequence of this decision, that such and such a being is to exist or such and such an event to take place. Exterminating Avicenna's necessitarianism at its very root, Duns Scotus proves, therefore, that, far from flowing from the first being in virtue of any natural law, creatures could not even have any distinct ideas in the first being without a free intervention of His will. That they may be determinately conceived, God must first of all will them ; the essence of the necessary being, therefore, necessarily moves His intellect, but the essence of the contingent being can move it only contingently, since, in order to move, it must first of all be, and its being depends on a will. But the converse holds no less, and thus we come in sight of our conclusion.

Nothing contingent can be a necessary and natural object for the divine intellect. Not its necessary object ; since the sole necessary object is the divine essence : nor its natural object, since it acquires the character of object only in virtue of a voluntary decision. But then also conversely, no created intellect can have God for natural object, for if there is no natural relation between the creature and God, neither can there be between God and the creature. That this should be otherwise would demand that the relation of our intellect to the divine essence should be the same as that of God's intellect to God's essence. Now we know very well that this is not so, since our intellect is contingent like our existence itself. That man should know God demands that the will which brought him from the state of a possible to a state of a real being should intervene anew in order to bring him across the abyss between the finite and infinite. The Infinite alone can do this. Thus a being who exists only in virtue of a free decision of God, because it has no necessary relation to God, will know God only in virtue of another divine decision equally free, because there is no natural relation between a contingent intellect and a necessary essence.[5] This decision God freely made in creating this visible world that manifests His glory, and by the consideration of which we raise our minds to Him ; and He makes it again in raising to Himself the blessed who rejoice to see Him face to face ; but in either case, whether indirectly through the creature or directly by His grace, the human intellect rises to the knowledge of God only because God Himself has willed it.[6]

In constructing this synthesis Duns Scotus not only cuts away the root of Arabian necessitarianism, but at the same time establishes the essential incompatibility of all ontologism, both with his own doctrine and with Christian philosophy in general. When we consider how often he has been accused of this, we might almost come to despair of any effort to be clear, since we see him saddled with that very theory of knowledge which most of all he detests, and which his own metaphysic shows to be radically false. Now here, on this plane of master theses and directive ideas, he arrives by his own path at the very positions that St. Thomas attains by his. For it is very evident that we have to do now with none of those questions of method where technical differences are always possible, it is the very essence of Christian thought that is concerned, and if St. Thomas were not already safe from ontologism in virtue of the all-sufficient barrier of his empiricism, he would have put up as many other safeguards as Duns Scotus did in his metaphysic. For both thinkers, in short, the vision of God's essence is altogether proper to God ; no creature can attain to this unless God raises him up to it : *nisi Deo hoc faciente.* In both doctrines the radical difference that divides necessary being from contingent being supposes a breach in the order of existence prolonged into the order of knowledge. The divine intellect alone, says St. Thomas, of itself sees the divine essence, and it sees it precisely because it is the divine essence itself : if our intellect, which is not the divine essence, is to know it, it must be rendered capable of knowing it by a divine action : *haec igitur visio non potest advenire intellectui creato nisi per actionem Dei.*[7] Thus, in both systems, a natural knowledge of the divine essence is a contradiction in terms : and that is why God cannot be the primary and natural object of our intellect.

This very firm conclusion does not make the second part of our problem any easier. We have now in fact to find a natural object of the intellect which, although not God, will yet put us on the way to God. In other words, if the knowledge of God is not our knowledge of God, the beatitude it brings will not be our beatitude ; now that this knowledge may one day be ours requires that our intellect should at least be able to be made capable of it, and consequently also, that there should already be found in it some foundation for that capacity. Let us ask St. Thomas and Duns Scotus in turn where they find this, and how they regard the relation of the human intellect to the most perfect of all objects.

St. Thomas' position is clearly marked out in advance by his theory of knowledge, and never for an instant do we find him trying to evade its necessary consequences. Our intellect forms all its concepts by the aid of sense intuition, wherefore, in its present state, it can have no object that is unattainable by means of such intuitions. But there is nothing at all to be alarmed about in this. St. Paul says that we can rise to the knowledge of God from our knowledge of creatures, and it is thus clear that an epistemology that binds itself to proceed by way of sense will always find at least one way open to God : that, namely, which finds its starting-point in the spectacle of creation. Sensible empiricism, therefore, entails no agnosticism in the field of natural theology ; but if it does not close the way to God we may still ask what knowledge of God it gives us ?

If we start from the sensible, God is not to be attained except as the creative cause of the world of bodies ; and all our natural knowledge of God will be limited to what we can know of such a cause through the medium of its effects. That is very far from nothing, nor is it even a little, but it certainly is not all. We can, in fact, reason from the existence of contingent beings and conclude to the existence of a necessary being. We know therefore that He is, and that He is the first cause of all the rest ; and that knowledge is enough, as we have seen, to effect a complete transformation in our philosophic interpretation of the universe. This term once attained, it is still possible to circumscribe, in a measure, the divine essence thus posited in existence ; for if God is the first cause in the order of being, we can be sure that He altogether transcends every given being and, indeed, every being conceivable by a created mind. In the intellectual effort to deny to God all those limits that we find inscribed in the sensible being that we know, we are led to posit the existence of a supereminent essence, entirely distinct from all the effects it causes. That said, all that man can say is said. The human intellect knows this essence as existing, it does not penetrate it as such, and as we have already seen, we know that by its own powers it will never do so. Dionysius had good reason then to say that the God Whom our reason reaches remains, so to speak, an unknown God : *Deo quasi ignoto conjungimur ;* for we know, indeed, that He is, and we know what He is not, but what He is remains wholly unknown to us : *de Deo quid non sit cognoscimus, quid vero sit penitus manet ignotum.*[8] Certainly the distance between the intellect and God is immense in Thomism ; it is, if one may say so, a " *distantia maxima* " ; nevertheless, it is not such that God Himself could not overcome.

It is to be noted in the first place, that however feeble the intellect may be, it is and remains an intellect ; that is to say a capacity to become, in a manner, all things by way of representation. It is due to its lowly estate among intellects that it can assimilate only the intelligible enclosed in the sensible ; but what it seeks in the sensible is precisely the intelligible, and nothing will quench its thirst for that as long as there remains any intelligibility to be assimilated. This is clear enough if we consider only the scientific or philosophic life of the mind. Limited as we are by the natural object of our investigation, at least we can never rest until we have attained its limits ; that is, until we have gathered up all the actually acquired material of knowledge under a small number of first intelligible principles. It is as if the whole life of the mind were animated from within by a natural desire for the fullest possible unification ; or rather, this natural desire is merely another name for the life of the mind itself ; it is, we might say, the contingent mind itself, striving to actualize its latent possibilities, to fulfil and realize itself. That, moreover, is why no natural desire can be in vain ; for the mere fact that it exists supposes an active possibility—one which is aware of itself in the case of an intellect—and the desire simply expresses its tendency to self-actualization. But note that we say " possibility," and nothing other than possibility, for the success of the effort does not depend only on the subject that puts it forth, it depends also on the accessibility of the object. Something more than an effort is required for success. But an effort arrested *en route* is no vain effort ; it fails of its object, but that does not prove that it has no object. Quite the contrary : even when it is proved that he who desires will never by his own powers attain the whole end of desire, and even if he has no reason to hope for any help that could make him capable of it, the desire experienced remains there still, not in the least extinguished, but rather exasperated into anguish by the very sense of its own impotence. This anguish St. Thomas knew ; and Alexander of Aphrodisias knew it, and Averroes, and Aristotle : *in quo satis apparet quantam angustiam patiebantur hinc inde eorum praeclara ingenia ;* for this is the anguish of the human intellect itself which has in itself the power to become all things, and, grasping the existence of Being from the starting point of sense, would become *that ;* and cannot.

Here, and only here, the familiar formula St. Thomas so often repeats in connection with this question attains its full meaning : *impossibile est naturale desiderium esse inane.* That the desire to see God is natural the whole history of philosophy proves, and, we may add, the personal experience of every man who, from the consideration of the world, rises to the consideration of its cause. The world of sense is given, and we would know the reason of its existence, and the answer is that things are because God is. Knowing that God is we then desire to know what He is ; and there philosophy is brought to a standstill. But far from being suppressed, the desire is only sharpened by its own repulse. Whether it can be satisfied or not, yet as many men as there are to know that God exists so many men will there be to desire to know His nature ; so many souls will there be to know neither rest nor beatitude while deprived of this knowledge, and even to suffer worse miseries than other men, because those who are unaware of their own ignorance are also unaware of what transcendent good they lack. The last word of philosophy is the assured affirmation of the existence of a supreme good and of our impotence to partake of it ; or rather, this would be the last word did not our knowledge of God's existence, and our desire to behold His essence, bear witness at the same time that the vision is possible. The anguish of the pagan soul measures the distance that divides the human intellect from the sole Object that can satiate it ; and that is why the divine promise delivers that soul from anguish, for now, knowing the nature of the intellect and its own immortality, it knows also that, by the grace of God, His beatitude may become our beatitude, because our truth will be perfected by His truth.[9] Let us be more precise on this last point.

Even in the philosophy of St. Thomas there is a point for the insertion of a divine gift which, since it is added as a grace to nature, by no means destroys nature, fulfils it rather, brings it to the highest perfection of which it is capable. That the intellect can receive such a perfection is quite evident, it results from its very nature ; which is a capacity for the apprehension of the intelligible. Between what it is, and what God is, there is no incompatibility as regards the order of knowledge, and for this reason St. Thomas can write that the divine substance is not entirely alien to the created intellect. Certainly it is entirely beyond its reach, but that is only because it is infinitely above it, not because it is formally unknowable : *divina substantia non sic est extra facultatem intellectus creati, quasi aliquid omnino extraneum ab ipso.*[10] The active *power* of the intellect in respect of its proper object, whereby it may be led to order all its knowledge with respect to Being, is thus accompanied by a complementary *possibility*, a possibility of apprehending even this Being itself which it posits as existing without in any way attaining to its nature. We say *possibility*, because, since the proper object of the intellect lies in the sensible, it has no power in itself to rise above that level. And, further, even were the pure intelligible its proper object, yet, since it is itself but a being by participation, it could never naturally attain to Being ; this would still infinitely transcend all its forces. But this possibility is no merely abstract one ; it is a real capacity of a subject, which, although altogether without the means to actualize itself, can nevertheless be actualized by God. It could never be actualized by Aristotle's God. The natures that strive to know him do not owe him their existence ; such as they are such they remain, that is to say in so far as they do remain before disappearing for ever. It can be actualized, nevertheless, by the Christian God, for since He created these natures He can also perfect them. He Who gave them being and made them immortal remains both able and willing to add to His gifts. He can give all there is to give—provided only that they are capable of receiving. Since, then, in itself, the intellect is capable of the total intelligibility, God can bestow it if He wills, and it is in this

sense that, while firmly maintaining the truth of the epistemology of Aristotle, St. Thomas leaves open perspectives
on to the supernatural that will profoundly modify its
significance.

His natural theology, in fact, legitimizes the whole ambit
of the Christian hope, but at the same time it remains as
modest as can possibly be. That of Duns Scotus is the only
one which, with similar firmness of thought, has ventured
to grant us a little more in this life without authorizing us
in the least to make God the primary object of the intellect.
It is a doctrine equally received by both philosophers, in
which for the rest they simply follow Aristotle, that being
is the first intelligible concept attained by our minds. We
can perceive nothing or conceive nothing otherwise than
as a being ; and it is only after we have done that that we
determine the nature of the object thus apprehended.
That is why St. Thomas affirms that being is the first intelligible, and the proper object of our intellect : *ens est
proprium objectum intellectus, et sic est primum intelligibile.*[11]
Only, when he expresses himself thus, St. Thomas always
considers that our concept is and remains an abstraction
from the sensible, so that if we try to apply it to a pure
intelligible, such as God, it is attributable only by way of
analogy, and becomes available only when corrected by
all the necessary negations.

The case is different in the philosophy of Duns Scotus.
I have already noted that, in his view, it is not on account
of its essence that the intellect is obliged to turn to sense, but
simply on account of the state in which in fact it finds
itself and the ultimate reason of which, whatever it be, is
certainly contingent. It is not in the least surprising, after
that, to see the Scotist intellect outrunning the data of
sense, leaving them much farther behind than the Thomist
intellect can possibly do.[12] By birthright, in virtue of its

proper nature, its proper object can only be a pure intelligible, exactly as in the case of the angelic intellect to which it is so closely allied. And, in point of fact, and notwithstanding its state of fallen nature, what it immediately attains is neither the essence of the singular sensible thing as such, nor yet the essence of this singular rendered universal by a logical operation, but the intelligible essence itself, neither singular nor universal, taken in its pure indetermination. And it is this that explains the famous Scotist doctrine of the univocity of being. This does not mean that the divine Being is of the same order as created being ; Duns Scotus is very well aware that they are but analogues. Nor does it mean that being is a universal concept logically attributable both to God and creatures ; for this everyone would admit, with the reservation, of course, that it has still to be determined in what sense the concept is thus attributable. What the doctrine really means is that the quiddity, the very essence of the act of existing, taken apart from the modalities which determine the different modes of existence, is apprehended by the intellect as identical, whatever in other respects the being in question may be. When Duns Scotus says that what first falls under the intellect is being, he no longer therefore understands with St. Thomas the nature of the sensible being as such, but existence in itself, without any determination whatsoever, and taken in its pure intelligibility.[13] To say, under these conditions, that being is univocal as regards both God and creatures, is simply to affirm that the content of the concept applied to them is the same in both cases, not because they are beings of the same order, or even of comparable orders, but because *being* is now regarded as signifying only the very act of existing, or the very existence of this act, independently of every other determination.

Evidently, then, Duns Scotus is able to carry the human

intellect much nearer to God than Thomism. If what he says is true, then at least one of our concepts, and precisely the highest of them all, namely, being, is something more than merely analogical. The great advantage he finds in this position—and he returns to the point with a certain complacency—is that the proper object of the intellect will coincide with its adequate object, and that this adequate object will point of itself to the true Christian God, Who is true, total, and infinite Being. Thus there will be an essential, although of course only virtual, continuity between what nature knows and what grace can give it power to know. Undoubtedly, grace is still necessary, and not less so than in Thomist doctrine. The very fact that being as conceived by the intellect is univocal, far from giving us the divine Being, proves that this still eludes our grasp. Inasmuch as what we think is the act of existing in its common applicability to all that exists, it is clear that it is not of God that we think ; did we do this we should have to conceive Him as the Infinite Being, the Pure Being ; that is to say under modalities that would break through the univocity and bring us back to analogy. The being that is thought as common to God and creatures is, by definition, the being neither of God nor creatures. And that, by the way, is why the Scotist proofs of God's existence are really proofs ; the very fact that they start from being supposes that this concept is not the concept of God ; for were it so there would be nothing to prove, we should have no need to look further. Univocity no more provides a starting point for the beatific vision than analogy, for the dividing line between man and God is situated on another plane ; it results from the fact of creation. Whether the conception of being be analogical or univocal, Being in either case eludes the grasp of beings, and it does so with equal necessity, for what divides these latter from Him is their radical contingence. If God Himself had never said it, how could man have ever known that he is destined to the possession of the infinite Being? But once he knows it, how clear everything becomes! The God Who offers Himself to us is Being; the natural object of our intellect is already being; its adequate object is this very same natural object, but seized this time in its complete indetermination, that is to say, it is still being. To reach God what do we need? That after having recognized the existence of the infinite Being, which natural reason is quite competent to do, God should now bestow on it the beatifying vision of His infinity; that is to say of the divine Being precisely as divine, and of this, as we have seen, every creature is of itself incapable. Thus the Scotist univocity is a radical negation of pantheism, since the common attribution of the concept of being to God and creatures requires precisely that it should not be extended to that which makes the being of God to be God; but at the same time it unifies the whole order of human knowledge in affirming the essential unity of its object throughout all the diversity of the states through which it may pass.

Let us now try to see the new aspect in which an epistemology appears by the mere fact of being integrated to a Christian conception of the universe. Here again there are few formulæ, whether Thomist or Scotist, for which no equivalent is to be found in the writings of Plato, of Aristotle, or even Avicenna; but when Christians repeat the formula there is always a doubt whether they are really saying the same thing. In the present case nothing could be more doubtful, and we have only to follow the mediæval thinkers to their final conclusions in order to realize it. Once it is known that the vision of Truth Itself is possible, and above all once it is actually hoped for, we shall look with quite another eye on the nature of each particular truth—not only as regards its value, but its very nature, for now it will have neither the same cause, nor the same object, nor the same end.

Things, as we have seen, have a truth of their own, because they are, and because all that is, is intelligible. Their truth, indeed, is their being, that is to say their fidelity to their own essence. But this fidelity, while measuring at once the degree of their being and that of their truth, is itself measured by the ideal type of the essence in question, that is to say by the divine idea itself. Thus, for a Christian thinker, the intrinsic truth of things is wholly suspended from the act whereby God thinks them and that whereby He creates them. In the same way the truth of our intellect is a real truth, and one that is truly ours since we make it. However, each time that a true judgment is formulated it stands between two distinct orders of divine relations, each of which conditions it at once, both as regards its content and its existence. Its content is the object, since a judgment is true when what it says concerning things corresponds to what things are. Thus every conceived truth consists in the intellectual apprehension of an essence which is in itself what it is in God, so that, by the interposition of the object, the divine thought rules our thought. And as to its existence, the judgment is something caused by our intellect, but our intellect could not cause it were it not itself caused by God. Every created intelligible light is a participation of the divine light ; everything of its own that the intellect adds to the raw material of experience, whether external or internal, it must of necessity have received from God. Mediæval philosophers may differ about the modes of divine illumination, but all are at one in teaching that God is the Creator and Ruler of intellects, inasmuch as He confers not merely existence but the power of forming the first principles, on which all the sciences, theoretical or practical, depend. In this sense we may rightly say with Albert the Great, that God is the first Mover in the order of knowledge just as He is the first Mover in the order of being,[14] or let us say rather, that He is the first in the order of knowledge because He is first in the order of being : and that is why the Christian intellect has no longer the same object as the Platonic or the Aristotelian.

Whether being is given it as univocal or as analogous, the intellect apprehends it as a participation of the divine Being, and we may say the same of the very act whereby it apprehends it. For us, no doubt, and in the order of acquisition of knowledge, the first object is the sensible thing ; but, absolutely speaking, the first intelligible is God, and so also is He for us in the order of knowledge once acquired ; for then we know that He is the first cause of knowledge itself, and we know also that if there is anything for us to know it is in virtue of the conformity of beings with the divine intellect. *Omnis apprehensio intellectus a Deo est* ; [15] there, at any rate, is something you will never find in Aristotle, any more than you will find the idea of creation it implies. *Veritas etiam rerum est, secundum conformantur suo principio, scilicet intellectui divino* [16]—there again is something not to be found in the same sense in Plato, unless, indeed, we gratuitously credit him with the Augustinian conception of the Ideas and the implied doctrine of the Word ; that is to say with Christianity itself. Thus the metaphysic of Exodus penetrates to the very heart of epistemology inasmuch as it suspends both the intellect and its object from God, to Whom both owe existence. What is new in this is the idea, unknown to the Ancients, of a created truth ordered of its own spontaneity towards Being which is at once its end and source, since by Him alone it exists, and by Him alone it can be perfected and fulfilled.

CHAPTER XIV

LOVE AND ITS OBJECT

For the idea of an intellect aware that it has a natural object but recognizing no other, the philosophers of the Middle Ages had but to turn to the Greeks. For the idea of a love aware that it has a natural object but recognizing no other, the men of the Middle Ages had but to turn to themselves, and in that respect they resembled the men of every age. If, for the rest, they wanted to see the thing in a book, they had Ovid's *De arte amatoria*, a very explicit and detailed manual which they did not omit to read, copy and imitate. That Platonic and Neo-Platonic love that so captivated the Renaissance was not yet known, or known only through the medium of Christian interpretations that profoundly modified its character. Aristotle himself, with the little he had to say on the subject, appeared on the scene only with the thirteenth century ; now already, from the twelfth century, St. Bernard and his disciples had built up a complete doctrine of Christian love. They started from the plain reality : the common experience of egoism and sensuality, in short from human desire in all its naked crudity.

The primary characteristic of this desire, as it came under their own interior observation, lies in the fact that it turns immediately to self and all that is related to self. Such were the terms in which it was described by St. Bernard and William of Saint-Thierry, the two masters of that Cistercian order, whose greatness, however unjustly neglected, was none the less of the purest : having built up the whole science of Christian love from its very basis it retired into silence to live it rather than discuss it. Wisdom less perfect has been often known, but even such wisdom as this had its disadvantages since its discretion has been taken for forgetfulness and neglect of self-justification for confession of error. I would therefore do simple justice to St. Bernard by recalling the exact meaning of his expressions. When he says, and when William of Saint-Thierry repeats, that human love necessarily begins in egoism and carnality he intends merely to state the case as it is in fact. What the state of humanity should be as of right is another matter, but one we shall be bound to consider in due course. Let us be content in the meantime to take the concrete reality just as it is. Man born of carnal desire, needs to live, and cannot live without setting up himself as the object of his own desire,[1] and since he could not desire himself without desiring all he needs as well, we must say that he loves himself in the first place, and then all the rest for pure self-love.

Let us add, however, that as soon as we look at it a little more closely this human desire presents a peculiar characteristic which does not seem in itself to be at all necessary. At first sight there seems to be no reason at all why intelligent beings like men, with all the resources of the world at their disposal, should not succeed in satisfying their desires. So little it seems is needed for the purpose. Epicurus remarked, and not without reason, that with a little bread and water the wise man is the equal of Jupiter himself. Let us rather say that he should be so ; and since the recipe for happiness is so simple we may reasonably ask why so few make use of it. The fact is, perhaps, that with a little bread and water a man ought to be happy but precisely is not ; and if he is not, it is not necessarily because he lacks wisdom, but simply because he is a man, and because all that is deepest in him perpetually gainsays the wisdom offered. It seems as though he could pursue no other end than his own proper happiness, but is quite incapable of attaining it because, although everything pleases, nothing contents. The owner of a great estate would still add field to field, the rich man would heap up more riches, the husband of a fair wife would have another still fairer, or possibly one less fair would serve, provided only she were fair in some other way. The experience is too common to be worth the trouble of many words ; but it is of some importance to recall it here because the great fact on which rests the whole Christian conception of love is this : that all human pleasure is desirable but none ever suffices.

This incessant pursuit of an ever fugitive satisfaction springs from troubled deeps in human nature ; the dull yet poignant restlessness of a creature who seeks happiness and to whom even peace is denied. *Pax* : a magical word for mediæval souls, sweetly and tranquilly announcing the most precious and yet most elusive of goods. For lack of this man wanders restlessly on from object to object, and not wholly without cause, for if what he already has is good, what he has not yet is good too. Hurried along by the stream, he must always let go one good thing to acquire another, drain one pleasure to the dregs to try the next, feel all the weariness of satiety with each as it passes and anticipate at the same time a new satiety with all that is to come. What he would really need would be the consummated sum of all possible pleasures at one blow and therewith also death—simply because such pleasure is not to be made eternal. Hence the madness that fails even to enjoy the passing pleasure for greed of the next that is about to follow it—and pass in its turn. The quest for pleasure is but a mockery of that unattained peace, even as movement is but the changing image of a changeless eternity.

Here, at this precise point, comes in the " conversion "

recommended by mediæval thought. With a little bread and water, a few friends, and such peace as the world gives, there is still no peace because desire for all the rest remains and is not satisfied : *dicentes: pax, et non est pax.* The whole ascesis of Epicurean and Stoic ethics is, in this sense, purely negative ; it demands renunciation of all and offers no compensation. Christian ascesis on the other hand is positive ; instead of frustrating desire by denying its object, it fulfils it by revealing its meaning, for if it can be satisfied with nothing in this world, that is perhaps because it is made for something greater than all the world. Man, then, must either make up his mind to be content with goods that cannot content him—and that would be to recommend a resignation that is first cousin to despair—or else he must renounce desire itself, for it is mere folly to wear oneself out in trying to appease a hunger reborn with every sop that is flung to it. But if he does this what shall he receive in return ? No less than all. The successive steps of the mediæval life of renunciation were but so many successful manœuvres that let slip the shadow indeed, but only to seize upon the prey. " Do you find it hard to lose this or that ? Then seek not to lose it ; for that is what you do when you would lay hold of what cannot be retained." Such is the way, a rough one, no doubt ; but less so than it seems for him who keeps the goal in view : " The way to God is easy ; for we have but to disburden ourselves and we advance. If we had to carry everything with us it would be hard enough. Ease thy load, therefore, let all things slip, then, lastly, renounce thyself." [2] To decide for this renunciation what do we require ?

We must understand in the first place that the very insatiability of human desire has a positive significance ; it means this : that we are attracted by an infinite good. Disgust with each particular good is but the reverse side

of our thirst for the total good ; weariness is but a presentiment of the infinite gulf that lies between the thing
loved and the thing within love's capacity. In this sense
the problem of love, as it arises in a Christian philosophy,
is a precise parallel to the problem of knowledge. By
intelligence the soul is capable of truth ; by love it is
capable of the Good ; its torment arises from the fact that
it seeks it without knowing what it is that it seeks and,
consequently, without knowing where to look for it. If we
regard it from this standpoint the problem of love is either
insoluble or already in fact resolved. On the plane of nature
it will never be solved, for the confused desire for the infinite
will never be satisfied with creatures. If we suppose that
satiety with the world is due to a love of God that is unaware
of itself, then the problem is solved already, for then we
understand both the love itself and the reason for its repulse,
and, moreover, that this very repulse attests the possibility
of victory. But the conditions of victory still remain to be
determined.

For this purpose we must recall what the Christian
universe really is : a sum total of creatures owing their
existence to an act of love. God willed all things, and particularly the intelligent creature man, for no other end
than that they might partake of His glory and His beatitude.
That is as much as to say that all created activity is essentially and necessarily referred to God as to its end. A thing
that is made for God, by the mere fact that it acts, tends of
its own spontaneity to God in virtue of a law written in
the substance of its very being. The quest for the supreme
end may very well be a quest without knowledge, and that
of course is the case for all things unendowed with intelligence, which therefore neither know their own acts nor
reflect on them so as to determine their end. It is none the
less true that even these things tend towards God, for they

act only in view of a certain good which is an analogue of
the supreme Good ; so that every one of their actions tends
to make them a little less dissimilar to the supreme Good.
This quest of God, which in other beings is merely blindly
lived, becomes conscious of itself in the case of man. He
has not sufficient intellectual light to be able to say whether
the Good is accessible, for a good has no more right to
Goodness than a being has to Being ; but he has quite
enough to know that it really is the supreme Good that he
seeks, and the reason why he seeks it. For if what we have
just said is true, then human love, in spite of all its ignorance,
blindness and even downright error, is never anything but
a finite participation in God's own love for Himself. Man's
misery lies in the fact that he can so easily deceive himself
as to the true object, and suffer accordingly, without even
suspecting that he does so ; but even in the midst of the
lowest pleasures, the most abandoned voluptuary is still
seeking God ; nay more, as far as regards what is positive
in his acts, that is to say in all that makes them an analogue
of the true Love, it is God Himself Who, in him and for
him, seeks Himself.

Thus, as we might indeed expect, the end of human love
is also its cause. In the pursuit of our own happiness God
sets out first and always precedes us, at one and the same
time it is for Him that we long and He that makes the
longing : *praevenit, sustinet, implet ; ipse fecit ut desideres,
ipse est quod desideras*.[3] In this quite ultimate sense we may
say that the cause of our love of God is God, for in creating
us He created our love.[4] But if that is true we must go
further still ; for if love is a seeking that would become
possession, to say that without God we should not be capable
of loving God, is to say that without Him we should not
even be able to seek Him. God, then, wills that man should
possess Him already by love so that he may be drawn thereby

to seek Him ; and He wills that man should seek Him by love so that he may possess ; he may seek and find ; what no man can ever do is to be beforehand with God, for none can seek unless he has already found. *Nemo te quaerere valet, nisi quod prius invenerit ;* [5] that word, of which an echo may be heard in a famous page of Pascal's, contains the true significance of the Christian conception of love.

We have only to transpose it into metaphysical terms and at once we see with what doctrinal implications it is charged. At the root of all lies St. John's phrase (i. 4, 16) : *Deus caritas est.* To say that God is love is not to say that He is not being ; on the contrary it is to affirm it a second time, for God's love is but the sovereign liberality of Being, Who, in superabundant plenitude, loves Himself, both in Himself and in all His possible participations. That is why creation is at once an act of love and an act creative of love : God is the cause of love, says Dionysius, *et sicut emissor et sicut progenitor,* and St. Thomas, in his comment, observes that God is the cause of love inasmuch as He generates love in Himself, and causes it in other beings as an image and likeness of Himself. As Being He is the sovereign good and the sovereignly desirable ; therefore He wills and loves Himself ; but since the good He loves is none other than His own Being, and since the love by which He loves this good is none other than His will, which is itself substantially identical with His being, God is His love. Now this love that God generates in Himself, and also substantially is, He causes in others by impressing on them a desire of His own perfection, analogous to the eternal act by which He loves Himself. Therefore we can, and, indeed, must, say that He moves His creatures to love Him ; but here, as always in a Christian philosophy, the first motive cause differs from Aristotle's inasmuch as it is creative. To make Himself loved by another, is to put that other in movement towards Himself, and to cause movement is, in the case of God, to create it. It is therefore altogether one and the same thing to say that God is worthy of love, that He moves beings, that He causes their movement towards Himself, or that He creates in them the very love by which they love Him : *ad Deum autem pertinet quod moveat et causet motum in aliis ; et ideo ad eum pertinere videtur quod sit amabilis, in aliis amorem creans.*[6] Now what is this but to say what St. Bernard had already said and Pascal will repeat : we, may seek God and find Him, but never can we forestall Him ? To confirm the unity of the whole Christian tradition on this point we have but to express in metaphysical terms, that to seek God is to have found Him already.

It is unnecessary to point out that our metaphysicians say it, and indeed how could they avoid saying it, seeing that they describe our love of God as a participation of God ? Eternally pre-existing in the Sovereign Good, flowing thence to things by an act of free generosity, love returns to the good whence it sprang. Here is no affair of a stream that ever gets farther and farther from its source and eventually loses itself. Born of love, the whole universe is penetrated, moved, vivified from within, by love that circulates through it like the life-giving blood through the body. There is therefore a circulation of love that starts from God and finds Him again : *quaedam enim circulatio apparet in amore, secundum quod est de bono ad bonum.*[7] The conclusion, then, is unavoidable that to love God is already to possess Him, and since he who seeks Him, loves Him, he who seeks possesses. Our quest of God is God's very love in us, but the love of God in us is our finite participation in the infinite love wherewith God loves Himself. Borne on the current of divine love which flows through us and returns thence to its source, we can say with St. Augustine that to love God is to possess God.

This new metaphysic of love, wholly based as it is on the metaphysic of being, raises problems equally new touching the nature and even the psychology of human love. The discussion of the difficulties thus raised is of such importance for the history of mediæval thought and literature that we must attempt some explanation of their meaning and point out the novel element their solution implied.

As long as we remain on the plane of humanity and of man's relations with finite goods, no metaphysical problem of love arises at all. There is a moral problem, certainly, and a highly important one : the very fact of human desire and its radical insatiability. Greek ethics might be considered, from this standpoint, as so many efforts to palliate an inevitable evil and render its consequences as harmless as may be, but without, for the rest, holding out any hope of suppressing the evil ; no matter by what route they proceed, they always end up with an act of resignation. The great difficulty that always faced them was to know what ought to be loved : as to what an act of love is in itself and how it stands to its object they felt no mystery at all. All that is, is good ; all that is good is desirable ; if, among all desirable things, we could determine those most useful to ourselves to desire, we should have said almost all that it is possible to say on the matter. But a Christian thinker is in a different position, for he feels little difficulty about what he ought to love, but there is an insistent, nay sometimes an agonizing question, as to whether it is possible to love it at all, and what this love really is. In such a system it is the very nature of love that becomes the difficult point, and it is therefore not in the least surprising that this problem offered mediæval thought one of the best opportunities it ever had of proving its originality.

When God is posited as the absolute Being an absolute Good is posited as well, and, since the good is the object of love, we are led thereby to posit an absolute love. This is expressed in the Great Commandment. Here at once we emerge from the categories of Greek philosophy, but nevertheless at the risk of getting into difficulties never felt by Plato and Aristotle. A finite being desires finite goods, a relative being desires relative goods, and its attitude may be quite simply explained. Lacking what it needs to fulfil or maintain its own being it desires all these things naturally, and desires them of course for itself. In this sense all human love is, spontaneously and normally, a more or less interested love. Not that the Greeks overlooked that disinterestedness demanded by every genuine friendship ; no one who remembers his Aristotle is likely to fall into that mistake ; but not even Aristotle ever dreamed of eliminating all love of concupiscence to make way for pure love of friendship. Greek man never owed more than finite debt to his gods. But the case is changed as soon as we admit the existence of an absolute good, such as that defined by Christianity. Human desire, henceforth, stands in the presence of an object such that it becomes quite impossible to desire it for anything other than itself. We stand once more between the horns of the same dilemma which always arises out of the relations of being to Being ; the only difference is that instead of encountering it in the field of existence, of causality, or of knowledge, we find ourselves at grips with it on the field of love. And just here, as it happens, it is particularly formidable. Man, having a will, naturally desires the good, the thing that is good for him—his good. On the other hand no Christian philosopher can ever forget that all human love is a love of God unaware of itself ; and indeed as we have shown, all human love is an analogical participation in God's own love for Himself. Now in virtue of His own perfection God loves Himself in view of nothing other than Himself ; therefore, as long as we love, not merely other things, but God Himself, in view of ourselves, for our own sake, our love is unfaithful to its own true essence. To love as we ought, we should, in the first place, have to love all things for God's sake just as He loves them, and then, next, we should have to love God for Himself just as He loves Himself. The difficulty lies precisely here : that it is by no means immediately evident that this demand contains no contradiction. A finite being's love for his own good is, and, it would seem, cannot but be, interested ; how then can we demand of him a disinterested love ? When God is said to love Himself for His own perfection alone, the case is simple enough ; since His perfection is such that nothing is left for Him to acquire, He can rejoice in it without power to complete it ; but that man who needs so many things, and needs God more than anything else, can or even ought to love his own supreme good otherwise than as a good to be acquired—is not that a mere impossibility ? There lies the whole knot of the Christian problem of love : an essentially interested participation in an essentially disinterested love, which should become disinterested in order to realize its own essence, and is quite unable to seek to perfect its essence without destroying it. How shall we overcome this difficulty ? By returning once more to our principles.

As the term of our enquiry we seek an act of love whereby man will love God as God loves Himself. Well, we may say once more that either the problem will remain for ever insoluble, or it is already solved. If the love of God were not already within us we should never succeed in putting it there for ourselves. But we know that it is, since we are all essentially created loves of God, and since all our acts, all our operations, are directed of their own spontaneity towards the being that is at once their end and source. The question, then, is not now how to acquire the love of

God, but rather how to make it fully aware of itself, of its object, and of the way it should bear itself towards this object. In this sense we might say that the only difficulty is that of the education, or, if you prefer it, the re-education of love. The whole effort of Cistercian mysticism is therefore brought to bear on this precise point, which is in truth the centre of the whole debate : is there any way by which man can continue to love himself while loving God only for God's sake ?

The first point to be cleared up, if the problem is to be solved, relates to the notion of love itself. Is it, after all, quite so certain that all disinterested love is impossible ? Would it not be much nearer the truth to say, on the contrary, that if it is to be real love all love must be disinterested ? What hides from us the authentic sense of the word love is our habit of more or less mixing it up with desire pure and simple.   Obviously, almost all our desires are interested ; but to say that we love a thing when we desire it for our own sake is a faulty mode of expression ; what we really love in such a case is precisely ourselves, and the other thing is simply desired for our own benefit. But to love is quite another thing : it is to will an object for itself, to rejoice in its beauty and goodness for themselves, and without respect to anything other than itself.

An attitude of this kind is, in itself, equally removed from the opposed extremes of utilitarianism and quietism. Love seeks no recompense : did it do so it would at once cease to be love.   But neither should it be asked to renounce joy in the possession of the thing loved, for this joy is co-essential with love ; love would no longer be love if it renounced its accompanying joy. Thus all true love is at once disinterested and rewarded, or let us rather say that it could not be rewarded unless it were disinterested, because disinterestedness is its very essence.   Who seeks nothing in love save love receives the joy that it brings ; who seeks in love something other than love, loses love and joy together. Love, then, can exist only if it seeks no reward, but once it exists it is rewarded. Thus the idea of a love at once disinterested and rewarded contains no contradiction, quite the reverse ; but even so, the difficulty is not wholly disposed of. Perhaps a man might love God with disinterested love if he could do so with complete forgetfulness of self ; but the question still stands whether such self-forgetfulness is possible, that is to say whether, in view of the absolute necessity of providing in the first place for his own needs, he can detach himself completely from self in order to attach himself wholly to God ?

Such a conversion, taking the word in its literal sense, is certainly not to be looked for on the plane of nature alone, and therefore it lies beyond the scope of philosophy ; but nature at least must be susceptible of conversion, it must have capacity to undergo the effect, and philosophy may properly consider the nature of this capacity. Now whatever the differences between the Cistercian and Thomist schools on this point, a profound unity of inspiration underlies the divergency of method and technique. St. Bernard and William of Saint-Thierry teach, as we have seen, that man of necessity loves first of all himself, and only by gradual steps rises to the love of God ; but we observed at the same time that this is a mere matter of fact, a consequence of the deflection of our natural inclinations by original sin. In himself, as the Creator willed him, man of his own spontaneity loved God more than he loved himself, and it is just for this reason that human love can still be re-educated, rectified, brought back to its proper object. There is no occasion therefore for surprise when the Cistercian masters go on to say that in spite of the downward deflection of our vitiated love, its proper object is still the divine perfection.

St. Thomas, on the other hand, does not overlook the fact
that human love too often stoops to objects unworthy of
itself; he is very well aware that fallen man, in point of fact,
spontaneously prefers himself to God, for that indeed
constitutes his very fall, but he does not on that account
forget that every created love is but a participation in the
uncreated love, and that this implies an identity of
object.

These two classical ways of dealing with the problem are
thus substantially the same, or rather they come to the same
thing. Before the fall man naturally knew that he ought to
love God, and how he ought to love God ; after the fall he
remembered it no longer, and has the lesson to re-learn :
" *Amor ergo, ut dictum est, ab auctore naturae naturaliter est
animae humanae inditus, sed postquam legem Dei amisit ab homine
est docendus.*" [8] That is why William of Saint-Thierry
would cast out this false teacher Ovid, and bring all human
love back to God as to its natural object. To adopt the
expression of an old author, he would write an *Anti-Naso.*
But St. Thomas has no other doctrine. As for the Cister-
cians, so for him, man naturally loves God more than himself. This love which puts God above all things is not yet
charity, but the natural dilection to be perfected and fulfilled by charity. To suppose the contrary, to admit that
man naturally loves himself more than God, would be to
admit that an inclination can be at once natural and perverse, according to nature and contrary to nature ; and,
furthermore, it would be to admit that grace, in order to
make the love of God prevail over love of self in the soul,
would have to destroy nature instead of perfecting it.[9] If
to love God above all things we need grace to-day, it is
therefore not because our nature is in itself incapable of
such love, but because it has become incapable without a
grace which shall first heal its wounds and then turn it

towards its true object.[10] The only question still left open therefore, is why man is naturally capable of loving God above all things ? To answer this question will be to dispose finally of the antinomy of which we seek to get rid, that is to say between the natural love of self and the natural love of God.

The problem is far from simple. It is in fact one of those in which some have taken certain pleasure in piling up the difficulties, by setting over against each other two essentially irreducible conceptions of love, the *physical*, namely, and the *ecstatic*. On the one side we have a love conceived in the Græco-Thomist manner, based, that is to say, upon the natural and necessary inclination of all beings to seek their own proper good before all else. For those who hold this *physical* conception there is a fundamental identity between love of self and love of God, as if it were altogether one and the same thing at bottom to love oneself and to love God, to love God and love oneself. The *ecstatic* conception, on the contrary, postulates self-forgetfulness as the necessary condition of all true love, of that which literally puts the lover " outside himself," " beside himself," and sets free our love of another from all that might seem to connect it with our egoistic inclinations. When, however, we go to the texts of the mediæval masters themselves it is not at all easy to find any such clear-cut distinction. If they suffered from any internal contradiction in this respect they were certainly not aware of the fact, and we may even ask whether, in their deepest thought, they did not always deny its legitimacy.

What has made it easy to go astray here is the common contemporary abuse of what, originally, was a mere metaphor, the first step in a *manuductio*, too often taken in a distressingly literal way. When St. Thomas would explain how it is that man naturally loves God above all things, he commonly points to the fact that anyone who naturally loves his own good must necessarily love that without which it would be impossible. Now every natural thing depends on another. If, then, being gifted with intelligence, it becomes aware of its dependence, it cannot fail to prefer that on which it depends to itself, since that on which it depends is a necessary condition of its own existence. This, for instance, is why a part is always ready to expose itself to danger for the preservation of the whole of which it forms a part. When the whole body is threatened the hand interposes to ward off the blow ; and does so with a quite natural and indeliberate movement. It is as if the hand knew that it cannot subsist apart from the whole body, so that to defend the body is equivalent to defending itself.[11] Hence the famous doctrine known as that of physical love, in which the relation of the divine to the human good is represented as the relation of a whole to its part. If man naturally loves God more than anything else, and first of all more than himself, is it not simply because God is the universal good, under which all other particular goods are contained, on which man's very existence depends, and which accordingly he spontaneously prefers before himself as the necessary condition of his own existence and his own perfection ?

There is some truth in this interpretation of the Thomist doctrine of love, but this something is exposed to falsification by something else that is not true at all. The hand is really and literally a part of the body ; and in this case it is quite true to say that the relation of the particular to the general good is the relation of a part to the whole. But as soon as we leave this biological example and seek one in the social order, then we can no longer hold to the same formula without risking a flagrant over-simplification. An individual man is certainly a part of the whole we call the City, but not in the same sense as that in which the hand is part of the body ; he is not a natural part, and so if he exposes himself for the welfare of the city this is no longer in virtue of a natural inclination, but the outcome of a rational decision. He knows the nature of his dependence on the City, he judges the case, decides that its interests prevail over his own, and freely exposes himself for the sake of the City. When next we turn to the still higher relation of dependence in which man stands to God, it becomes quite obvious that our comparison no longer holds. It is still true to say that God is the universal Good under which all particular goods are contained, but the relation of dependence in which man stands to God is no longer that of a part to its whole. God is not a whole of which man is a part ; man is not a part of which God is the whole ; the universal here in question embraces the particular in quite another manner than that in which the body contains the hand that exposes itself for its defence, and, as a necessary consequence, the love by which man naturally loves God more than himself is very different from the mere brute instinct that moves the hand to protect the body ; and it differs even from the rational process which prompts the citizen to sacrifice himself for the City. Would we know in what this love consists we must first of all ascertain in what sense it is true to say that God is the " universal " good of which man is but a particular case.

Happily, as it turns out, in view of all we have said above, it needs only to be recalled. In a Christian universe, in which beings are created by Being, every creature is a good and an analogue of the Good. At the root of all this order of relations there lies, therefore, a fundamental relation of analogy which rules every derived relation subsequently set up between creature and Creator. If, for example, we say that God is the universal good, we can only mean that

God is the sovereign good, the cause of all good. If we say that each good is but a particular good we can only mean, not that these particular goods are detached parts of a whole which would be Goodness, but that they are analogues of the creative Good that gave them birth. In this sense, then, it is true that to love any good whatsoever is always to love its resemblance to the divine goodness, and, since it is this resemblance to God that makes this good to be a good, we can say that what is loved in it is the Sovereign Good.[12] In other words, it is impossible to love the image without at the same time loving the original, and if we know, as we do know, that the image is only an image, it is impossible to love it without preferring the original. What holds of the whole totality of creatures holds much more of man in particular. To will any object is to will an image of God, that is, to will God ; to love oneself, then, will be to love an analogue of God, and that is to love God.

If the case stands thus, the antinomy that troubled us so much is disposed of. Moved and directed by intelligent substances, beings deprived of knowledge act nevertheless in view of ends and of their own spontaneity tend to their own good. Now, in their case, to tend to their own good is indifferently to seek their own perfection or to seek the divine likeness, for their perfection precisely consists in this divine likeness. If, then, such and such a particular good is desirable only on account of its resemblance to the supreme Good, it is impossible to desire the supreme Good merely in view of this particular good ; on the contrary, the particular good must always be desired for the supreme Good. Consequently, for all purely physical beings, to perfect themselves is to make themselves more like to God.[13] All the more must this be so when we have to do with an intelligent being such as man, for it is his intelligence above all that confers on him both his proper perfection and his

proper analogy to God. If we add to this that he has
received the promise of the beatific vision, that is to say a
state in which his intellect will know God as God knows
Himself, it will be seen without difficulty that man is
destined to reach simultaneously, and by one and the same
act, both the summit of his own perfection and all the fullness
of divine resemblance that is open to him.[14] What is this
but to say that the key to the problem of love lies in the
doctrine of analogy : *ratione similitudine analogiae principia-
torum ad suum principium*, and that for its resolution we must
return to the basic principles laid down by the Cistercian
masters ? For there, where St. Thomas speaks of similitude
and analogy, St. Bernard and William of Saint-Thierry speak
of resemblance and image. Here, then, does not the physical
conception of love rejoin the ecstatic ? Or rather are they not
but one and the same fundamental conception developing
into its consequences according to two different techniques ?

The Word alone is the Image of God ; man, for his part,
is only made to the image of God. And that, no doubt, is
already a very high prerogative, since it makes him capable
of participation in the divine majesty and beatitude ; it
is a greatness again which is inseparable from man's nature,
because bestowed by the same creative act that bestowed
existence ; but a greatness nevertheless not unaccompanied
by possibilities of misery, for although man could not lose
his capacity for the divine without ceasing to be man, he
could lose, and in fact has lost, the original rectitude of will
that made him love the divine. And when rectitude of will
was lost, then, at the same blow, the soul lost the perfection
of the divine likeness. Now by a fatal consequence—since
to be an image of God is of his essence—man could not
cease to resemble God without ceasing to resemble himself ;
but then, conversely and not less necessarily, he has only to
turn to God, by the aid of grace, to recover at once the

divine likeness and that conformity with his own nature which he had lost by sin. Here, then, by the classical ways of the Cistercian school, we arrive at the conclusion which will also be that of St. Thomas Aquinas : *et haec hominis est perfectio, similitudo Dei.*[15] At the same time we see how it assures the unity of the two forms of Christian love which it was proposed to distinguish. For if man is an image of God, the more like God he makes himself the more he fulfils his own essence. Now God is the perfection of being, Who knows Himself integrally, and loves Himself totally. If man is fully to realize his virtualities and become integrally himself he must become this perfect image of God : a love of God for God's sake. For him whose thought moves on the plane of likeness and analogy, which is that of creation, the supposed opposition between love of self and love of God has no *raison d'être.* To say that if man of necessity loves himself he cannot love God with disinterested love, is to forget that to love God with disinterested love is man's true way of loving himself. Whatever of *amour propre* he retains, makes him so far forth different from that love of God which is God ; and all love of self for the sake of self that he abandons, makes him, on the contrary, like to God. But thereby also it makes himself like himself.[16] As image, the less he resembles the less he is himself ; the more he resembles the more he is himself ; wherefore *to be* is, for him, to distinguish himself as little as possible, to love himself is to forget himself as much as possible. And he attains his last perfection when, remaining substantially distinct from his original, he has become no more than a subject carrying God's image.—It is time, however, to descend from these heights. Our natural moral life develops on a humbler level. We know now the transcendent end that Christianity sets before man ; let us then see what resources it finds in man, whereby he may dispose himself to receive it.

APPENDIX

NOTE ON THE COHERENCE OF CISTERCIAN MYSTICISM

THE coherence of St. Bernard's doctrine of love has been formally placed in doubt by an historian whose views on these matters are not lightly to be disregarded.* To bring such a charge against the author of the *De diligendo Deo* is to bring it against the mystical doctrine of the greatest of twelfth-century mystics. If we venture to dispute the conclusion here it is for no lack of admiration for the deeply regretted Père Rousselot, whom to know is to love, but the question raised concerns the unity and intelligibility of a doctrine without which the Middle Ages would not be what it is in our eyes. He who became Dante's guide towards mystical ecstasy—was he, or was he not, clear on the basis of his own mystical doctrine? Such is the question raised by P. Rousselot's critique. Let us first see, however, in what that critique consists.

The " problem of love " might be thus formulated in abstract terms : " Can we possibly have a non-egoistic love ? If so, how does this pure love for another stand to that love of self which would seem to lie at the root of all natural tendencies ? " All mediæval thinkers agree in taking the love of God as the typical case concerned in the problem, for all hold that God alone is man's beatifying end ; but as to how love bears upon the end to which it tends, is a point on which these mystics seem on the contrary to hesitate between two possible views. Two different conceptions of love, in fact, divide their minds : the *phy-*

P. Rousselot, "Pour l'histoire de l'amour au moyen âge," in *Beitrage-Baeumker*, VI, 6, Münster, 1908.

sical conception, which we may characterize as Graeco-Thomist, and the *ecstatic*.

The *physical*, that is to say the " natural " conception of love, is characteristic of all those thinkers, " who base all real or possible love on the necessary inclination of natural beings to seek their own proper good. Between love of God and love of self there is, for these authors, a deep but hidden identity ; they constitute a two-fold expression of one and the same appetite, the deepest and most natural of all, or better still, the only natural one." The *ecstatic* conception of love, on the contrary, " is the more marked according as an author is the more careful to cut all cords that seem to connect our love of another with our egoistic inclinations ; love, for those who hold with this school, is so much the more perfect, so much the more *love*, in the degree in which it puts the subject " out of himself," " beside himself." *
Here, as will be seen, is no question of two opposed systems, but of two opposite tendencies, and, in fact, when P. Rousselot asks himself which of the two predominates in St. Bernard, he has to admit both. The physical conception prevails in the *De diligendo Deo*, but the ecstatic conception in the Sermons on the Canticles,† and that, moreover, is why " his doctrine is not perfectly coherent."‡ Now is this quite certain ?

It seems far from easy, in the first place, to distribute St. Bernard's texts into two groups, and find in each a particular solution of the problem of love. Keeping to P. Rousselot's terminology, it is incontestable that the physical conception is to be found in the *De diligendo Deo*, and even plays an essential part, but this part is not to be understood without grasping the exact sense of St. Bernard's terminology, nor without considering the doctrine in the

---

* P. Rousselot, *op. cit.*, pp. 1–4.
† P. Rousselot, *op. cit.*, p. 5, note 1.
‡ P. Rousselot, *op. cit.*, pp. 49–50.

light of its relation to the whole treatise. To deal at once with the second point, it is a fact that the ecstatic conception of love is already developed (A.D. 1126) in this same *De diligendo Deo*, in which also the physical conception is set forth in its severest form. To be more exact, it is in the oldest of all the mystical writings St. Bernard has left us, the *Epistola de Caritate* (1125), that both are found together. In the same chapter, only a few lines apart, St. Bernard affirms that man's love " begins necessarily with himself," and again that the upshot of this love of self is to enter in the joy of the Lord, to tend wholly towards God, " as if in wondrous wise he should forget himself and, as if delivered from self, should be wholly God's." * How maintain, after that, that the ecstatic conception in St. Bernard is only a late development ? † And how can we avoid asking whether the " intuitions " he is supposed to have missed, and the " illogicalities " charged against him, are not rather chargeable on his historian than on St. Bernard himself ? That is the question for consideration : and we will begin by defining the meaning of the terms St. Bernard uses in the *De diligendo Deo*.

What is meant by *natural love* in a doctrine like St. Bernard's ? For him, as for St. Augustine, *man's nature* always designates man in his concrete state ; that is, man as God created him. For him, then, the word " *nature* " is not opposed to *grace*, but rather itself evokes the idea. The reality is the same for St. Augustine and St. Bernard as for St. Thomas ; none of them believed in the real existence of a *state of pure nature* in which man would have subsisted, were it only for a moment, without the gifts of grace ;

---

* *De diligendo Deo*, cap. XV, reproducing part of the *Epistola*.
† It would be easy to show that conversely, in the Sermon on the Canticles, St. Bernard has by no means forgotten *physical* love ; it will appear clearly further on.

but the Thomist terminology is not the same as the Augustinian. With St. Thomas, the concepts serve to analyse the concrete, and so by the *nature* of man he understands the very essence of man conceived as rational animal, with all that appertains to the essence and nothing that does not. When, under these conditions, a Thomist speaks of a natural love in man he understands a human love as such, without intervention of grace. For St. Augustine and St. Bernard the concepts serve to designate the reality in its concrete complexity, as yet unanalysed ; if, then, they speak of a natural love, the word " nature " does not necessarily exclude grace, since God created man in a state of grace, and since, even when man lost this grace, he could still recover it. In short, St. Bernard never speaks of nature save as in possession of grace or in capacity to receive it.

Apply this principle to the analysis of our texts. Is it *natural* for man to begin by loving himself ? Certainly ; the *De diligendo Deo* affirms it. But in what sense is it natural ? Not on account of any precept imposed by God, but as forced on us by the necessary conditions of our infirm and feeble nature. What St. Bernard calls carnal love, *amor carnalis*, that is, love of ourselves for our own sake, is therefore the necessary starting point for all ulterior evolution of love, not because God commands it, nor on account of any excellence of its own, but because, without self-love, we could not even subsist. To love God we must live, and to live we must love ourselves.* Thus, in the first place, it is a fact, and, to begin with, nothing more.

Man, then, finds himself in the following situation :

---

* " Sed quoniam natura fragilior atque infirmior est, ipsi primum imperante necessitate, compelletur inservire ; et est amor carnalis, quo ante omnia homo diligit seipsum propter seipsum." *De diligendo Deo*, cap. VIII. In so far as this love is limited to the acquisition of the necessaries of life, it remains " in necessitatis alveo " it has nothing to reproach itself with. If, as is indeed its natural bent, it tends to the superfluous, it becomes dangerous and bad. Its legitimacy is thus based on its necessity.

God asks him to love God above all things, and, on account of his frailty he cannot help loving himself first. To grasp the reason for this contradiction we must take the cause of human weakness into consideration. Now on this point St. Bernard expresses himself with a clearness that leaves no room for misconception. Carnal love is not prescribed by God, nor yet is it a fault, but it is the result of a fault. It is because we are carnal, and born of the lust of the flesh, that our love, or our cupidity—for they are all one *— begins necessarily with the flesh. This love or cupidity may afterwards be rectified by grace and duly ordered towards the highest spiritual end, but man, born of sin, does not begin with spirituality, but with the animal and the carnal.† In other words, the purely factual necessity we are under to turn our love first to self results from the fact that we are born of concupiscence, which is itself a sequel of original sin.‡ *Amor carnalis* is thus that love by which a flesh already delivered up to sin, of necessity begins ; *natural* self-love is the love that has become natural for fallen man ; but if we take nature in this concrete, and, so

---

* St. Bernard makes no distinction between *amor* and *cupiditas*. He follows a classical terminology, St. Augustine's namely, for whom cupidity is merely love itself in its aspiration to the object loved : " Amor ergo inhians habere quod amatur, cupiditas est . . . proinde mala sunt ista, si malus est amor ; bona, si bonus." St. Augustine, *De civit.* Dei, XIV, 7. If we look at the whole chapter we shall see that Augustine refuses to oppose *caritas, dilectio* and *amor*. Love may be good, and then it is charity or dilection ; what puts love in opposition to charity is not that it is love, but that it is bad. And so for *cupiditas*. If the love is good, the cupidity is good. There is a cupidity of charity, namely the very aspiration of charity to God.

† " Verum tamen quia carnales sumus, et de carnis concupiscentia nascimur, necesse est cupiditas vel amor noster a carne incipiat, quae si recto ordine dirigitur, quibusdam suis gradibus duce gratia proficiens, spiritu tandem consummabitur, quia non prius quod spirituale, sed quod animale, deinde quod spirituale." *De diligendo Deo*, cap. XV.

‡ Not merely does the expression " *de carnis concupiscentia* " leave no room for doubt, but the " *quia carnales sumus* " takes us back naturally to St. Paul : " Scimus enim quia lex spiritualis est, ego autem carnalis sum, venumdatus sub peccato " (Rom. vii. 14). This text opens the famous development on man's incapacity to do what he would. The sense both of *carnalis* and of *concupiscentia* is thus beyond all doubt, and we are certainly concerned with a sequel of sin.

to speak, historical sense, it will not do to consider its fall alone ; for this fall is measured with respect to a grace, and since what remains of this grace also forms part of nature, the *natural* love of man is not to be completely apprehended without taking account both of man's misery and his possibilities of recovery.

Man is not the image of God ; there is but one Image of God, namely the Word. But man is an image of this Image, and that is why the Bible says that he is made *ad imaginem.* This expression signifies that man was created in a state of high dignity, and capable of participation in the divine majesty : *celsa creatura in capacitate majestatis.* Therein lies his greatness, and since it was bestowed on him by the creative act itself, it is altogether inseparable from his being. True, the soul's greatness is not to be identified with the soul ; but it is in it as its form ; now, on the one hand, *nulla forma est id cujus est forma,* so that the soul remains distinct from that which constitutes its greatness, but, on the other hand, it cannot lose its form without ceasing to be itself, so that any separation is inconceivable.* But on the other hand the soul was not only created capable, but also desirous, of heavenly things : *appetens supernorum,* and thus it was created in a state of rectitude. Sin was committed, rectitude lost, and now the soul stoops to terrestrial things ; from *recta,* therefore, it has become *curva ;* but even while thus losing its rectitude it has not therefore lost its greatness, for it retains its form, and even in its abasement remains *etiam sic aeternitatis capax.* Were not that the case, were the soul's native greatness suppressed by sin, no hope of salvation would remain to it : for with its greatness would go the capacity for divine things that precisely constitutes it :

---

* St. Bernard : *In Cant. Cant.,* sermo 80, art. 5.

non superesset spes salutis ; nam si desinat magna esse, et capax. *
There, then, we see what it means for us to have been made
to the image of God ; what now is the bearing of this fact
on the question of mystical love ?

It helps us first to understand what is vitiated in us,
therefore also what must be reformed if our primitive human
nature is to be restored, and, *by that very fact*, enabled to
surrender itself to the mystical embrace of the Word :
ut ad amplexus Verbi fidenter accedat.† What retains its likeness
to God, even after sin, is the soul's greatness, its form.
What is now unlike is the soul's stooping, its declination
earthward, its loss of rectitude. Standing upright, the soul
loved God and obeyed no other law than that of divine love ;
as soon as it turned to earthly things it enslaved itself to the
law of fear, and lost the liberty of love. Not that fear
destroys its power of free choice, for that is identical with
the will, and is and always will be, indestructible ; still
less, then, does it destroy the will, but clothes it over in a
way as with another vesture, and hides for a time the
spiritual liberty with which love had endowed it : *non
tamen ut libertate propria nudaretur, sed superindueretur.* The ills
the soul now suffers do not replace that native goodness
which is the original gift of the Creator, but they are super-
induced on that goodness, and disturb it, deforming an
order they can in no wise destroy.‡

Now this disorder bears with it an all-important consequence : the soul, becoming unlike God, the soul of whose

* *In Cant. Cant.*, 80, 3. Thus we have the following schema :

Anima	Magna	Recta	Curva
ad	quo	quo	quo
imaginem	capax aeternorum	appetens super-	appetens
	(forma animae)	norum	terrestrium

The quality *magna* is inseparable : those of *recta* and *curva* separable. St.
Bernard's doctrine is here very close to St. Anselm's.

† *In Cant. Cant.*, 81, 1.

‡ *In Cant. Cant.*, 82, 4–5 : " Itaque boni naturae mala adventitia dum non
succedunt, sed accidunt, turpant ubique ea, non exterminant, conturbant,
non deturbant."

very nature is to be like God, becomes unlike itself. False
to its own nature, which is to be a divine analogue, it ceases
at one and the same time to resemble God and to resemble
itself : *unde anima dissimilis Deo, unde dissimilis et sibi.** Now,
conscious of what it is in itself, it can ignore neither its own
remaining and inherent capacity for greatness, nor the
cruel loss of that greatness of which it is naturally capable.
In other words it feels itself both like to God and faithful
to itself inasmuch as its aptitude for divine things subsists,
but at the same time false both to God and its own true
nature ; and hence it is rent in twain and feeling itself
still like and seeing itself in part unlike, it conceives that
horror of self which is the inner tragedy of the sinner's life. †
Then the soul longs to recover the fullness of its resemblance
to God and itself, by casting off the dissimilitude that divides
it from either. It cannot do so save by charity and grace.
Now to recover charity is not only to become once more like
God, and therefore like itself, but also, on account of the
soul's intimate knowledge and sight of itself, it is to see God
in His image at last restored by grace, and to rejoice in its
own recovered splendour. Hence the following magnificent
passage, in which all the leading ideas of Cistercian mysticism
are closely knotted together : " Admirable and astounding
that likeness assuredly is which brings with it the vision of
God, or rather *which in this vision itself*, [thus, the *similitudo*
participated by the soul is identified here with *visio*, as if it
were the same thing to see God and to become like Him.]
But I speak of the likeness and vision which are one and
the same with charity. For charity is that vision : *caritas
illa visio :* and charity too is that likeness. Who does not
stand amazed at beholding the charity of a God despised
and yet recalling the soul that spurned it ? Well therefore

* *In Cant. Cant.*, 82, 5.
† *In Cant. Cant.*, 82, 6.

did that wicked one, to whom I referred a while ago, deserve the reproach addressed to him for claiming a likeness to God, since, loving iniquity, he loved neither himself nor God. For it is written : ' He that loveth iniquity hateth his own soul.' But let him remove from his soul the iniquity which forms in her a partial unlikeness to the Word [for the soul's greatness and capacity for God remains] and then there shall be perfect unity of spirit, mutual vision, and reciprocal love. ' When that which is perfect shall come, that which is in part shall be done away,' and then between God and the soul shall be pure and perfect love [*castus*—totally disinterested], they shall know each other fully, behold each other clearly, they shall be united to each other firmly, they shall live together inseparably, they shall be like each other absolutely. Then shall the soul know God even as she is known ; then shall she love as she is loved ; and over His bride shall rejoice the Bridegroom, knowing and known, loving and beloved, Jesus Christ Our Lord Who is over all things, God blessed for ever." *

Seize this central point and Cistercian mysticism will once again appear coherent, and its supposed contradictions will vanish. How reconcile love of self with love of God ? There is no difficulty. Grant that a being is an image, and then the more it resembles its Original the more faithful it is to itself. But what is God ? He is Love : *Deus caritas est* (1 John iv. 8) ; that is to say, being charity by essence He lives by charity. His charity is Himself, and therefore it is His life, and in a certain sense we may say it is His law. Cistercian mysticism is altogether suspended from a theology of the Trinity of which the central idea would seem to be that God Himself lives by a law, and that the law that rules His intimate life is love. The Father generates the Son, and the bond that unites the Son to

* *In Cant. Cant.*, 82, 8.

the Father and the Father to the Son, is the Spirit Who is
their mutual love. Charity is thus, so to speak, the bond
that assures the unity of the divine life, and thus also the
divine peace, and the divine beatitude ; not a charity
that would be something superadded to God, but one which
is *substantiam illam divinam :* the divine substance itself.
As participated in us, charity is no longer God's substance
but God's gift.* It is as if God Himself lived by a substantial law—*nec absurdum videatur quod dixi etiam Deum
vivere ex lege*—and as if this law of love, participated by
things, were the eternal law, the creative and directive law
of the universe, but more particularly of man in whom it
reigns as charity. " *Cum ipsa quoque lex omnium sine lege non
sit, non tamen alia quam seipsa, qua et seipsam, etsi non creavit,
regit tamen* " ; † that is the theological keystone of the whole ;
for the slave lives also by a law, the law of fear which is not
the law of God, and the hireling lives also by a law which is
indeed a law of love, but, since this love is self-love, he is
ruled by a law himself has made, which therefore is not
the law of God. But slave or hireling may do what they will,
the divine law will still be there ; and no matter what laws
they give themselves, it will always overrule them. In
preferring his own will to God's man wills, with a perverse
will, to imitate his Creator ; that is to say to rule himself
by his own law, but in rejecting the divine law of charity
he transforms it into a penal law that crushes him. The just
man, on the contrary, taking upon him the yoke of divine
love, is no longer *under* a law, but neither is he *without* law :
non sub lege, nec sine lege. But the law that he makes his own
is now God's law ; it is in a very deep sense that he daily

* " Dicitur ergo recte caritas et Deus et Dei donum. Itaque caritas dat
caritatem, substantiva accidentalem. Ubi dantem significat, nomen substantiae est ; ubi donum qualitatis." *De gradibus humilitatis,* cap. XII. Cf.
St. Augustine, *De Trinitate,* XV, 19, 37.
    † *De gradibus humilitatis,* cap. XII. Cf. *In Cant. Cant.,* VIII, 3.

prays that God's will be done on earth as it is in heaven ; that is to say that as God lives eternally by love of His own perfection, so man may desire nothing here below save that very perfection of God. But to love God as He loves Himself, that truly is to be one with Him in will, to reproduce the divine life in the human soul, to live like God, to become like God, in a word, to be deified. The marvel is that in thus becoming God man also becomes or re-becomes himself, he realizes his very essence as man in realizing its end, plucks up by the roots the miserable dissimilitude that divided the soul from its own true nature. Losing that whereby it is but partially itself, it finds once more the fullness of its own being, as it was when it came from the hands of God. Where then is the supposed opposition between love of God and love of self ? Man is so much the more fully himself as he becomes more fully a love of God for God's sake.

The second antinomy survives examination no better than the first. It is very true that Cistercian ecstasy is at once annihilation and fulfilment, but this simultaneous double assertion indicates no deviation or internal embarrassment in St. Bernard's thought. In the first place the Cistercian texts never speak of annihilation, but only of an all-but annihilation ; and we must also notice what it is that is annihilated : dissimilitude. Iron reddened in the fire seems to have become but fire ; the air filled full of light seems to have become but light ; but iron and air, now no more than subjects carrying fire and light, are still there to carry them. These comparisons, taken from Maximus the Confessor, and deriving indeed from still older sources, rather assert the persistence of man under the influence of charity than exclude it. If we weigh St. Bernard's expressions we shall easily see the limits beyond which he would not have us go. A drop of water mingled into wine *deficere a se tota videtur ;* it seems to be dissolved. So also under the action of grace does man's will seem to be dissolved, to have passed into the divine will : but that takes place in a mysterious manner : *quodam ineffabili modo.* So, finally, even when nothing of man remains in man, his substance nevertheless remains, and will remain even in the beatific vision, *manebit quidem substantia, sed in alia forma.** In short the soul's substance is indestructible ; to be annihilated in God, is to surrender its own will, that is, the separative will that made the man at once different from God and from himself ; it is, therefore, to become at one and the same time, both a perfect image of God, and a human plenitude. Charity begins the work of restoration ; ecstasy realizes it as far as it can be realized in this life ; it is consummated in the beatific vision.

The essential character of Cistercian mysticism is now plain : it rests wholly upon a conscious effort to perfect the natural likeness of the soul to God, by means of a conformity, ever more fully realized, between the human will and the divine will. To love God is, in a way, to make God love Himself in us, as He loves Himself in Himself. That is the true meaning of the mystic marriage : " *Talis conformitas maritat animam Verbo, cum cui videlicet similis est per naturam, similem nihilominus se exhibet per voluntatem, diligens sicut dilecta est. Ergo si perfecte diligit, nupsit.*" The mutual embrace of God and the soul lies in the very union of their wills : " *Complexus, plane, ubi idem velle et nolle idem unum facit spiritum de duobus.*" Love stronger than fear, love that is its own cause and its own fruit, because it is the love of Love Itself ; unequal love, however, since between man and God is no equality, yet able at least to will itself to be a total love, and even so to be, since God gives it, we

---

* *De diligendo Deo*, X.

but give it back to Him.* The whole doctrine of disinterested (*castus*) yet rewarded love depends on this :
that charity is not merely the soul's love for a being that
loves it in return, but its love for the very substance of love,
the end beyond which no other end exists.

I have described St. Bernard's mysticism as Cistercian
mysticism because it is to be found, in essence, in the other
representatives of the same school. William of Saint-
Thierry, whose greatness is but little appreciated, has
forcibly expressed the *rôle* of grace in the work of restoring
the image of God, and the profound identity of the act
whereby man becomes himself in becoming like to God.
" When Thou lovest us, Thou lovest us only for Thyself,
wherefore the perfect law of justice forbids us too to love
anything beyond Thyself. And certainly, if but great grace
be granted, he who loves God may go so far that he loves
neither Thee nor himself for himself, but both Thee and
himself for naught but Thee. And so he is reformed in
Thine image, to which indeed Thou didst create him,
Thou, who in the truth of Thy sovereign nature, and in the
nature of Thy truth, canst love nothing, neither man, nor
angel, nor Thyself, but for Thyself." † Now for man to
become like God is to fulfil the desire of his true nature :
Ipse enim imago Dei est. Et per hoc quod imago Dei est, intel-
ligibile ei fit, et se posse et debere inhaerere ei cujus imago est." ‡
But to become like to God is to will what God wills : " *Velle*
autem quod Deus vult, hoc est jam similem Deo esse. Non posse
velle nisi quod Deus vult, hoc est jam esse quod Deus est, cui velle
et esse idipsum est." § One more step and we see that an
image that attains to perfect resemblance attains, by
definition, its own perfection. The author of the *Epistola*

* *In Cant. Cant.*, 83, 2-3.
† William of Saint-Thierry, *De contemplando Deo*, IV, 9.
‡ *Epist. ad fratres de Monte Dei*, II, 2, 5.
§ *Op. cit.*, II, 3, 15.

does not delay to take it : "*Et haec est hominis perfectio, similitudo Dei :* to be like to God, that is man's perfection. Not to wish to be perfect is a fault. We should always therefore nourish the will to this perfection, we should encourage love and oblige the will, lest it dissipate itself on alien things. Love must at all costs be preserved from corruption, since for love we were made, for love we live ; and to become like to God who were made in God's image." *
What wonder, after that, if the divinization of man is effected by the union of his will with the divine will in the embrace of charity, and if clearer sight of God is accompanied by an always closer conformity of the image with its original : "*Similitudine ei appropinquans, a quo longe factus est per dissimilitudinem ; et sic expressiorem visionem expressior semper similitudo comitatur.*" †

If indeed it is thus, then not only is the internal coherence of Cistercian mysticism assured, but perhaps also in a measure, the historical and doctrinal continuity of all mediæval mysticism. The "Graeco-Thomist" conception of love is not inappropriately called "physical," on condition, however, that by this term we understand a nature in the Christian sense, made to the image of its Creator ; that is to say a nature but little Greek and very strongly Thomist. The well-known comparison between man's love for God and the love of the part for the whole has been pushed rather too far. Man is not a part of which God would be the whole, he is an analogue, a similitude, of his Principle : and that is why, according to the famous passage of the *Contra Gentes*, III, 24 : "*Propter hoc igitur tendit in proprium bonum, quia tendit in divinam similitudinem.*" If this is so, the conception of image, ruling even the philosophy of nature and orienting the desire of all created things, should secure

* *Op. cit.*, II, 3, 16.
† *Op. cit.*, II, 3, 18.

in Thomism, as in the Cistercian mysticism, that the perfection proper to man, and his complete submission to the divine will, shall converge to the same point. When the fragments of history are better understood they tend of themselves perhaps to re-constitute a unity.

CHAPTER XV

FREE-WILL AND CHRISTIAN LIBERTY

THE assertion of man's freedom is as old as Christian Thought. Not that Christianity invented the idea of liberty [1] ; it would even repudiate any such claim if the need arose. As early as the second century Irenaeus reminds us that, if it was judged necessary to " manifest " liberty in the Scriptures, the law itself, promulgated by revelation, is nevertheless coeval with the human race : *veterem legem libertatis hominis manifestavit*. But what first claims our attention is the emphatic way in which the Fathers of the Church insisted on the importance of the concept of freedom, and the very special nature of the terms in which they did it.

God, in creating man, prescribed him laws, but left him nevertheless free to prescribe his own, in the sense : that the divine law does not constrain the human will. One may say that from the first awakening of Christian thought a whole series of philosophical terms, whose mere equivalences are instructive, passed into general currency. God created man with a rational soul and a will, that is to say with a power of choice analogous to that of the angels, since both angels and men are beings gifted with reason. At once it was accepted on all hands that freedom consists in an absence of absolute constraint, even in respect of the divine law ; that it appertains to man in virtue of his reason and is expressed in the power of his will to choose : *liber, rationabilis, potestas electionis*, are terms henceforth inseparable from each other, inseparable, moreover from the central thesis which made them acceptable to the Christian thinkers, forced them on them we might say. God created man free because He would leave him responsible for attaining his last end. It is entirely for him to choose between the way that leads to happiness and the way that leads to eternal misery ; man is a fighter who has to rely on his own forces, but one also who ought to rely on them—master of himself, gifted with a genuine independence—τὸ αὐτεξούσιον—he effectually collaborates in his own destiny.[2]

As soon as the idea of liberty was submitted to analysis its extreme complexity became apparent. Much that would aid its elucidation was to be found in the philosophy of Aristotle. Natures, as he conceived them, are internal principles of operation in beings, so that every natural being has a true spontaneity, at least in the sense that the principles of their acts are within themselves. Determined to one mode of action only in beings not gifted with know-ledge, the nature acquires a certain indetermination in the animals who are offered a plurality of possible objects by sensation ; and then it gives birth to what we call appetite or desire. The field of this indetermination is considerably extended in man owing to the fact that he is gifted with an intellect. Capable, in a way, of becoming all things by knowledge, he has a vast multiplicity of things at his dis-posal, and by means of his will he has to make a choice. In the first place, then, he has a spontaneous will to his natural end, that is to say to happiness ; and then the power of rational deliberation on the means to be adopted to attain this end ; and then finally an act of will choosing one of these means in preference to the others. The act of choice, προαίρεσις, *electio*, occupies a central place in Aristotle's ethics [3] ; and the Christian thinkers early understood that this idea was indispensable for their purpose, albeit they felt that in the end it would not suffice to satisfy

them completely. Let us therefore try to see what they added
to it and what they retained.

The very conception of voluntary choice, as Aristotle
defines it, implies the idea of reason ; but it remains true
to say that if will makes its appearance in the series of beings
only when reason comes to enlighten desire, it is nevertheless essentially a kind of desire. Choice then would be impossible without knowledge, but remains above all the act of
an appetite fixing on one of its possible objects. Now
appetite itself, according to Aristotle, is only the expression
of the internal dynamism of a nature. The intervention
of reason so profoundly modifies the conditions under
which this dynamism energises, that will, in a sense, may be
opposed to nature as that which chooses to that which does
not. But man is also a natural being, and his will is only
the particular form that natural desire assumes in a rational
creature. Just as, then, prior to the choice of means, there
must be the will to the end, so also, prior to this will, there
must be the actuality of the human being. Here again the
first act is the root of the second act, being is the cause of
operation. The will, therefore, is simply the organ of that
efficient causality which is proper to man, and voluntary
choice, first and foremost, expresses the spontaneity of a
nature which contains in itself, or rather which is, the
principle of its own operations.[4]

The Fathers of the Church and the mediæval philosophers carefully noted this conclusion ; and, indeed, their
own conception of created nature and the causal efficacy
they attributed to it as a participation of the divine power,
led them even to accentuate its importance. Unless we
remember that the dynamism of second causes was regarded
by the Christians as an analogue of the creative fecundity,
the whole evolution of the problem of liberty in the Middle
Age becomes unintelligible. People sometimes speak as if

it were something surprising that Christians should attribute a real efficacy to the determinations of the will. But why not ? They had no reason at all to hesitate in accepting the result of Aristotle's analysis, and everything invited them to go beyond it.

However elaborate was the Aristotelian theory of choice conceived as a decision of will following upon a rational deliberation, yet it remains a fact that Aristotle spoke neither of liberty nor of free will. What we now call psychological liberty was certainly present to his thought ; when he speaks of ἐχούσιον, the voluntary, he is indeed thinking of an action that springs from the very depths of the being, but it is not easy to disentangle the idea of liberty from analyses that nevertheless imply it, and the term itself is lacking. Among Christians, on the contrary, and especially among the Latins, liberty at once comes to the forefront. The very complexity of the formula *liberum arbitrium* led them to put the question : where lies the element which makes the choice, that is to say *arbitrium*, a free choice, that is to say *liberum* ? And all agree in situating it, to begin with, in the will's power to determine itself from within. It might even be said that if St. Augustine deals so freely with grace it is, to a certain extent, because he regards the incoercibility of the act of will as altogether obvious. Why should he go out of his way to say that the will's choice comes from the will, seeing that the will is a power of choice by definition ? But it is important, on the other hand, to insist that to will is to be free. St. Augustine therefore sees a clear attestation of this *free* will in all those Scriptural texts—and they are innumerable—in which God commands or forbids this or that action or this or that desire. And this truth, moreover, is no merely religious one, something indispensable in the economy of salvation,[5] but a fact that our daily internal experience perpetually attests. The will " is its own mistress," it is always " within its power " to will or not to will ; nothing is " more immediately at the disposal of the will than itself " ; formulæ that express the natural inseparability of the will and its act ; it is because it is born of the will and expresses the will that the act of will is always free.

Here we stand at the source of one of the most important of mediæval ideas, the idea of liberty of exercise. A will may be said to be free in many various senses, but first of all in this, that it can will or not will, put forth or not put forth its act ; and this primary freedom is absolutely essential. Christian philosophers express the fact positively when they identify free-will with the will, or rather with the act of choosing put forth by the will ; for when it chooses, it wills ; if it wills, it is certainly the will that does so and is also able not to do so.[6] A very lively sense of moral responsibility drew Christian attention to the fact that the subject that wills is the real cause of its own acts, for it is just on that account that its acts are imputable. This moral preoccupation, to which we shall soon have to return, thus led them to place in the will the root of a liberty that nothing can take away—unless indeed the will itself be extirpated along with it. This they habitually express, but now negatively, by setting over against each other, in irreducible opposition, the two terms *necessity* and *will*. *Libertas a necessitate*, or *libertas a coactione* signifies for them, before everything else, the total impermeability of the act of will by any kind of constraint. You can compel a man to do this or that, you cannot compel him to will it. Either there is will and thus no violence, or there is violence and thus no act of will. To affirm that free-will is " free from constraint " or " free from necessity " is therefore to affirm, first and foremost, the natural spontaneity of the will, the indissoluble bond between the act of choosing and the causal efficacy of the reasonable being that puts it forth.

We shall have to see later on how these conceptions entered into the Thomist synthesis. In the meantime they may be observed so to speak in their nude state in several theologians and philosophers of the Middle Ages. Nowhere is their terminology more fluid, and there is constant interchange of meaning between the two elements of the complex term *liberum arbitrium*. The importance of the spontaneity of the act of will is such that some of them went so far as to identify spontaneity with *arbitrium ;* thus returning, after a long lapse of years, to the primitive sense of Aristotle's προαίρεσις. Such at least seems to me to have been the thought of St. Anselm, for whom will, spontaneity, and absence of necessity are equivalent terms. Such seems also to have been the thought of Hugh of St. Victor, in so far at least as it can be divined through the medium of his rather obscure formulæ. For him, the spontaneous movement by which free-will fixes on its object pertains to *appetitus*, and since it is desire that chooses, it is desire that is *arbitrium*, freedom depending on the rationality of the will enlightened by knowledge.[7] But it is undoubtedly in Duns Scotus that the spontaneity of the act of will, already so strongly marked in St. Anselm, attains its definitive expression. Passing beyond even Aristotle's dynamism, he makes a radical opposition between the order of natures, which is that of necessity, and the order of wills, which is that of liberty. Every nature is essentially determined and a principle of determination ; every will is essentially undetermined and a principle of indetermination. But this indetermination must not be taken for a mark of insufficiency ; on the contrary, it attests the excellence of a faculty not tied down to any determinate act. Duns Scotus goes so far as to treat reason itself as a nature, so that all determination may be put down to the account of knowledge, and all freedom, on the other hand, to that of the will. The contingency of the choice does not depend upon rational judgments proposing alternative possibilities of action, but on the spontaneity of the act of will by which the alternative is resolved. Freedom is therefore wholly centred in the radical indetermination of the will, whose unforeseeable decisions spring from within as from a source of determinations wholly undetermined by anything else.[8]

Not all the mediævals went so far as that ; many even hesitated to rely on the will alone to assure the free exercise of its spontaneity. In Aristotle's philosophy it is true to say that choice is essentially voluntary, but Aristotle himself indicates, and his successors still more strongly emphasize, that if rational deliberation did not precede the will's decision, this act would not really be an act of choice. Deprived of the light of the intellect, the will would fall back into the rank of animal appetite ; without sense knowledge, which determines the animal's reactions, it would be no more than a necessarily determined natural inclination. The example of Stoicism, moreover, always stood as a warning of the importance of the question and the danger of over-simple solutions. Placed in a universe ruled throughout by the most rigorous determination, the Stoic sage can find liberty only in voluntary acceptance of the universal order. If we object to him that in this case the will itself is subject to the necessity of this order, he replies that the thing is without importance, since in spite of that the will necessarily remains a will. Boethius very clearly showed that a pure spontaneity is easily compatible with an absolute determinism and with all the necessity that results from it. If, in order to be free, we are content to be no more than an internal source of voluntary decisions, it matters little that these decisions are as rigorously predetermined by fate as any exterior event ; inasmuch as they are made by us, they are ours and we are free. It was this that led Boethius to go wholly over to the other side. The insufficiency of a liberty content at need to be merely a necessarily determined spontaneity was strongly borne in upon him, and it is to remedy this that he insists, with all his well-known force, on the importance of the rational element in the free act. Not only is it insufficient to say that the act of will is spontaneous—since in that case animals, too, would be free—but he expressly puts the root of will in the reason. What makes a choice a free choice is the rational knowledge that precedes it. Thus for Boethius and all who follow him *arbitrium* has ceased to mean the spontaneous option of will, but means now the free movement of reason. In defining free will as a *liberum nobis de voluntate judicium* it is the judgment that he really regards as free, and the will itself as free only inasmuch as it is judged by reason.[9]

But Boethius introduced no novel element thereby into the description of free will. The part played by knowledge in the production of the free act had never been over-looked. St. Augustine, for instance, penned some famous pages on the mysterious premonitions whereby the will is always guided to its object. The will, in itself, taken as a simple appetition, is blind, or rather non-existent, for in this sense it does not deserve the name of will at all. Every-body also saw that because reason offers will so many objects, and various appreciations of those objects, it expands practically to infinity the field open to its spon-taneity. Given an object, the will is free either to act or not to act ; given several objects proposed by reason, it becomes free to choose one or the other indifferently. So far there was no difficulty ; the world did not have to wait for Boethius to see this. What was really new in his doctrine, its most disputable and yet most philosophically fruitful point, lay in his integral rationalism.

This will be best appreciated when set over against the voluntarism of Duns Scotus. As we have already indicated, the Scotist doctrine of liberty sets out especially to show that in no case and in no sense can reason be the total cause of the free act. And it will not be sufficient to understand this in the sense that the will concurs in the production of the free act ; that would be altogether too evident. What Duns Scotus means is that when reason has judged, when all possible alternatives have been defined, weighed, criticized, when the moment for choice has come, then, in so far as choice is not yet effected, the will remains essentially undetermined to fix on this or that. The Scotist liberty of indifference is therefore one with the spontaneity of willing, which thus remains the sole possible element of contingence in face of the determinations of reason. For Duns Scotus, in short, no number of contradictory determinations can amount to one freedom. But for Boethius it is quite otherwise ; the option of appetite is a mere blind spontaneity. What makes it free is the critique of reason that judges it, compares the various possible options, and declares one to be better than the others. If he can still appeal to Aristotle it is in fact because the will, as Aristotle conceived it, is not a simple appetite that follows knowledge, but, precisely owing to this knowledge, is a power *ad utrum-libet*, capable, that is, of making a true choice, a προαίρεσις or real *preference ;* we might say of it once more that it is in a state of *indifference*, since it is due to the interval that separates two or more possible actions that its spontaneity will take on the character of a liberty.

The way in which results converge is no less remarkable here than the divergence of methods. Proceeding directly from appetite to will, we are led to put the will's own indetermination at the root of free-will ; following Boethius along the road that goes *via* the understanding, we situate free-will in the indetermination left open to the will by rational judgments. St. Thomas, seeking as always, the point of equilibrium between opposed tendencies, endeavours to allot, both to understanding and to will, the place their nature assigns them in the production of the free act. Faithful to one of the deeper exigencies of Aristotelianism, he maintains that choice is essentially an act of the will, that free-will appertains directly to the will, or rather is the will itself : *liberum arbitrium nihil aliud est quam voluntas.* It is will that wills or does not will, and chooses this rather than that ; as a faculty it stands to the multiple choices of its freedom as the simplicity of the intellect stands to the discursive movement of reason. On the other hand, it is equally true to say that without the judgment, will would not be will. An act of free choice is therefore not to be exhaustively described without the voluntary decision endorsing a rational judgment, nor without the judgment thus endorsed by the will. Hence his very flexible mode of expression, which, however, should never mislead us as to his true thought. He stands midway between the older intellectualism of Boethius and the future voluntarism of Duns Scotus. Like Boethius he will say that an act of free will is a free judgment, but then adds " so to speak," since it is essentially an act of will willing, not of reason judging. Like Duns Scotus he would readily admit that free-will is to be referred essentially to the will, and, moreover says it, but refuses so to define it without taking into account the judgment of the practical reason, which issues in the voluntary choice as its conclusion : materially, free-will is voluntary ; formally, it is rational.[10]

However it be understood, the power of free-will is as wholly inadmissible as will itself ; man is as inconceivable without the former as without the latter. On this account all Christian thinkers from St. Augustine to St. Thomas and Duns Scotus agree that after the original transgression free-will remained what it was before it was committed. St. Bernard himself insists forcibly on the " integrity " of the will in the state of fallen nature, and St. Thomas finally clinches the matter when he calls this freedom, which nothing can take away from man, " natural." [11] It can in fact disappear only with the will, that is to say, with the man himself. However, as a necessary counterpart to this " physical " character, so to speak, of free-will, it is now deprived of all moral qualification. Without freedom no morality, but no moral element is implied in its essence. Since the will can always choose, it can choose good or evil, and the goodness or badness of its choice in no wise affects the freedom of the act. All choice, precisely as choice, is at once both psychologically indetermined and morally indifferent. Now it was quite impossible for Christians not to take a lively interest in the moral qualification of the free act ; and considerations of this second order were bound to have important repercussions on their psychology of liberty.

In comparing Christian free-will with Aristotle's " choice " I noted that the latter speaks of *arbitrium* rather than of liberty. When he uses the word ἐλευθερία, which we usually translate as " liberty," he gives it a political meaning. Aristotelian liberty is more especially independence, the state of a person who, socially or politically, is independent of any other—the ideal of the democrat, as Aristotle himself puts it. The notion of liberty, which nowadays we take as something eminently psychological and metaphysical, was, in the first instance, merely political or social. But when the Stoics began to consider the freedom of actions with respect to the necessity of the laws of nature, they were led to integrate this social conception with ethics. Each of us now stands to the universe as the citizen stands to the city : laws exist and bear upon us, and the problem is to accommodate ourselves to them so well that we no longer feel their constraining force. The Stoic sage is one who succeeds in this ; making himself independent of destiny, he achieves freedom. It must be added that he considerably complicated the question of liberty for Christians.

For the Christian too finds himself in a universe ruled by laws. Not that he has any destiny to fear, but he is now aware that he is under providential guidance, subject to the will of God as Sovereign Legislator, and more particularly to the divine laws that now rule the state of fallen nature. In a celebrated passage St. Paul defines the various states of Christian liberty with respect to the various laws that rule it (Rom. vi. 20–23) ; in servitude to sin man is free from justice ; in the service of justice he is free from sin. Thus is brought into the foreground a conception of liberty and servitude analogous, in the supernatural order, to the condition of freeman and slave in the social order. Essentially religious as it was, this Christian conception of liberty could not fail to have a profound influence on the philosophic and moral problem of the freedom of the will. Already St. Augustine found himself at grips with the multiplicity of meanings which the word " liberty " may convey to the Christian mind. Sometimes it means simply " freedom of the will," sometimes liberty as opposed to servitude, and since servitude may be servitude to sin or servitude to death, we can call " liberty " either the sanctity that delivers us from the one, or the resurrection that rescues us from the other. Accordingly the mediæval doctors, carefully classifying these various senses, will add to the psychological liberty proper to free-will, that is, *libertas a necessitate*, these other properly religious or supernatural liberties, namely, that which delivers us from sin : *libertas a peccato*, and that which delivers us from suffering and death : *libertas a miseria*. The temptation seems to have been strong in certain theologians to reduce this multiplicity to unity by absorbing free-will into these religious liberties which strike off its fetters and allow it to put forth its powers without restraint—but this simplification brought with it a merely apparent facility, and although the sole possible issue was certainly more complex we shall see that philosophically it proved more fruitful.

To reduce psychological freedom to its liberation by grace was a convenient way of unifying the meanings of the word *liberty*. Taking the spontaneity of the act of will for granted, nothing then stood in the way of saying that the only liberty is *true* liberty, that is to say liberty to do right. Peter Lombard attests the existence of this view in the twelfth century, but he prefers another and maintains the existence of a twofold liberty, that of free-will in a state of moral indifference, and that of free-will liberated. The reasons for this preference are easily understood : free-will, as we have said, is a physical power and therefore inamissable, whereas freedom to do right has been lost ; how, then, can we fail to distinguish things so different ? To maintain the difference under the same name, on the other hand, would lead to perpetual misunderstandings. Hence the remarkable effort made by the mediæval philosophers to disassociate, within the idea of liberty, elements which had been always there implied, but up to that time had been more or less confused.

The starting-point is found once more in St. Paul and St. Augustine. The Epistle to the Romans makes a clear distinction between the act of will and its efficacy, for these two do not always go together : " *Nam velle adjacet mihi ; perficere autem bonum non invenio* " (Rom. vii. 18). St. Augustine puts·the distinction more precisely : *velle* as against *posse*.[12] Henceforth no ambiguity is possible, for, absolutely speaking, will is not power nor is power will. But another difficulty remains. If we made a radical distinction between power and will, we should end by withdrawing the *potestas* from *voluntas*, so that the new liberty we are here concerned with, so far as defined as a power to do what one wills, would be entirely out of the reach of the will. And this would not only be out of accord with all ordinary use of language— since of an efficacious will we say that it is free, not that it is powerful—but also with the very just feeling for the close ties between the freedom of the act of will and its efficacy. Something else, then, was required, and St. Augustine himself pointed the way. Grace, as he conceived it, was not without a profound effect upon the play of our free-will. It is not sufficient to picture it as taken into partnership as a complementary power ; it has a direct effect on the state of the will in that it confirms and heals it. In possession of grace we have something more than free-will plus the power of the grace, it is the free-will itself which, by grace, becomes power and achieves its liberty.[13] The problem then was so to dispose the constituent elements of the free act that freedom was really attributed to free-will, while at the same time distinguishing between a will in servitude and a will liberated.

Perhaps it was St. Anselm who most clearly grasped the meaning and scope of the question. " Power," for him, is the aptitude to do what one wills. Now, in a sense, the will is a kind of power ; it is a power of willing. The more apt it is to will, the more fully it will be itself. Apply this principle to the problem in hand. A will that wills but is not able, is not merely a will without efficacy, but a diminished will. The strength it lacks is its own strength, that which, as a power of willing, it ought to have. Free-will, therefore, is always in a position to will the good or will the evil—how otherwise could it choose ?—but since the will is essentially power, the bad choice is not to be confused with its liberty. Man is free, and by his own choice he does evil, but not by that which makes his choice free. Let us go further. Created free, able to sin, man sinned by his power of sinning ; but this power was no part of his true liberty which is a liberty of not sinning, of not serving sin.[14] In other words, man's liberty was the liberty of a will created free from servitude to sin ; his free-will, therefore, was not only free, but was, at the same time, an efficacious power. This free-will abdicated its power in sinning ; but are we then to say that this abdication was constitutive of its liberty ? A liberty that enslaves itself, even freely, is unfaithful to its own essence ; the free act by which it makes itself less free betrays its own freedom. For this reason, precisely because every will is a power, all diminution of the power of will diminishes the liberty of free-will. The real power is the power of efficaciously willing the good ; having done evil, the will remains free to will the good, but not to do it ; it is therefore but a wounded liberty ; in restoring the lost power grace restores to free-will something of its first efficacy ; far from diminishing, it liberates it ; to the spontaneity of the *arbitrium* it adds that *liberty* which is its efficacy ; a true *liberum arbitrium* is a *libertas arbitrii*.[15]

Once the power of sinning ceased to be regarded as a constitutive element of liberty as such, the philosophic problem of free-will had of necessity a changed aspect. Whoever sought to determine all the conditions of liberty was led to distinguish, with St. Thomas, three different standpoints : liberty as regards the act (in that the will is free to act or not to act) ; liberty as regards the object (in that it can will this or its contrary) ; and liberty as regards the end (in that it can will good or evil). There is no difficulty about the liberty of the act, since the will is a spontaneous master of its own determinations. Whether we are concerned with will before the original transgression, or after it, or even in the blessed confirmed in grace, in each case will always wills what it wills, and is consequently free. Nor is there any difficulty when we consider the will as choosing the means to the end ; the choice is free beyond all doubt. Man, indeed, does not choose his end ; necessarily he wants to be happy merely on account of his nature as man ; but various routes to happiness lie open, and he is free to choose what seems to him the best for the purpose.

It is otherwise, however, when we consider the choice of means in their relation to the end in view. Absolutely speaking, it is because man is free that he is able to deceive himself about the nature of his end, or the means to be adopted to achieve it. Apart from intellectual error he would always know what he has to do, and apart from defect of will he would never refuse to do it ; and the errors, like the defects, are indices of a free-will. It is not these, however, that constitute liberty. Even where they are no longer possible, as, for instance, in the blessed, liberty nevertheless exists whole and entire, for if to do ill is to be free, to do always right is not less so.[16] It will be granted, then, on all hands that the free decision of a fallible will owes its freedom solely to the voluntary character of the act, not at all to its fallibility. But are we not forced to go further ? If the power of choosing ill is but a defective use of freedom, does it not mark a diminution, or, so to speak, a mutilation of freedom itself? If it seems difficult to deny this, are we not bound further to affirm that the less fallible the will the more free it is ?

At this precise point the Christian doctrine of the liberation of the will by grace penetrated the analysis of free-will and profoundly modified its structure. Although Thomism did not push the matter to its final conclusion, it prompted the attempt. First of all, taken precisely as nature, the will is necessarily determined to will the good. We have seen, and must maintain, that freedom resides in the will as in its root, but it is none the less true that its cause lies in the reason. The will is drawn to the good of necessity, the necessity of the nature in which it is rooted, and it is therefore only on account of the diverse conceptions of the good proposed by reason, that the will can be undetermined.[17] But if reason is the true cause that makes our natural will a free-will, why should we not ask ourselves what becomes of free-will when the will stands face to face with a perfect determination of the reason ? The case of God and that of the angels and the blessed are very different to be sure from ours ; nevertheless, they may serve to throw light on the matter. Now the reason of God is certainly infallible ; yet none the less St. Thomas affirms that His will is free as regards everything other than His own proper perfection. Likewise he attributes free-will to Jesus Christ and even to the blessed,[18] for the will of Christ, although determined to the good, was not determined to this or that particular good ; and the will of the blessed, although confirmed in grace, has none the less to order its acts in view of the end.[19] The case of the angels is a parallel one, and may help us perhaps to understand that of the blessed. They see God face to face, and not only therefore cannot but desire Him, but cannot even be deceived as to the means to be adopted for the service and praise of God. Now, far from prejudicing free-will, the infallibility of its acts of choice merely attests its perfection. It is therefore more perfect in the angels, who cannot sin, than it is in us, who can ; and it is more perfect, not because more free from necessity or from constraint, since freedom as such does not admit of more or less, but more perfect in its cause, which is the intelligence itself, for wherever there is intelligence there is free-will, and the more intelligence there is by so much is there the more liberty.[20]

St. Thomas' theology was always big with philosophic consequences, and time for their development has not been wanting. There were two possible interpretations open here, according as circumstances suggested that the emphasis should be put on the *root* of free will, that is, the spontaneity of willing, or on its *cause*, that is, the reason. At the end of the Middle Ages the sudden appearance on the scene of Wycliffe, Luther and Calvin brought matters abruptly to a head. Taking up the question solely in the religious sphere, the Reformers were interested only in the power of free will ; and since the will can no longer do anything without grace, there was nothing left for them to do but deny free will altogether. Luther's *De servo arbitrio* is the most complete expression of this attitude : a will that has lost all its power has lost all its liberty. To combat the Reform the Catholic theologians had to effect the necessary rectification. They therefore appealed to those elements of the patristic and mediæval tradition which were in direct opposition to the doctrine of the slave-will, that is to say they appealed to will's radical indetermination, and its faculty of choice. Deliberately exalting the power *ad utrumlibet*, Molina built up a doctrine in which indifference became the essential character of liberty. Duns Scotus was in a position to furnish him with arms, and it is from him, in all probability, that he drew inspiration ; but St. Thomas too, had texts to offer, and it is very certain that in maintaining indifference of choice at the root of free will, Molina remained faithful to one of the permanent exigencies of Christian thought in this matter. But the power of free will also claimed its rights. It could not possibly be granted to Luther that this constituted the whole question ; but the Thomists, at least, owed it to themselves to admit that it was a part. That is why, without in the least letting go of the *root* of liberty, they recalled attention to the nature of its *cause*, and those who followed them down this path developed a close critique of Molina's liberty of indifference. I have pointed out elsewhere how the controversy supplied Descartes with material for his doctrine of liberty.[21] I conceive that his Fourth Meditation would never have seen the light had it not been for the long effort of mediæval theology that lay between him and the Greeks. Through Père Gibieuf of the Oratory, to whom he appeals, he owes to St. Thomas the conception of free will as a power of choosing with a simple and indivisible liberty from constraint, but whose exemption from error and evil increases as its indifference decreases. In Cartesianism, as in Thomism, the infallibility of the judgment, far from destroying free will, only exalts its liberty.

Thus understood, these doctrines of free will stand in direct connection with that Christian naturalism, the exigencies of which were felt throughout all mediæval philosophy. When confidence in the indestructibility of nature and in the efficacy of second causes as issue of the creative fecundity was lost, the world was ripe for the Reform ; wherever, and in whatever measure, this confidence was retained, the Reform was held in check. For this reason all Christian ethics of the Middle Ages, like those of the Fathers that inspired them, have their necessary basis in the doctrine of an indestructible free will. If, as some are pleased to insist, the Reform inaugurated the era of modern thought, then this era was inaugurated with an abdication of free will. To men of whom even God Himself can no longer make coadjutors and co-operators, Christian philosophy had no message. By the mere fact, however, that it was Christian, it was bound to emphasize the importance of the will's power and of the place it fills in the definition of the free act. To this philosophy therefore we owe a richer and more comprehensive idea than is anywhere to be found among the Ancients, of the spontaneity of an act of will which, in virtue of its efficacy for the good and true, becomes liberty.

CHAPTER XVI

CHRISTIAN LAW AND MORALITY

THE vocabulary of ancient ethics is not at all easy to translate. When Aristotle wants a comprehensive term for actions which he approves and recommends in his Ethics, he calls them καλὰ καὶ σπουδαῖα, that is to say beautiful and noble things, also things of some real moment, arising not from any merely utilitarian, servile or playful activity, but from one that has a value in and for itself.[1] It is not easy to recognize here our plain and simple term : the good. The Latins were already in a difficulty about the rendering of τὸ καλόν. Cicero, who gave the matter some attention, fixed upon *honestum*, which we usually render into English as " morality " or " virtue," but which Cicero uses to mean essentially something laudable for its own sake, apart from all utility or reward.[2] Christians could not possibly remain indifferent to ideas which, in spite of all their imprecision, or even perhaps on account of their very imprecision, were so rich in substance. In all moral good there is a certain beauty, the Greeks were quite in the right there ; and so for Christians too the beautiful, the *decorum*, will remain one of the characteristics of the good : it invests the soul with the splendour of intelligible beauty, as the Greeks say, or of spiritual beauty, as St. Augustine prefers to put it. What is beautiful and good deserves honour and praise ; Cicero was certainly not in the wrong ; but first we must recollect (for it is a point to which we shall have to return) that before everything else it is to the interior act of the will that all the praise is due,[3] and especially that this will itself is beautiful and praiseworthy only on account of its self-orientation towards an overruling end.

As far as this point is concerned the evolution of ethics may be summed up by saying that the Christian moralists sought first to attach all moral worth to the voluntary act as its root ; that at the same time they gathered up the concepts of the beauty and honour of human acts into a concept still more comprehensive, that, namely, of the good ; then referred the good to a transcendent principle worthy of all honour in itself and absolutely, more truly even than virtue, which is only honourable on account of this.[4] They regarded the soul of a just man as beautiful and worthy of honour because virtuous, but virtue itself as honourable only because it leads man to God. It is therefore not the supreme good, the *nec plus ultra* that it was to the Greeks, the all-sufficient unconditioned condition of all morality. To understand how the change of outlook was effected the best way will be to define, first of all, what it is that gives the moral act its good or bad quality ; and it will be simplest to bring our effort to bear on the bad act ; on what Christians call sin.

The word " sin " and its equivalents has a moral and religious connotation in modern ears, and it is difficult to realize that this was not always so. The mediæval theologians were so careful to preserve all that is true in Greek moral doctrine that it is not easy to distinguish their own additions in this matter, and even in a case which so directly concerns Christian morals they do not always make it clear at what exact point they abandon their predecessors. Let us therefore try to follow the path indicated by the mediæval thinkers, that is, to go along with the Greeks as far as their thought can take us and abandon them only when they abandon us. No one ever carried out the experiment more systematically, or pushed it further, than St.

Thomas Aquinas, and if we take him for guide we shall be sure of going to the extreme limit that it is possible to reach in this direction.

Sin, above all, is a vicious act ; if, then, we would understand the nature of sin, we must understand that of vice whence it flows, and, since vice is opposed to virtue we shall have to recall first of all the nature of virtue. Essentially it is a habit, that is, an acquired and durable disposition inhering in the subject, which enables him to act in accordance with his nature. The definition is Aristotle's ; [5] and thus it is on an Hellenic basis that the whole edifice is to be built. A thing is said to be good when it is as it ought to be in order to fulfil its own essence, and the exigencies of its nature ; to acquire the habit of acting as one ought (it being granted what one is) is therefore to acquire a morally good quality, and to accomplish the act that spontaneously proceeds from such a habit is to act well, or, in other words, to do right. An act is morally good or virtuous when it accords with the true exigencies of the nature of the agent.

Three things, consequently, are opposed to virtue— sin, malice, and vice. Sin, by definition, is a disordered act, that is to say an act contrary to the order demanded by the nature of the agent. As such, it is in direct opposition to the good act, as we have just said, and is therefore necessarily bad ; and precisely on that account it must be considered as the result of a vice, since the vice of a nature is precisely the lack of a perfection demanded by that nature, that is to say the contrary of virtue. To define virtue and vice in this way, by their accord or disaccord with nature, is not merely to rest in the position of Aristotle, but to accept what St. Augustine himself had expressly wished to retain from Greek naturalism : *omne vitium eo ipso quod vitium est, contra naturam est.* [6] Here, then, is no question of a tardy invasion of Christian theology by Hellenism, but a position common to the Patristic era and the Middle Ages, and thus essential to Christian thought itself. Nothing could be more purely classical ; and yet, in bringing in his technical elucidations borrowed from the Aristotelian ethic, St. Thomas was preparing himself to pass beyond the master who inspired him.

For what indeed is a nature ? That which puts a thing into its proper species, and, consequently, it is its form. But what is the form of the human composite ? As we have seen, it is the rational soul. It is from reason then that our nature derives its properly human character. If this is so, then, to define moral good and virtue as that which accords with our nature, is equivalent to defining them as that which accords with reason. Conversely, moral evil, sin, and the vice whence sin arises, can be conceived only as a lack of rationality in act or habit. In whatever way we look at the matter we shall always have to return to Cicero's famous definition of morality as the habit of acting as reason and nature demand.[7] The mysticism of Dionysius the Areopagite was quite sufficiently Hellenized to take up these notions of moral good and evil : " Good, for man, consists in following reason, and evil in forsaking reason." Hence the ultimate determination of the concept of virtue to which St. Thomas leads us : that which makes the man who has it good, and makes the thing he does good, by making him apt to act according to his nature, that is to say according to reason.[8]

At first sight, nevertheless, this definition would seem to be insufficient for a Christian ; or rather it is even evidently insufficient, since it makes no mention of God, nor of those relations, good or bad, which virtue or vice cannot fail to establish between the will of man and the will of his Author. Accordingly other definitions are to be found in the Fathers or mediæval theologians which seem at once to be more direct and more Christian ; as, for instance, St. Augustine's, so often cited : sin is any word, deed or desire contrary to the eternal law.[9] For is not the divine law above nature— and does it not therefore seem better to define sin by its opposition to the law that rules nature, rather than by its opposition to the nature ruled ? But since natures are natures because God made them so, to deviate from their own essence is the same thing as to contravene the rule laid down by God in the creative act.[10] The rectitude of the human will is therefore measured at once by its accord with the divine will and by its accord with reason. In Anselm's vigorous phrase there is, in this sense, a truth of the will just as there is a truth of the judgment, and at bottom it is the same truth, since in either case man's rectitude consists in ruling himself according to the divine law, so as to think and will what he ought. There is even such a thing as a truth of actions—*qui facit veritatem, venit ad lucem* [11]—for to *do* the truth is simply to do right, that is, to bring our act into conformity with the divinely prescribed rule, just as to do ill is to swerve from it. If we remember, then, that God is the Author of reason, we see at once that to neglect what reason prescribes is to neglect what God prescribes. To act against its dictate, that is against nature, touches also its Author : *unde ejusdem rationis est quod vitium et peccatum sit contra ordinem rationis humanae et quod sit contra legem aeternam.*[12] There are therefore not two definitions of sin, but only one, since to violate the rule of reason, the proximate rule of human action, is to violate its first and most fundamental rule, the eternal law, which is, in a way, nothing but the reason of God : *quasi ratio Dei.*

In reducing thus to unity the two regulative principles of moral action, the Christian thinkers were enabled to link up the results of Greek speculation with their own metaphysic of creation. Clearly, however, this was to subordinate the former to the latter, for human reason becomes legislative in matter of morals only in so far as it is " informed " by the divine law at each of its degrees (*e.g.*, by natural information, by sacred doctrine, or by infusion). As natural reason it follows the first principles of the practical reason, which is itself subject to the divine illumination, the rule of the moral conscience ; as directed by God to the supernatural end man's reason is subject to the prescriptions of revelation or to the hidden action of grace. Although therefore it is true that the natural law of reason is reducible to the eternal law of God, it is not therefore true that the eternal law is reducible to the prescriptions of natural reason. In other words, there are more things commanded or forbidden to man than are commanded or forbidden by his natural reason ; the two orders could be purely and simply coincident only if the natural and supernatural ends of man were rigorously identical. We know on the contrary that this is not the case ; the eternal law therefore envelops human reason, provides a basis and sanction for its prescriptions, but for much more besides ; and on this account the philosopher may well define sin as something opposed to reason, but the theologian, nevertheless, will regard it as above all an offence against God.[13] And although the theological order is one thing and the philosophical order another, we might reasonably ask whether the philosophical order itself would here be what it is without the theological order that crowns it. Sin, as a breach of the divine law, as disobedience to God, is more a matter for the theologian than for the philosopher ; very true, but then it seems inevitable that even the Christian philosopher himself, when he comes to consider why acts against reason are wrong, will in the end be led to the same sort of answer. The divine law, taken in its strict sense, signifies only the commandments of God, and it is with reference to this divine legislation that moral theology defines sin ; but, taken in a wide sense, it comprises all the various " informations " of human reason by divine reason, including the moral conscience itself. That, then, is why Christian moral rationalism ends by integrating itself with a metaphysic of the divine law ; disobedience to reason is disobedience to God Himself : all sin is a breach of the divine law (*praevaricatio legis*).[14]

We should look in vain for any such conception in Aristotle's moral doctrine. Attentive to all that conditions morality, he classifies with minute care the various errors that go to vitiate our acts, but his analysis never takes him higher than the human reason. At the basis of all his analyses and all his conclusions lies Socrates' fundamental principle : all wickedness is ignorance. Corrected, completed, made full and flexible with all the necessary precision of detail, this principle still rules Aristotle's ethics : " The wicked man is ignorant of what is to be done or not done ; and it is this kind of error that makes men unjust and, speaking generally, bad." [15] At the root of the evil, therefore, lies an ἁμαρτία or an ἁμάρτημα, that is to say an initial error of judgment, which in the upshot causes the action to miss the mark it aims at. Aristotle, in fact, regards man as engaged in the pursuit of a happiness—for the rest a purely relative happiness—the achievement of which is the term of the moral life. Actions are good in so far as they tend towards this goal, bad in so far as they tend away from it ; in any event his conception of moral good or evil is closely allied with the ideas of success or failure.

There are errors of conduct which do not depend on ourselves, mere ignorances due to exterior causes, for which, therefore, we are not responsible ; but none the less they make us miss our mark, they are defeats properly so called, failures in any case, misfortunes if you will. When the cause

of the failure lies in ourselves, then there is no longer a
mere mischance but a fault. Thus the only difference
between a failure and a moral fault lies in the cause of the
final frustration of the human act. And nothing is changed
if to these causes we add impulses of desire or wrath which
hurry us into evil ; the resulting act is unjust, but the
injustice is still due to a deep-seated ignorance of which
the bad act is but the indirect consequence. Even if the
act is consciously willed, deliberate, chosen after a consideration and evaluation of motives, its malice expresses
the injustice of the agent—not merely the act but the man
is bad ; but the man's own malice, the radical injustice
which vitiates his nature, always presupposes the same
initial error in the reason.[16] Virtue is the rational habit
that puts us in the way of attaining happiness, just as vice
is the irrational habit which condemns us to miss it. In all
this there is no hint of any law higher than that of the human
being, we get the benefit of our address and suffer the consequences of our awkwardness ; absorbed in the contemplation of his own thought, the First Unmoved Mover makes
no attempt to legislate for man ; not only is it difficult to
picture him as the author of a supernatural revelation, or
as the law-giver of a Decalogue, but since he is not the
creator of consciences he lives his own divine life without
creating any interior law to direct us or light to enlighten
us ; our errors enslave us, and there is no Christ there to
suffer on that account ; the just man perseveres and the
sinner repents, and there is no joy in heaven at his per-
severance or welcome for his repentance ; it is for man to
live out his moral life and gather up the fruits ; God is not
interested.

It seems to be otherwise with Plato ; indeed, it is necessarily otherwise in a philosophy where the gods are the
authors of nature and rule it providentially by laws. In

Platonism, as in Christianity, there is a divine order which rules and defines the moral order. Expressing himself in the language of myth, the philosopher recalls the old tradition which relates how Saturn, realizing man's incapacity to govern other men with suitable authority, set certain higher intelligences over the cities, and made these Demons shepherds of men, as men are the shepherds of their flocks. When we follow the precepts of reason it is really to these divine guides that we submit, committing to them thus the conduct of our private life, of our families or of our cities. There is therefore no difference at all between obeying the gods and obeying " the immortal part of our being." [17] If now we simply identify Saturn with the Demiurge of the *Timaeus*, have we not at once an equivalent of the Christian scheme, where the natural law of reason is but an expression of the divine law, and where the ultimate foundation of morality lies in this latter ?

That Plato is as near to Christianity here as one can very well get without actually being a Christian is altogether evident, and I say it merely to avoid the accusation of forgetting it. But there are important differences, nevertheless, of which history is bound to take account ; and they all arise from the fundamental difference between the Greek and Christian worlds. Subordinated to a plurality of gods, organized but not created by one amongst them, Plato's universe is not wholly penetrated with intelligibility because it is not wholly dependent upon its author. Hence the divergence between the Christian and Platonic conceptions of providence which I have tried to explain in its proper place. The same divergence appears here. All moral injustice is a violation of the law, but this law depends on complex causes, much more complex than that which imposes the law on the Christian world. According to Plato it is set up in the first place by the god who is interpreted by the art of the human legislator, but the laws that rule States depend also on chance and circumstance, which, along with the god, govern all human affairs.[18] The intervention of this μετὰ θεοῦ τύχη χαὶ χαιρός not merely presupposes a universe at any rate partially left to itself—as indeed it must be in so far as it is uncreated—but also a human order over which the divine law reigns quite otherwise than it does in the Christian world.

It reigns, no doubt, just as providence does, of which, after all, it is merely a form ; and moreover, like providence, it surrounds itself with sanctions in order to ensure that its decrees are respected ; but in neither case does violation of the law involve a violation of the order of creation, a thing, for the rest, that Plato never even conceived. The god who, according to Orphic tradition, is the beginning, middle and end of all things, goes always straight forward ; Justice follows him with rewards for the good and punishments for the wicked, but this Justice administers the world in an altogether impersonal manner. Whoever follows the footsteps of the deity becomes thereby like him and will have the reward of living justly with the just ; whoever thinks himself able to stand alone is abandoned by the god : after some delusive appearance of success Justice will turn it into vengeance in the end and the man is lost, himself, his family and his country.[19] The "ancient law" to which Plato constantly appeals is the law that like loves like ; it is sufficient, both for himself and his gods, to secure the reign of order on earth, but this reign of order exists only for the sake of the beings ruled, their lapses of conduct hurt no one but themselves, they touch neither Justice nor the god that goes before him ; he is insufficiently responsible for man to take offence at his conduct.

The position is altogether different in the Christian scheme. "The eternal law," says St. Augustine, "is the divine reason, or the will of God, ordaining the preservation of the order of nature, forbidding its disturbance." Nothing could be clearer than his formula ; but consider its implications. If the eternal law is one with the divine will or reason, it is closely allied with the Ideas ; in fact, each of these is an eternal and immutable law dwelling in the Divine Wisdom ; that is to say in the Word. Now we know that God's ideas are God. The divine law, then, thus identified with God's reason, is itself identified with God. It is, according to St. Augustine's familiar expression, often taken up by St. Thomas, the divine art whereby all things are created and governed. Although they are only comparisons, these expressions are to be given their full weight. Art is the rule which guides the thought of the artificer, and determines what the work is to be. When God is in question it has to be remembered, in the first place, that His art is not a perfection superimposed on His nature, but is His nature itself, and we must add, moreover, that its efficacy does not go to a mere production but to a creation. To take due account of both these aspects of the matter we must therefore say that the eternal law is none other than God Himself, Whose reason governs and moves all things inasmuch as it has created them. If we lose sight of the intimate connection between the concept of law and that of creative providence, we shall find it impossible to discern the special character that distinguishes the Christian moral law from Plato's.

Between the two there interposes once more the capital fact of creation, and all the consequences it carries with it in both orders ; that is to say in the order of action and the order of knowledge. The eternal law is one with the Wisdom of God which moves and directs to their end all things it has created. And so we may say with St. Augustine that God " concreated " the natural law along with all things that He called into existence ; just as, in virtue of their existence, they participate analogically in the divine Being, so, in virtue of the fact that the law of their activity is inscribed in their essence, in the intimate structure of their being, they participate analogically in God's eternal law. How could they receive one without the other ? The natural law is to the eternal law what being is to Being, and the principle holds for all orders of creatures without distinction.

This is a point on which the Fathers of the Church and the mediæval philosophers, whatever their divergencies in the detail of technical discussion, are entirely at one. It would be hard indeed to imagine it otherwise, seeing how freely they all avail themselves of Biblical texts as soon as they take up the question. *Praeceptum posuit, et non praeteribit* (Ps. cxlviii. 6). *Quando circumdabat mari terminum suum, et legem ponebat aquis ne transierent fines suos* (Prov. viii. 49). *Per me reges regnant et legum conditores justa decernunt* (Prov. viii. 15). The creative God of the Scriptures affirms Himself, therefore, as the Source and Cause of all legislation, natural, moral and social. The laws of the physical world are the work of a Sovereign Legislator, prescribing the rules that nature blindly follows. Man, being gifted with intelligence, must obey them.

The eternal law, then, may be said to be " written " in our hearts. If reason is the rule whereby the goodness or the malice of our will is measured, it owes it to this supreme rule, which is itself but a ray of the divine reason, shining in us by way of participation. *Multi dicunt : quis ostendit nobis bona ? Signatum est super nos lumen vultus tui Domine* (Ps. iv. 6–7). The classic text which forms the basis of every doctrine of intellectual illumination is equally applicable to moral illumination, since the rules of action, like the rules of knowledge, also derive from God. They are, indeed, themselves cognitions : the first principles of the practical reason whence by way of rational consequence flows the whole series of particular duties. " The light of reason within us is able to show us good things, and guides our will, in so far as it is the light of His countenance, that is, derived from His countenance. It is therefore evident that the goodness of the human will depends on the eternal law much more than on human law ; and when therefore human reason fails we must have recourse to the Eternal Reason." [20] The moral law that was divinely revealed to the people of Israel merely put plainly before men's eyes what they had refused to read in their conscience, where, nevertheless, they might have found it already written ; what God reveals for man's good when He deems it necessary, He has, in the act of creation, already inscribed in their conscience ; to consult the revealed moral law, to consult conscience, or to consult reason, is, in any event, to consult God.

It is easy enough to see from all this where the basis of all legitimate political or social legislation is to be found ; [21] but let us be content for the present to note the consequences of the principle as far as they concern the idea of sin. Man, as he here appears, is a being who participates in the eternal law by way of rational knowledge, and knowing the law, directs his footsteps by its light instead of simply submitting to it like the rest of nature. To see the law directly in itself would be to see God, and is therefore out of the question in this life, but since all knowledge of the truth is an irradiation of the divine light in us, every infraction of the prescriptions of the practical reason, is *ipso facto* an infraction of the eternal law which rules it ; that is to say an opposition set up between the human will and the divine reason. Furthermore, since the order of nature set up by creation is merely a real participation in the divine law, all that sets itself in opposition to the divine law, which is followed by this natural order, is, by that very fact, a vice ; *omne illud quod contrariatur ordini naturali est vitiosum.*[22] All virtues, says John Damascene, are natural. They are natural because rational, for our nature is defined by rationality ; but they are rational only because they conform to the sovereign prescriptions of the divine reason, which, through the medium of human reason, legislates for human wills. We must understand therefore, conversely, that all that contradicts a natural inclination, unless indeed it does so in the name of a higher natural inclination, is sin. Thus, to violate a natural law and rational prescription may be, assuredly will be, to endanger our own happiness, for it is a violation of an order that always retains its rights and reasserts them by sanctions ; but it is something much more at the same time, namely an offence against God, Who is Creator both of the order and the legislative reason, and is Himself the Supreme Legislator. The sacred character with which all order, even natural, is invested by the mere fact of its creation, carries ineluctable consequences. Here is no mere question of disturbing, at our own risk, an order established in matter for our own benefit, by some more or less benevolent Demiurge ; when man sets himself up against it he denies and subverts as far as in him lies, the end which God Himself proposed in creating him. All refusal of the natural order is tainted with sacrilege, and, when a grave sin is in question, with sacrilegious madness. When a human will revolts against the divine will, the sin consummates the suicide of a moral person created for beatitude—and rejecting it. There lies the true heart of its malice ; and nothing then remains but the inevitable consequences, or the atoning remedy.

It is hardly possible to come to this conclusion without recalling numerous Biblical expressions that represent God as offended, angered, avenging or appeased ; images in which no one sees any attribution of human passions. The Judeo-Christian God is certainly very different from the gods of Greek mythology ; He feels neither anger nor sorrow ; His intimate life is no more troubled at our offences than gladdened at our praises. It is Aristotle rather than Homer who is in the right here. Nevertheless, the Bible makes use of these metaphors only in order to suggest, by way of sensible comparisons, a profound metaphysical truth which is an integral part of Christian philosophy and of which Aristotle, for lack of the conception of the natural law as a creation of God, never so much as dreamt. Even in the world of Plato, sin does not violate God's work as it does in the Christian world. Of course nothing that we can do can touch God formally and in Himself, but the sinner does all he possibly can against a God who lies beyond his reach, by lifting up a destructive hand against His creation. Even there, neither man nor angel can boast of being able to set a term to the divine power or obstruct the fulfilment of the divine will. What God wills to be will be ; what He wills to be done will be done, and will be done, moreover, by us, or, if necessary, in spite of us. Man disturbs the prescribed order only in so far as God permits ; to this extent, however, it is only too true that free-will can and does disturb it. That is why sin, as the Christian thinkers conceive it, makes the sinner an adversary, a rebel against God : *facit hominum rebellem et contrarium Deo ;* an enemy that fights against Him and resists Him : *peccatum facit hominem inimicum Dei, et pugnare adversus Deum et resistere.*[23] This enmity, which makes the punitive sanction inevitable, is, at bottom, nothing but a will to deny justice, and to refuse God that obedience which is His due. Sin often springs from mere ignorance or weakness, and then, naturally, its gravity is much less ; but when its essence is fully realized, when there is a deliberate denial and refusal of the order willed by God, then it becomes a sheer contempt for God ; without power to diminish His glory it refuses to recognize it. When man acts thus he excludes himself from his own share in the glory to which he was originally destined. God retains the whole perfection of His beatitude, but the sinner loses his, for it is precisely to this divine effect in himself that the sinner sets up an obstacle, and destroys by the revolt of his will.

Hence we may understand how, in the Christian conception of the world, the idea of sin is to be linked up with the idea of a sanction on the one hand and of grace on the other. Both good and bad are subject to the divine law ; the former inasmuch as they fulfil it, the latter, who refuse to fulfil it, inasmuch as they must submit to it. On the purely natural and philosophic plane the upshot of the moral life is merely this life's happiness, gained or lost according as to whether human acts are in accordance or otherwise with nature and reason. Now, however, it is the divine will and reason which, ensuring the operation of the natural law, determine the fate of human beings to which that law applies ; but simply because this natural law is a participation of the divine law, every grave infraction of its prescriptions, if voluntary and fully advertent, has a character quite other than it could possibly have in any Greek ethic. By expressly promulgating in the Decalogue what the moral consciousness of man already prescribed, God made manifest how man's reason participates in His own. Warning him that in violating his own law he violates the law it derives from, the divine Legislator makes it henceforth impossible to misconceive the real nature of the moral law, and the real bearing of its infringement. The crime would be irremediable were it not that God takes it upon Himself to provide the remedy.

Sin, in fact, as Christians conceive it, is none other than a destruction of the divine order. When he sins, therefore, man destroys what he could not create, and, moreover, cannot recreate. The rectitude of his will was itself a grace ; he lost it by his fault ; and how, after that, could he possibly recover it by his own efforts ? But what God has given, God, if He so wills, can restore. And further, in sinning against God in his works, man sins against an infinite Being ; how, then, by what personal efforts, in virtue of what merits, can he bridge the infinite gulf between his own injustice and the divine justice ? Once more, if human merit is to be restored there is none but God Who can restore it. Thus the Christian finds himself placed in an order in which natural morality itself calls for the supernatural as its necessary complement. He can appeal no longer to his own virtue, righteousness and merit, in the first place, but to those only that he acquires by grace. Is there any real need to insist how the changed conception of sin in the Judeo-Christian tradition, which links it up with the idea of creation, profoundly transforms the whole conception of the moral life ? Doubtless we touch here on the supernatural order which no one would dream of integrating with philosophy properly so called ; it transcends it, but it would not even find anything to transcend in a Greek philosophy because there the conception of moral evil is no longer what it is in the Christian scheme. St. Bonaventure puts this capital point with the greatest clearness : " The Greek philosophers did not know that sin is an affront to the divine majesty, nor yet that it deprives our faculties of their power. They asserted, therefore, that in the performance of just acts a man might restore that justice which, by forsaking the order of right reason, he had lost. Wherefore Aristotle said that in striving to act better, the wicked man will greatly improve, and may even effect a full restoration of the habit of well-doing. . . . But Catholics on the contrary, who consider the matter in the light of faith, and subject to Scriptural authority, know well that sin offends God—*Dei est offensivum*—and also in a way, is the cause of damnation—*et etiam, quodam modo, damnificativum*—inasmuch as it withdraws man from the service of God, and subjects him to the service of the devil ; and finally deforms the divine image in us—*et etiam imaginis deformativum.* And they concluded from all this that if free will is to be saved from slavery to sin, grace is altogether necessary." [24] Here all lines of Christian thought converge ; when we realize how sin effaces the image of God in our souls, we realize what it defaces in the work of creation, what consequences it entails, and what sacrifice of infinite price will be necessary for our redemption.

Far as we are in all this from Greek philosophy, we are farther away still from modern thought with its insistent demand for complete autonomy of the will. It is by no means certain, however, that even this idea of the autonomy of the will is a complete stranger to Christian influence. For it presupposes that man is a person : and who, if not the Gospel, has taught us to consider the human person as a member of a realm of ends ? The citizen of the realm of ends is a person, because he is a rational being ; his practical reason is legislative inasmuch as it prescribes action from a maxim universally valid for reason in general ; when it obeys the moral law the will submits only to its own law, which is the law of reason. If that is what is meant by autonomy of the will, who will maintain that the will of the Christian is not autonomous ? Of course, it is subject to the law of God ; but is it not so even in Kant's ethics ? The individual man, even there, is no more than a member of the realm of ends, for even if he is legislator he remains subject to the legislative reason. One Being only enters into that realm as Chief, *als Oberhaupt,* and that is God, for He is the Sovereign Legislator and alone depends on the will of no other. To say, with the Christian thinkers, that human reason is not " homogeneous " with God's reason,[25] is therefore merely to recall that the creative and created reasons do not belong to the same category ; it is not in the least to deny that what the second imposes as universally valid is also imposed by the first. Quite the contrary, as a participation of the divine reason and law human reason is legislative in its turn, since in thinking truly it can always will rightly, that is to say determine its actions according to a maxim that can be erected into a universal law of nature. Christian morality, understood in this sense, is not opposed, like Kant's, to the empirical standpoint of happiness just as it is not opposed to the rational standpoint of perfection ; it envelops and justifies both. Perhaps Kant's ethics are but a Christian ethic cut loose from the Christian metaphysic that justifies it, the still imposing ruins of a temple with undermined foundations.

CHAPTER XVII

INTENTION, CONSCIENCE AND OBLIGATION

WE have just seen the very considerable importance of ancient Greek ethics in the eyes of the mediæval thinkers, and the respect in which they held them even when they felt their insufficiency. But now we arrive at a point where, whatever their desire to preserve all that could be preserved of Greek moral philosophy, they were nevertheless thrown altogether on their own resources and obliged to invent almost everything. The Ancients can never be accused of neglecting the study of virtues and vices : their ethic was altogether made up of it. The isolated act, whether good or bad, if it were connected with no permanent quality of the subject, no stable habit affecting it durably and having a right to a place in its definition—such isolated act hardly interested them as moralists at all. No human action deserves discussion or moral appreciation save in so far as it expresses, not merely the particular act of will whence it flows, but the whole moral being such as it has become by the patient efforts of a lifetime. As long as the Greek influence endured, this problem of virtue and vice retained its preponderating importance in the field of morals ; and now in our day it has almost completely vanished. No moralist any longer considers anything but the individual act, taken precisely in its individuality, as if the whole totality of our moral life were thereby brought into play. Since Kant, especially, it seems that the analysis of duty tends more and more to concentrate the whole essence of morality in the quality of the immediate decision of the will.

Slowly and with infinite pains the Ancients set out to model their interior statue of a Man, they judged of a life only by its totality ; consummate artists as they were, the details had no interest for them save in function of the whole. Modern man, it seems, feels differently. We might say that he throws himself wholly into each one of his more important actions, and with the essence of his will, whether good or evil, he engages his whole personality. A single act may be a final triumph or an irremediable disaster, redeeming all vices, annulling all virtues. Perhaps between the Ancients and ourselves there has been a gradual displacement of an old and familiar outlook ; or possibly even modern thought, in search of a basis for its ethics, has only begun to believe itself able to dispense with Christianity at the very moment when the teachings of Christianity have become so thoroughly engrained that they are mistaken for the laws of its own reason.

Among the innumerable consequences for philosophy which flow from the Bible and the Gospel, there are none more important than what we may call the interiorization of morality. God is Being, and creates beings ; not only as to their body but, if they have one, as to the soul. Permanently conserving, continuously creating all being, and providentially watching over all He conserves, He knows all because He makes all. Without Him no soul subsists, and without His knowledge no thought can possibly pass through that soul. No refuge, henceforth, is left for those who cry peace and meditate evil in their hearts : *qui loquntur pacem cum proximo suo, mala autem in cordibus eorum* (Ps. xxvii. 3). For the sin once committed there is no hope of remission save by the forgiveness of the God Whom it offends, to Whom it is confessed, Who is free not to impute it to the contrite of heart (Ps. xxxi.). And since God already knows it, how shall it not be confessed ? If we forget the

name of the Lord to call upon false gods, He calls us to
account Who knows all the secrets of hearts : *ipse enim
novit abscondita cordis* (Ps. xliii. 22). For He made us and not
we ourselves (Ps. xcix. 3), and He therefore knows His
handiwork, its righteousness as well as its sin : *scrutans
corda et renes Deus* (Ps. vii. 10).

The God of the Bible thus immediately lays claim to the
whole of that moral jurisdiction to which His creative power
entitles Him. Because He knows all, man owes Him an
account of all, even of his inner thoughts : *omnia enim corda
scrutatur Dominus, et universas mentium cogitationes intelligit*
(Paral. i. 28, 9) ; since nothing within us escapes His
glance His judgments are just : *qui judicas juste, et probas
renes et corda* (Jer. xi. 20). Wicked and inscrutable is the
heart of man, says Jeremiah, who shall discern it ? And
always comes the same reply : *Ego Dominus scrutans cor et
probans renes* (Jer. xvii. 10). There is therefore no reason at
all to be surprised at the insistence with which the Gospel
reminds us that sin is anterior to the act that outwardly
manifests it, and, in many cases, independent of it. The
interior consent is already an act, as manifest to God as the
exterior act is to man, so that the internal accord or disaccord of the will with the divine law suffices to define a
fully determined order of moral obedience or transgression.
Exterior acts, of course, retain a considerable importance,
but now there is a whole series of acts anterior to these,
which do not count at all in man's eyes, but are all-important
in God's. *Ut quid cogitatis mala in cordibus vestris ?* (Matt. ix.
4). It is there, in the heart, before any deeds are done or
even any words are uttered, that evil is accomplished and
crime committed ; for " The things which proceed out of
the mouth, come forth from the heart, and those things
defile a man. For from the heart come forth evil thoughts,
murders, adulteries, fornications, thefts, false testimonies,

blasphemies " (Matt. xv. 18–20). The law forbids adultery, but God forbids that it be committed even in the heart (Matt. v. 28).

This doctrine, so clearly stated in the Gospel, was bound to be taken up at once by the Fathers of the Church. Justin teaches it in the second century, relying on this last text from St. Matthew, and for all commentary contents himself with a reference to its ultimate basis. The law forbids us to contract a double marriage, the Master of Christians forbids us even to desire it, for even the desire of defilement defiles, " since not our acts only but even our thoughts are all manifest to God." [1] The feeling that we are not justified by a mere material observance of the law, no matter how scrupulous, was often expressed by the early apologists both as against pagan legalism and Jewish pharisaism. " Whatever we say or whatever we think, whether by day or whether by night, is said or thought in God's presence, and we know, since He is altogether light, that He sees all that we think to hide in our hearts." [2] To the sin committed in deed, and the sin committed in word, is henceforth added the sin committed in thought ; and it is easy to guess that it will assume a special importance as the root and source of the other two. The sin of the heart is the beginning and, so to speak, the root of all sin ; the word and deed that spring from it merely bring it to its full development.

We might say that from this moment the essence of moral good and evil was in the way to be transferred from the outward act to the will ; for it is by the will, as St. Augustine says, that a man's life is made righteous or sinful : *voluntas quippe est qua et peccatur et recte vivitur.* [3] And by *will* we must here understand the secret impulse of the heart whereby it turns to a certain object or a certain end in preference to others. So constantly in the language of the Psalms does the " cry " or " clamour " of the human heart go up to God that with the Fathers it almost passes into a technical expression. The thief who watches in the darkness of night for his hour to strike already cries aloud his crime to God ; for he has decided on the deed and God knows it, the crime is committed already. Perhaps a man hardly knows himself from what a heart his actions spring, but God knows ; and thus the moral act is often—one might be tempted to say almost always—transparent to the eye of God alone, opaque and hidden from ours who carry it out. This spontaneous self-declaration of the will to its Creator, of the will that is often unaware of its own motive, this perfect openness before God of all its most secret movements, of all that we conceal from ourselves but cannot dissimulate before Him—all this is what the Psalmist means by *clamor*. When we come to give it a name in philosophy then, as a real movement of a will tending to its end, we call it simply " intention." [4]

Intention, so defined, is evidently an act of the will. Reason, doubtless, is directly involved, since it is impossible to tend to any end without knowledge of the end ; but the appetite, or desire, thus lit up by reason, is precisely what we call will ; and hence it is simply self-evident that intention is essentially voluntary. Now it is enough to reflect on the place it occupies in the complex structure of the voluntary act to see the determining part it there plays. Man tends to his ends by simple acts, but to attain them he has further to choose, that is to say to will, the means. This choice, in its turn, presupposes a rational deliberation, and this deliberation brings our intellectual and moral virtues into play, and lasts until the will, now judging itself sufficiently enlightened, decides upon this means rather than on that. The operation is thus analysed into a number of distinct elements, but, in reality, a single movement runs through it from end to end : that of the intention. It is by one and the same movement that we tend to the end and will the requisite means to the end. And we can go further. Since the intention of the end is the root, the source, and, to put all in one word, the cause of the choice of means, it is clear that the moral qualification of the intention will affect, and in large measure determine, that of the whole act. If our " eye " is evil, as the Scripture puts it, that is, if our intention is bad, then the whole series of voluntary choices which determine the means to be adopted to accomplish it will itself be bad : nothing that we do is good if we do it for the sake of something evil. Conversely, it may so happen that the choice of means is not so good as the will to the end that inspires it, for with the best intentions in the world we can easily miscalculate our acts, or fail in execution ; but even so, the good intention is felt throughout the whole series of acts it sustains ; it redeems in a measure their defects, retrieves their mediocrity, so that something good at bottom remains even in that which we do ill, provided only it be done with right intention.

All this was so obvious to the Christian thinkers that when they sought the technical definition of moral good and evil, they went at once as far as they possibly could in the direction of a morality of intention. Peter Lombard, in the twelfth century, asking himself what is properly to be understood by the word " sin," cites those for whom it is constituted by the evil will alone, and not in the least by the exterior act.[5] When we recollect the influence exerted on this theologian by Abelard, we can have no doubt about the master he is thinking of. Abelard's *Scito te ipsum* is no less important for the history of ethics than his commentary on Porphyry for the history of logic ; and what at once comes to the forefront in his moral philosophy is the preponderating importance of the internal consent as compared with the external act that follows it. To sin is one thing ; to put the sin into execution is another. So far does he carry this distinction, that not only does he regard the evil deed as only improperly called sin, but even as adding nothing to the gravity of the sin. What causes a certain hesitation to admit this is the common sight of punishments falling on bad exterior acts rather than on sins ; but if we are going to look at the matter from the standpoint of social law we shall have to look further. The law often punishes actions that are not bad at all, and tolerates others that are. It is not concerned with moral good and evil, but rather with the maintenance of social order ; and hence the extreme importance it attaches to the execution or non-execution of the wrongful act. It is not so with God ; He, and He alone, takes account not so much of what we do as of the spirit in which it is done, and, in all truth, He weighs our guilt by the intention : *veraciter in intentione nostra reatum pensat ;* and, adds Abelard, that is precisely why it is written in Jeremiah that He searches the reins and the hearts : " Seeing, in a most wonderful manner, what none other sees, He takes no account of actions when He punishes sin, but the intention only, while we, on the contrary, take no account of the intention which quite escapes us, but punish the action which we see." [6] There are, therefore, in reality, only sins of the soul, and sin is essentially concerned only with what may be called the soul of the bad act, that is, with the spirit that animates it ; just as on the other hand, there is no moral good save in the spirit that bodies itself forth in a good action, and alone qualifies it as good. And what then makes the intention itself good ? Is it our conviction that we do well, that in acting as we do act we please God ? These are certainly excellent and even necessary dispositions, but if we recollect the existence of a divine law ruling all intentions and actions, we shall see that they do not suffice. A further condition is required before a complete determination of the idea of morality can be effected.

This cannot be done without bringing the concept of intention into relation with that of the moral conscience ; and indeed they are closely allied. Having put all the emphasis on the part played by intention, the Christian thinkers found themselves face to face with a considerable difficulty. The human reason is one and the same faculty, whether in its theoretical or practical exercise. On the one hand the divine illumination makes it capable of thought by means of necessary first principles ; on the other internal or sensible experience supplies it with materials to which these principles are to be applied. Whether we are concerned to know what things are and to achieve a science, or to know what our acts ought to be and to set up a system of morals, in either case we have to make use of the same principles of construction, and the same materials. But when there is question of determining the moral good we have to take account of a special element which interposes between the rational promulgation of principles and the formulation of the particular judgments which flow from them. This is conscience. By this word we do not mean any new faculty distinct from will and reason, but an act, or rather certain acts, which bring a rational judgment to bear on our conduct. When we simply recognize that something has been done or not done we are said to be conscious of the act or omission ; when we judge that something ought to be done or not done, we say that conscience commands or forbids ; and when, finally, the judgment bears upon an act already done we say that conscience approves or disapproves, and in the latter case its voice is the voice of remorse.[7] Of these three functions, the second is specially important in the determination of moral good. Since the intention qualifies the act, what is to be the attitude of the will when faced with rational judgments ? The answer would be simple enough were the human reason infallible, which, unfortunately, it is not as we all know. Suppose, then, that the moral conscience errs in the application of principles to the detail of actions, what, in that case, ought the will to do ? If it obeys conscience it wills evil, but if it disobeys it sets aside the very thing that reason presents as good ; the intention, then, will be bad, and thus also the whole act will be bad. How is this difficulty to be overcome ?

Christian moralists are all agreed that every prescription of conscience lays an obligation on the will to conform to it ; and this was inevitable once they had decided that the moral worth of the act depends on the intention. Here, then, the Greek objectivism, particularly as it was expressed in the moral philosophy of Aristotle, had to give way all along the line. The will is now qualified, not by the object as it is in itself, but by the object as reason presents it ; in itself indifferent, it becomes good or bad according as reason proposes it as a good to be done or an evil to be avoided ; what is good in itself becomes evil, and what is bad in itself becomes good, if the judgment of the practical reason represents it as such to the will. For example— and let us take what in Christian eyes would be extreme examples—" To believe in Christ is good in itself and neces- sary for salvation : but the will does not tend thereto except inasmuch as it is proposed by reason. Consequently, if it be proposed by the reason as something evil the will tends to it as to something evil ; not as if it were evil in itself, but because it is evil accidentally, on account of the way in which reason apprehends it." [8] St. Thomas Aquinas holds, then, that a will that tends to something really good as though it were an evil, by the very fact that it forsakes reason, even an erroneous reason, is an evil will. Con- versely, to persecute Christ was clearly wrong, nothing

could make such action good : nevertheless, if His per-
secutors, or those of His disciples, merely acted in accordance with their own conscience, then they sinned only by
ignorance ; their fault would have been much more
serious if, against the voice of conscience, they had spared
Him.[9] This affirmation of Abelard's is quite in accordance
with what the Thomists were soon to be saying : *voluntas
discordans a ratione errante est contra conscientiam, ergo voluntas
discordans a ratione errante est mala.*

We are as near as possible here to a morality of intention
in the strictest sense of the term. If the mediæval moralists
did not get as far as the Kantian position, it was not because
they failed to conceive the possibility of such a moral philosophy, but because, having conceived it, they rejected it.
Abelard seems to have come nearer to it than anyone else,
since, as we have seen, the performance or non-performance
of the act, does not, in his view, affect the morality of the
voluntary decision. But nobody followed him here ; the
realism and robust good sense of the Christian philosophers always refused to admit that it would be just as
good to will to give an alms and not give it, as to will to
give it and to give, or just as bad to will a murder and not
kill the man, as to will the murder and carry it out. The
adultery that is committed in the heart is a true adultery,
but that is no reason at all for adding the other. Moreover—
and Abelard is here in accord with all later Christian
moralists—however necessary it may be to put the intention
of conscience at the heart of morality, it does not in itself
suffice to define morality. Above and beyond what seems
to us to be good or bad there is the thing that really *is*
good or bad. An erroneous conscience obliges, no doubt,
but an act prompted by a right conscience is very different
from an act prompted by an erroneous conscience, and only
the former is truly good. Besides being bound to obey our

conscience we are also bound, wherever an error of judgment is to be feared, to criticize it, and to replace a bad conscience by a better.

If we are to take proper account of this new factor, the preponderant part played in the economy of Christian morals by the determination of ends will have to be made clear. The Ancients were far from overlooking the importance of this ; the mere title of *De finibus bonorum et malorum* is enough to prove it. But the very difficulty we feel in translating this title is due perhaps to the fact that we are so penetrated with the Christian spirit that we can hardly conceive of the ends of goods and bads, precisely because in our eyes the goods and bads are defined by the ends themselves. Up to a point it was already the same in Aristotle's ethics ; absolutely, we might say, were it not that Christian ethics are there to show that, however advanced are Aristotle's conclusions, they yet fall short of the mark. The whole purpose of his ethics is to teach mankind the means to the last end, that is to say, happiness ; which consists in living out a whole life according to the best and most complete of human virtues. The difference lies in the fact that his conception of the very relation of the means to the end is other than that of the Christian moralists owing to their different conception of good and evil. Aristotle assuredly has a deontology ; there are things that *must* be done, and the only reason why they must be done is a certain end to be attained. The man who does not do them is like an awkward archer, who aims at a mark and misses it. But the end which in Aristotle's doctrine qualifies all moral actions is not proposed to the will as a term laid down by the divine law, imposed by a Creator on His creatures. The means therefore may, and indeed they do, stand in a certain relation to the end ; a certain relation, however—and much more one of expediency than of obligation. Only a law obliges, for only a law binds ; and that is why the Christian conscience, as an expression of the divine legislative reason, always prescribes the act to be done as a *moral obligation*. The idea has become so familiar to-day that we forget how novel it once was, and by whom it was invented.

The Christian thinkers nevertheless had a very clear sense of what they were here adding to ancient Greek ethics. The formidable problem presented by the virtues of Pagans and how they should be valued, made it quite impossible for them to overlook it. Here are not merely individuals, but peoples, almost the whole human race, living in total ignorance of the divine law. Probably that world was not always wholly given up to vice throughout the whole of this long history. We know indeed beyond a doubt that there were virtues ; but we know also that since these men knew nothing of the Gospel and therefore nothing of their true end, not a single one of their actions could have been directed naturally to the end that it should have been. Shall we then stigmatize all their intentions as bad, and consequently all their actions as bad ? Faced with this question Christian philosophers were forced to examine the nature of the bond which, in their morals, links up the act with its end. Theological research was once more acting like a ferment at the heart of philosophy and lending powerful aid to its development.

We propose to consider the problem only in so far as it concerns the idea of moral good. The first point to notice is that the Greeks and Latins, simply as pagans, lacked the conditions requisite for the exercise of a fully determined morality. So much is evident. When St. Paul, speaking of God, says, To Him be glory for ever (Rom. xi. 36), " he does not say *to them*, for there is but one God. And what is meant by : To him be glory, save : To Him be the best

and highest and most widespread praise? For the more
His praise is spread abroad the more ardently is He loved.
Then with a sure and steady step man goes forward in
the best and happiest life. For what we chiefly wish to
know in moral matters, is what is man's sovereign good,
that good to which all else in life must be subordinated,
and beyond this there is little else to be sought. We have
shown by reason as far as we are able, and by divine
authority that enlightens our reason, that this sovereign
good is none other than God Himself. For what could
possibly be better for man than to cleave to what really
makes him blessed? Now this is God alone, to Whom we
cleave in no other way than by affection, love and
charity." [10] The interest of this passage for our present
purpose lies less in the thesis affirmed—since it is evident
for the least exacting of historians—than in the kind of
relation set up between the Christian's moral act and his
end. What the Ancients knew nothing about was not merely
the unicity of the term of human life, but, by that very fact
and for the same reason, the mere existence of such a term.
It is not enough to say that they were mistaken as to the
nature of the end, on which, as they say themselves, all
morals depend, it must even be questioned whether they
ever conceived the idea of an end in the same sense as the
mediæval doctors.

When Aristotle places happiness in a good that shall be
coextensive with a complete human life : ἐν βίῳ τελείῳ—for
one swallow does not make a summer—he clearly enough
includes the end of life in the life itself. He conceives happiness, to be sure, as something other than the joy that
accompanies all moral perfection ; it is this very perfection
that he regards as the constitutive and substantial element
in happiness itself. But precisely on this account happiness
is not regarded as a reality that transcends the moral life

and crowns it. The sole end to be looked for, the true sovereign good, is the moral life itself, the act or series of acts which will naturally result in happiness. In an ethic of this type there will always be a very strong tendency to emphasize the importance of the means. Doubtless it is only because it leads us to the end that a means is good ; but it is not good because it is adopted for the purpose of attaining the end. If man is always careful to live as nature and reason demand he will find thereby his beatitude, but the value of his moral acts does not depend on any intentionality. It is easy to guess the preponderant importance which the notion of " appropriate act " (*convenable : officium*) was soon, thanks to the Stoics, to assume in Greek moral philosophy. " The first principle being laid down," says Cicero, " that things according to nature are to be chosen for their own sake and those which are contrary in like manner rejected, the first of ' appropriate acts ' (for thus I render καθῆκον) is for a man to preserve himself in his natural constitution." This initial choice being once made, it becomes, as in Aristotle, a virtue, that is a habit, the habit of acting spontaneously and constantly as nature demands. The upshot of this attitude, as Cicero very clearly saw, is that we might be almost tempted to believe in two sovereign goods : the end contemplated by the act, and the act itself as looking to the end. Having to eliminate one of them, the Greek eliminates the end. " When in fact we speak of a last end in the series of goods, we are in the position of one who makes it his purpose to take a true aim with an arrow at a mark. The archer, in the comparison, would have to do all he could to hit the mark, but nevertheless his last end, corresponding to what in treating of life we call the Sovereign Good, will be to do all he can to realize his purpose ; as for actually hitting the mark, that is something ' to be chosen,' not something ' to be desired ' for its own sake." [11] The καθῆκον, the *officium*, tends therefore more and more to supplant the end, and the means at last usurps its place. Nothing could be more natural in Hellenic morals where the end is immanent in life itself; and nothing could be more unsatisfactory for a Christian, for whom man's end transcends man. At one and the same stroke he invests the end with its true supremacy, and subordinates the means far more thoroughly than had ever been done before.

St. Augustine therefore made an admirable choice of the point at which to direct his attack on the morals of Pelagianism, that is to say of Hellenism, when he condemned it for admitting that genuine virtues could exist at all when the end of all virtue worthy of the name remained unknown. " You know very well that it is not by ' appropriate acts,' but ends, that virtues and vices are to be distinguished. The ' appropriate act ' is the thing to be done, the end is that for which it is to be done. When therefore a man does something that seems no sin on the face of it, if that for which it is done is other than that for which it ought to be done, then is he convicted of sin. But paying no attention to this, you have separated ends from ' appropriate acts ' and pretend that ' appropriate acts ' without true ends can be called virtues." [12] Augustine had already passed quite beyond Hellenizing Pelagianism, when, from the outset of his conversion, he made all morality depend, not on any perishable thing, but on the internal movement of a will firmly fixed in God, Who is man's sole end ; and in so doing he fixed for ever the nature of moral finality as Christian thinkers understand it. To say that the goodness of the will depends on the intention of the end, is, in their case, to say that it should order itself from within, if not constantly, at any rate really and intentionally, towards the supreme and transcendent good, objective and altogether distinct from ourselves, God, that is to say the divine Will, with which man should have the intention of conforming his own. Whenever, materially, we will some particular good, we ought, formally, to will that the choice be in conformity with the divine will ; for on such conformity its whole moral worth depends. *Rectum cor habet, qui vult quod Deus vult.*[13]

This principle assures, in the first place, the complete unification of human life. At the heart of all lies this inclination of will towards the end that reason reveals, this gravitational force, this *pondus* in a word, as St. Augustine and St. Thomas both call it, which by attraction of love unites man to God. If this intention be firmly and durably fixed on its end, then, since it is from this love that all other affections derive, there will be no place left in man's heart save for joys, sadnesses, fears or hopes that are oriented to God. And if, furthermore, it lifts itself up, or rather is lifted up to the order of charity, this love will blossom into supernatural virtues, springing from it as flowers and fruits that are borne on their common root.[14] Henceforth, as Augustine says, man seeks no more to " manufacture " a happy life ; rather will he ask it of God Who alone can give it ; it is only with respect to Him that there are any " appropriate acts " at all, and, what is more important, necessary ones. *Neque enim facit beatum hominem, nisi qui fecit hominem :* He Who made man alone can make man happy.

This insistent, inevitable appeal to the metaphysical principle whereon rests the whole Judeo-Christian moral system, brings us back to our starting point and sets its nature clearly before our eyes. In a sense it is something wholly and integrally interior, since the quality of the human act depends essentially on the intention that directs it, and since the intention itself comes only from the will. The will is free ; it is always within its power to will or not to will ; and here to will is to obtain, for what is required of it is precisely will : *ut nihil aliud ei quam ipsum velle sit habere quod voluit.*[15] Moral life is henceforth completely interiorized ; set free from all exterior conditions, and even from interior conditions when these do not lie within our power, it is free in a far deeper sense than in Stoic doctrine, since the will is free not only of the world but of itself. The wise man of the Stoics is not wise unless he is free, but the wise man of the Christians is wise if he knows that it does not suffice him to free himself, but desires to become free. Wherefore the " good-will " of the Gospel announced along with the good news itself (Luke ii. 14) is something so new, of such incomparable value. This good-will not only suffices ; but it suffices to itself because it suffices for God : *nihil tam facile est bonae voluntati quam ipsa sibi ; et haec sufficit Deo.*[16] At every instant man is, in the eyes of God, what he really is in himself ; but, conversely, he is at every instant really in himself what he is in the eyes of God : and there we have the other aspect of Christian morals. We have nothing that we have not received. In the universal circulation of love spoken of by St. Thomas and Dionysius, it is God Himself Who turns our will towards Himself, and in consequence makes it good ; in freely turning ourselves away we can but make it bad. In this sense all our moral activity is therefore ruled, directed ; its measure lies in God, and nevertheless it remains true to say that its source is altogether interior, since the divine law that rules us, expresses itself in us by the organ of our reason.

Thus between Greek ethics and Kantian moralism stand the morals of Christianity transcending and reconciling both. Perhaps it would be better to say that apart from Christian morals the second of these two terms would not exist ; it is a mere product of their decomposition. Apart from the philosophy of the Middle Ages neither Kantian morals nor Cartesian metaphysics would ever have seen the light of day. A legislative will, subject to the will of a supreme Legislator, autonomous however, since every reason is an expression of the supreme Reason—there, as we have already said, is the whole essence of Christian moral doctrine. It may be added, now, that a morality of intention, wholly based on good-will, that is to say on the will to act always out of respect to the moral law, is not out of accord with the doctrine which prescribes not only respect, but ardent love for the moral law ; which binds us, obliges us to morality, by means of a reason not homogeneous with God's reason, certainly, but not heteronomous either, since the natural law is the created analogue of the divine law whence it derives. In uniting his intention daily to God, in praying that His will be done upon earth as it is in heaven, the Christian legitimately gathers up the heritage of Greek naturalism, but he marks its transcendent condition, and, by one of those apparent paradoxes which are so often found at the heart of truth, he interiorizes it in the very act of submission : *Interior intimo meo et superior summo meo :* " There is One within me who is more myself than myself." [17]

Such profound differences within the bosom of so close a continuity are rather embarrassing for the historian. All that Christianity preserves of Greek ethics takes on a new meaning ; all that it adds seems a natural growth, inevitable, almost necessary. It is easily therefore understood that historians who are otherwise of very different spirit, agree that ancient and modern morals, at once divided and linked together by mediæval morals, differ *toto coelo*,[18] and yet are in no wise contradictory, but may, on the contrary, give each other mutual support. Agreement ceases on the nature of the support. According to one view, Greek moral philosophy is essentially a rational Eudaemonism, and, as such, an entire stranger to the ideas of moral law, obligation, duty, responsibility and merit ; and must so remain. All these ideas are of Judeo-Christian origin and essentially religious ; they should never have been introduced into philosophy, and it was precisely Kant's error to have introduced them. The first condition of any return to Greek ethics, that is to say to reason, is to expel them, and restore them to religion where they belong. According to the other view, it is incorrect to call these notions essentially religious, they are essentially rational and belong therefore of full right to philosophy. True, the Greeks never clearly conceived them, but they might well have done so on the basis of their own principles, or at least by purely rationally perfecting their own principles. Why, then, not integrate them with the body of moral doctrine, where now they play so useful a part ? That would save all that is worth saving in Greek ethics, without sacrificing all that later philosophic reflection has developed out of it.

Many minds favour a return to the Greeks, but the truth is that it would be almost as impossible to return to them simply as to pass them over altogether. The attempt might be conceivable if the introduction of the idea of duty and law into moral philosophy had been the work of Kant and had taken place in the eighteenth century ; but Kant himself only conceived them as rational because Christian philosophy had rationalized them. For such long centuries and so intimately and deeply have they become entwined in our moral conscience, that a moral philosophy claiming to be rational and pretending to ignore them would be regarded by everybody as ignoring the very essence of morality. To throw duty and conscience wholly over to religion, would be to throw morality wholly over too ; a thing that no one dreams of. Schopenhauer, on the other hand, was not altogether wrong when he reproached Kant for having gone to the Decalogue for the idea of duty ; the religious affinities of the notion are of the strongest, and it is inconceivable how anyone could write its history without taking account of its Judeo-Christian sources. For my own part I have no sort of doubt that the idea is rational, and should be maintained at the very heart of moral philosophy. I would by no means deny that as a matter of right the Greeks could well have conceived it ; I simply state it as a fact that they lacked the strength to do so.

To rationalize the notion of morality and its corollaries it is enough, we are told, to conceive " God as the total and conscious cause of nature and man ; once the dependence of our being on His will is established, we are free to construct a moral philosophy based on the idea of duty." Very true ; but, " to connect up human reason with God's reason " we must first of all admit the concept of creation, and conceive God as a unique and almighty Being, the Disposer of all beings and Legislator of the universe. When, with V. Brochard, we admit that the idea of an omnipotent and infinite God was not merely a " revolution," but a " great advance " with respect to the finite god of the Greeks, we can hardly deny that the idea of moral obligation belongs to the philosophical order, since it follows as a necessary consequence ; but when we are asked to retain Greek moral philosophy and complete it " with the only notion that it lacks, that of the divine absolute," it would be well to note expressly that Greek morals would thus become Christian morals. To suggest that moral philosophy cannot be brought into any true connection with religious thought, on the plea that it belongs to the philosophical order, would be to suggest something tenable if by " connection with " we mean " deduction from " ; but there are other links between ideas besides that of logical deduction. Duty is not to be deduced from a revelation, but from a doctrine of creation which is itself deduced from a metaphysic of Being. And this metaphysic of Being is deduced from nothing but the exigencies of rational thought. It is altogether natural, therefore, that a morality of obligation should find itself in no conflict with Greek ethics ; the Greeks themselves would have attained to it had they but pushed their metaphysics further. It is equally natural, too, that such a morality should differ *toto coelo* from Greek ethics ; the Greeks failed to attain it because they failed to push their metaphysics further.

And why did they fail in this ? Is it because mankind moves, as St. Thomas says, only *pedetentim* on the road to truth ? Undoubtedly. Nevertheless, there are steps and steps. In particular there is a step that metaphysics itself took, but did not, however, take alone : in pushing the problem of being to the plane of existence it opened a new road for ethics. That was accomplished for it by the Fathers of the Church and the mediæval philosophers. They would have been very much surprised to hear that their moral philosophy was not based on reason ; perhaps, even, with the help of a few explanations, they might have been got to admit that it was " laic." But we shall find it very difficult to believe, for our own part, that if this laicized ethic was so very different from the ancient, the fact that its authors were Christian priests had nothing at all to do with it. The study of mediæval thought, in helping us to restore or, if need be, to construct the conception of Christian philosophy, will perhaps help us at one and the same time to bridge a yawning gap in history and to reconcile philosophical positions which each contain some portion of the truth.

CHAPTER XVIII

THE MIDDLE AGES AND NATURE

BEHIND every one of the problems we have hitherto been considering we seem to feel the presence of another, and its discussion and solution may seem to have been unduly deferred. At least it has not been unintentionally deferred. Everywhere in mediæval philosophy the natural order leans on a supernatural order, depends on it as for its origin and end. Man is an image of God ; the beatitude he seeks is a divine beatitude, the adequate object both of his intellect and his will lies in a being transcendent to himself, before Whom his whole moral life is played out, and by Whom it is judged. The very physical world, created as it is for God's glory, tends with a kind of blind love towards its Author ; and each being, each operation of each being, depends momentarily, for existence and efficacy, on an omnipotent conserving will. And how, if the case stands thus, do we dare to talk of nature at all in a Christian philosophy ? Would it not be better to say with Malebranche, that the idea of nature is essentially anti-Christian, a pagan relic preserved by the indiscreet zeal of theologians ? The mediævals hardly thought so, as we may guess, seeing that it is against these that Malebranche's criticism is directed. They believed in nature like the Greeks ; yet did not conceive quite as the Greeks did, and I would therefore ask : what transformation did the idea of nature undergo at their hands ?

Here it is important to make a careful choice of witnesses. It would be of little use to consult the mystics ;

they are not interested in what nature *is*, but rather in what it *signifies*. To go to the authors of Lapidaries and Bestiaries, as some well-known philologists have done, would be like going to an almanack-maker for the contemporary scientific astronomy. The only possible witnesses to mediæval philosophy are, once more, the philosophers themselves ; if we attempt a short cut to it through the literature of the Middle Ages, or even through its history, we shall expose ourselves to grievous misconceptions ; such by-paths are dangerous and better avoided. It is a plain fact that nature existed for the mediævals, but it was not the " nature " of Greek philosophy nor yet the " nature " of modern science ; albeit it retained much that was characteristic of the first, and a certain amount of it has perhaps been retained by the second. In what then precisely did it consist ?

In the mediæval philosophies, as in those of antiquity, a natural being is an active substance, with operations flowing from its essence, and necessarily determined by that essence. As for Nature, it is simply the sum-total of natures ; and its characteristic attributes are therefore the same, that is to say fecundity and necessity. And so true is this that the mediæval thinkers always rely on the observation of a necessity in order to infer a nature. Whenever it is possible to recognize a certain regularity, whenever something happens *ut in pluribus*, we can be sure that this regularity has a cause, and this cause can be none other than the presence of an essence or nature, which, by its operation, always produces the same phenomena. For the same reason the operation of this nature is necessary, for it is only to account for the observed regularity that it is posited at all. The connection between the conceptions of nature and necessity is so much the closer, inasmuch as, strictly speaking, the existence of natures is not to be demonstrated ; sense perception reveals the existence of things acting from an internal principle, hence from a nature, and to want to reason it out further would be to set out to prove the known by the unknown. It is hardly an exaggeration therefore to say that the nature is perceived in its necessity itself, for it is revealed precisely in a general rule and this generality is based on the necessity : *hoc enim est naturale quod similiter se habet in omnibus, quia natura semper eodem modo operatur.*[1] It is thus easy to understand that when the problem of the basis of induction was raised, Duns Scotus had no need to look further than the principle ; all that happens regularly in virtue of a cause, is the natural effect of that cause.[2] The regularity could be broken only if the cause were not a nature : there is no medium between the necessity of natures and the freedom of wills.

It is quite easy therefore to conceive the mediæval universe as a possible object of scientific explanation, in the sense in which scientific explanation is understood to-day. I trust I have no illusions as to the extent and quality of mediæval science, but it seems to me necessary to draw a clear distinction between scientific knowledge of the world on the one hand and the general conception of the world that science interprets on the other. The universe was not very well known in those days and progress was made the more difficult because of the obstacle that the Aristotelian qualitative physics opposed to the birth of a physics of quantity. Much time was wasted in the fourteenth century on the calculation of qualitative intensities : which have to be quantified before they can be measured. That said, it must be added that what retarded mediæval progress in science was no backwardness in believing in universal determinism. Quite the contrary ; putting man's free will on one side, philosophers and theologians all agreed in a universal determinism of an astrological kind. St. Thomas considers that the movements of lower bodies are caused by

those of the heavenly bodies, and that all the phenomena of
the sublunary world are ruled by the movements of the
stars. Albert the Great and Roger Bacon went still further ;
in fact, the latter did not hesitate to cast the horoscope of
religions, not even excluding the Christian religion. Since,
then, the determination of natural phenomena was generally
admitted, it is not surprising that the idea of a true scientific
study of nature was conceived. Robert Grosseteste went far
beyond the Aristotelian physic of forms when he reduced
them all to light, the science of which belongs to optics and,
ultimately, geometry. Roger Bacon followed down the
same road and his insistent call for an experimental science,
albeit he did little to forward it, shows a very just appreciation of what a true scientific demonstration should be.
Although, then, the mediævals knew little enough about
nature they were not deceived as to the essential characters
that make it an object of rational knowledge : and it can
even be said in a sense that if their conception of nature
differed from the Greek it was not because it contained less
determinism, but rather because it contained more.

When St. Thomas puts the question whether astrological
determinism imposes an absolute necessity on terrestrial
phenomena he replies in the negative on the ground that
besides all that is determined by the movements of the stars
there exists a vast field for chance. His whole argumentation
falls back upon Aristotle, and not without reason, considering how important a part is played by chance in Peripateticism. It is by no means there conceived as a pure indetermination, that is to say as something that happens
without cause, and, in this respect, it makes no breach in
the universal determinism ; nevertheless, it is incompletely
determined, it is accidental with respect to the efficient
cause because not produced thereby in view of an end, or
because the thing produced is other than the end for which

the cause acts. In nature, then, the fortuitous is that which lacks an end. It might be said, in a sense, that since the final cause is the true cause, the fortuitous is that which has no cause ; but this would merely indicate certain lacunæ in the teleological order, not at all in the order of efficiency. Two series of causes, for instance, equally determined both as to their efficiency and their end, may cross each other's path without the meeting being either foreseen or willed : the intersection is accidental. This is what happens when two men, each proceeding to one point to which both intend to go, meet inevitably without intending to meet at all. It happens again when the matter on which a man is working fails, on account of its own proper necessities, to lend itself to the work : the point of conjunction between what the artist wishes to do and what he finds himself able to do, although determined by his art on the one hand, and the structure of the matter on the other, may nevertheless be in itself an accident. Now art merely imitates nature ; like the artist, nature may fail of its due effect ; the form encounters the blind necessities of matter and produces only a bungled being, a monstrosity. Such monstrosities are accidental failures of nature frustrated of its due end by an unforeseeable concourse of circumstances.

The remarkable thing is that even this dose of purely relative indetermination disappears in the mediæval conception of nature. Absolutely speaking, there can be neither chance nor monstrosities for a Christian thinker, for although on the relative plane of human experience these conceptions may and ought to be upheld, they lose all meaning when we set out to describe the universe from the standpoint of God. The ancient conception of the fortuitous is quite familiar to St. Augustine ; it is all that which is produced without cause, which depends on no rational order : *qui ea dicunt esse fortuita, quae vel nullas causas habent, vel non ex aliquo rationabili ordine venientes.*[3] Now in the Christian universe nothing ever happens save in the name of a rational order, nothing exists save as depending on it. In everyday conversation we may be permitted to speak of chance, but since the world is God's work and nothing it contains is withdrawn from His providence, it is impossible to regard anything as absolutely fortuitous. *Nihil igitur casu fit in mundo ;* nothing happens by chance : that is the ultimate Christian attitude to the universal order.

The mediæval philosophers therefore could accept the Aristotelian standpoint and accord it a conditioned validity. Boethius, in his prison, calls upon the Lady Philosophy to console him, and she replies in the language of the Stagirite. There is, in a sense, such a thing as chance. A man who, digging his field, comes across a buried treasure, may be said to have found it by chance ; he who hid it and he who finds it have both gone to the same spot but without in the least intending the discovery. Chance, then, is still here the accidental intersection of two chains of causality that meet, without being determined to meet by any end. That is to say no *human* end : but how about divine ends ? Nothing ever happens outside God's providence ; even matter, since matter too is created, introduces no element of blind necessity into the universe, nor can it play the part of accidental cause as it does in the uncreated world of Aristotle. On the divine plane there is no more chance ; for even the apparently accidental convergence of two different chains of causality depends on the unchangeable order established " by that admirable Providence that disposes all things wisely and brings each event to pass in its due time and place." [4] And to what St. Augustine and Boethius here conclude let us add with St. Thomas that, absolutely speaking, monstrosities are just as impossible as chance. Nature is the work of God and makes no mistakes ; matter lends itself to form just in so far as its Author wills, no more, no less. When these *defectus naturae* occur they must have been willed by God with some end in view ; human monstrosities, for instance, are born in accordance with the laws that govern fallen nature, but unfortunately philosophers prefer to deny God's designs than to admit that they know nothing about them ; they accuse nature of irrationality when it does no more than merely follow the higher laws imposed on it by God.

Thus, where Greek thought tolerates an indetermination resulting from a certain lack of rationality, Christian thought tightens the bonds of natural determinism by reducing the apparent disorder of nature to the laws of a higher reason. And the converse is equally true. Where Greek thought admits an anti-rational necessity, Christian philosophy casts off the bond of this necessity precisely because it is anti-rational. In the act of bringing chance under law it frees nature from Fate ; for everything has a sufficient reason, but this can be none other than Reason Itself.

St. Augustine more than once lifted up his voice against those who spoke of fatalities, that is to say events brought about by some blind necessity independent alike of man's will and God's will. There is, indeed, in all rigour, a sense in which we may speak of *fatum*. If the word be used to express the very will of God as prescribing laws to nature which it can do no other than obey, there is nothing to object to in the doctrine ; all that needs to be done is to correct the expression.[5] The mediæval doctors often allowed themselves this licence ; the ancient Fate had weighed too heavily on men's minds to be too summarily dismissed. Boethius took the trouble to put up some rather complicated architecture in order to ensure it a niche in the Christian temple. Providence is then the divine intelligence itself comprehending all things in the world ; that is to say, their natures and the laws of their development. As reunited therefore in the divine ideas the universal order is one with Providence ; as particularized, broken up, and, so to speak, incorporated with the things it rules, this providential order may be called Fate. All that is subject to Fate is thus subject to Providence, since fate depends on Providence as a consequence on its principle ; and we can even go on to add that many things depend on Providence that do not depend on Fate : " For as the innermost of several circles revolving round the same centre approaches the simplicity of the mid-most point . . . while the outermost, whirled in ampler orbit takes in a wider and wider sweep of space— even so, whatever departs more widely from the primal mind is involved more deeply in the meshes of fate, and things are free from fate in proportion as they seek to approach the centre ; while, if aught cleaves close to the supreme mind in its absolute fixity this too, being free from move-ment, rises above fate's necessity. Therefore, as is reasoning to pure intelligence, as that which is generated to that which is, time to eternity, a circle to its centre, so is the shifting series of fate to the steadfastness and simplicity of provi-dence."[6]

This great idea so splendidly expressed by Boethius proved most attractive to generations of thinkers and poets after him ; but found no favour from the metaphysical rigour of St. Thomas. Owing to its false connotations he could not approve of the use of the word Fate. Those of the Ancients who admitted an absolute chance, Aristotle for instance, were evidently forced to reject Fate. Those who rejected chance and wished to subject everything, even apparently fortuitous things, to the determining influence of the heavenly bodies, readily gave the name Fate to astronomical laws. But, clearly, human acts are free and therefore not subject to these laws ; and it is no less clear that all this apparent chance, that no Fate can ever explain, finds its sufficient reason in divine Providence. In such measure therefore as Fate may seem useful to account for fortuitous events, it is in fact nothing but the effect of this Providence ; but since in that case the word has quite another meaning for the Christian than it had for pagans, it is better avoided. It would be well to take Augustine's advice and refrain from using it altogether.[7]

How profoundly the Greek conception of nature was transformed by the Christian doctrine of creation is very apparent when the preceding conclusions are brought together and applied to the problem of future contingents. Since, in Aristotle's system, there is chance, there are also future contingents. What is essentially accidental must, by the very fact that it escapes the order of necessity, fall within the order of contingency. Now science is knowledge by causes ; and since the fortuitous is such because it has no cause it cannot be regarded as an object of science, still less of foresight. To be foreseen it would have to be determined, and would cease therefore to be contingent. In the Stoic system it is the other way round ; it is possible to foresee the future—and we know the importance the Stoics attached to divination—but they based the foresight of the future on the doctrine of Fate, whose function precisely it was to eliminate all contingence from the universe. Either, then, we grant contingence and deny that the contingent can be foreseen, or we admit the possibility of foresight, and deny contingence. But Christian philosophy surmounts both difficulties and affirms simultaneously both the contingence and its foreseeability, because it disassociates the notion of contingence from that of chance, and the notion of determination from that of Fate.

St. Augustine knew his Cicero too well not to be aware of the embarrassment of the Ancients on this question.

Cicero was unwilling to admit Fate ; he therefore attacks the Stoic notion of divination at its root, and the more surely to make away with it he goes so far as to maintain that all science of the future is impossible, either to man or even to God. That was a rather high price to pay for the vindication of freedom. Here in fact are two opposite follies ; the affirmation of Fate on the one hand, and the denial of the divine prescience on the other ; and it was on the latter that Cicero thought to find a foundation for our freewill. But a Christian mind, on the contrary, accepts both the freedom and the prescience ; for God is Creator and Providence ; it was He Who created the causes ; He knows therefore what they are and what they will do. If He created free causes He knows also what these free causes will do. Thus in the physical order, all that results from a concurrence of causes which to us looks accidental, falls under the prescience of God Who disposed the concurrence ; in the voluntary order, not only does the fact that God foresees our free acts still leave them free, but, on the contrary, it is just because He foresaw that we should perform these free acts, that we do in fact perform them. His prescience is His providence ; as provident for our freedom God does not destroy it, He establishes it : *profecto et illo praesciente est aliquid in nostra voluntate*. Nowhere, it seems to me, is the specific character of Christian philosophy more plainly visible. Aristotle's chance, belonging to the irrational order, was unforeseeable ; it became rational and foreseeable. The Stoic Fate was foreseeable, but eliminated chance and contingency. Providence, like Fate, foresees, but respects contingency.[8] Everything is taken up into a rational order, but without altering its essence. " From the fact that the order of causes is known with certainty to God, it does not follow that no freedom of choice is left to our will. Our wills themselves, in fact, form part of this order of causes, certainly known to God and comprehended in His prescience, for men's wills are the causes of what they do. Thus He who foreknew all the causes of things would certainly, among these causes, not have been ignorant of our wills, for He knew in advance that they would be the causes of what we do." [9] All is known to God according to its kind because all is the work of His creative intelligence, which willed the necessary as necessary, the contingent as contingent, and the free as free.[10]

These principles bring us to the threshold of an idea still stranger than any that have gone before, of which the writings of the Christian thinkers are full, and of which the Ancients, in full accord therein with many modern minds, would have refused to admit into their philosophy : the idea of miracle. That the Middle Ages was an age of miracles must surely be evident, since even the historians are aware of it. To be sure they often hit upon false miracles ; we know there were such from Salimbene, since at Parma in 1238, Franciscans and Dominicans *intromittebant se de miraculis faciundis*.[11] There were naïve miracles too, in plenty, amounting merely to the extraordinary ; everything surprising was taken to indicate the immediate intervention of God. What the historians miss is that mediæval miracles, when considered in their philosophic setting, rather attest the presence of a nature than deny it ; but of course it is a specifically mediæval and Christian nature, that same nature in short which is related to God in the way we have just indicated.

 For a Father of the Church, such as St. Augustine, there was no particular difficulty about the idea of miracle. In a way everything is a miracle. At the marriage of Cana Jesus made water into wine and everybody was astounded ; but rain becomes wine in our vines every day, and we take it all as a matter of course. Nevertheless, it is God Who creates the rain and the vine and the wine ; but He does it regularly, and we get so accustomed to it that we cease to wonder. Again, He speaks, and one rises from the dead and the whole countryside flock to see ; but men are born every day in the usual manner and we enter the birth in the civil register as if it were the most natural thing in the world. Thus in a created universe miracle remains supernatural, but still philosophically possible : " The very God, the Father of Our Lord Jesus Christ, makes and rules all things by His Word ; these primary miracles are effected by the Word as God ; the secondary and later ones are effected by the same Word, but now Incarnate and made man for us. Since we wonder at the mighty works of the man Jesus, let us also wonder at what He has done as God." [12] Between the *priora miracula* and the *posteriora miracula* there is no essential metaphysical difference : the divine omnipotence accounts equally for both.

Naturally, however, no Christian thinker ever dreamt of putting the events of the marriage at Cana on the same footing as events we call natural ; it is only in a certain sense that all things are said to be miracles. Our legitimate surprise at the true miracle is due to its happening out of the ordinary course of nature : *praeter usitatum cursum ordinemque naturae.* Miraculous phenomena are not necessarily more admirable in themselves than the daily spectacle of nature ; the government of the world, at once as a whole and in all its least details, is a much more wonderful thing than the feeding of the five thousand with five loaves. What strikes us about the multiplication of loaves is not so much the grandeur as the unusualness of the fact : *illud mirantur hominem non quia majus est, sed quia rarum est.*[13] Elaborating this idea a little further, St. Augustine dis-

tinguishes two superposed and co-ordinated orders of nature
that created by God when He created the seminal virtues,
these seeds, as it were, of all beings and natural future
events ; and that which is known only to the wisdom of
God, to which miracles, properly so-called, belong. From
this standpoint every creation which is added to the first
is miraculous,[14] but even so the miracle is only such for us,
not for God. If it appears contrary to the established order
of nature it is not so from the standpoint of the God who
established it, *cui hoc est natura quod fecerit.*[15] All that He
makes will always be nature for Him.

Thus, bound by the rigour of its principles, Christian
philosophy dismisses all these portents and marvels so dear
to the hearts of the Ancients, consigns them, along with the
monstrosities, into an irrational order it has long left behind
it. A Christian miracle is not a portent any more than a
defect of nature is a monstrosity. How can anything that
flows from the divine will be against nature, seeing that
nature is defined precisely as that which flows from the
creative will ? *Omnia portenta contra naturam dicimus esse, sed
non sunt.*[16] For the full definition of the concept of miracle,
therefore, we simply have to make it precisely clear that if
nature is reducible to the will of God, it is really to a will
that it is reducible, that is to say to something the reverse of
arbitrary. Plastic to excess in the hands of its Creator, to
such a point that sometimes, in reading the mediæval
Augustinians we feel inclined to ask whether any metaphysical necessity of essences is left at all, nature, nevertheless, gradually acquires the character of an intelligible
created order, until at last with St. Thomas and Duns
Scotus the doctrinal development attains its term. Nature is
henceforth defined as the order of second causes willed by
God ; if He had so willed, another and a different order of
nature would have been possible, and if He so wills the

established order can be completed with another, for God is not to be bound by an order of second causes that owes all its existence to Him. Since nature does not proceed from God by way of a necessary emanation, but by an act of His free-will, He can always produce the effects of second causes without the help of the second causes themselves, or even other effects that lie beyond the power of these causes. But even when miracle is thus rigorously defined as that which wholly transcends the power of second causes, therefore also as something altogether mysterious for human reason, it remains wholly rational from God's point of view. What does not belong to our order belongs nevertheless to His, on which ours depends. In derogating from the natural law, God but follows a higher law, against which He can in no case act, since it is one with Himself.[17]

To express this distinctive characteristic of the Christian nature the mediæval theologians invented the famous phrase " obediential power." So ill has this *potentia obedientialis* been understood that it has sometimes been regarded as a mere artificial afterthought invented *ad hoc* to patch up embarrassed theological positions ; but the fact is that it expresses one of the profounder aspects of the Christian natural order. A good deal of dialectical virtuosity was doubtless expended in cavils on the meaning of the term, but no mediæval philosopher could reject what it stands for without abandoning the Christian conception of the world. Henceforth, the idea of " *possibility* " will always have a double meaning. One may look at it, in the first place, from the standpoint of natural causes ; active second created causes exist, and passive subjects also, capable of receiving their action, and the order of natural possibility is defined by all that can happen within this order of second created causes. But that is not all. More things can happen in a created universe than created things themselves can effect. Above the special order of nature, as St. Bonaventure puts it, there is a general order depending only on the divine intellect and will. What only God can make of nature is impossible from the point of view of nature, but possible, nevertheless, from the point of view of God. The obediential power, then, is the possibility inherent in created nature of becoming what God can will, and does will, that it shall become. It is a purely passive possibility, excluding as by definition all aptitude for self-realization, but a real possibility, nevertheless, since it corresponds to all that God makes of nature, and the power He retains of actualizing it.[18]

Thus by throwing the obediential power into a general and divine order we reach a technical definition of the place of miracle in nature. Applying this idea to the problem of grace, mediæval theology achieved the synthesis that has been forming through so many centuries, and completed the systematic picture of the Christian universe. The conception of it was coeval with Christian thought. Just as for a Christian all, in a certain sense, is miracle, so, in a certain sense, all is grace. The error of Pelagius was at bottom only an exaggeration of this truth. It is doubtless true, as is often said, that Pelagius was a Greek in spirit, but he was not purely Greek, and his heresy itself would have had no meaning on the plane of the ancient naturalism. He is divided from the Christians by his extreme attenuation of the notion of sin and his effacement of the grace of the redemption ; but he is divided from the Greeks inasmuch as he is so intoxicated with grace that he absorbs nature almost wholly into grace. Incessantly he repeats that to merit by freewill is to merit by grace, and indeed from the first moment of its creation freewill itself is and always remains a grace. St. Augustine showed the greatest perspicacity when he discerned the partial truth that blinded his adversary to everything else. Pelagius is a thorough-going anti-Manichean ; and original sin, in his eyes, is a relic of Manicheanism ; created nature is so wholly good that nothing can be supposed capable of corrupting it to such a point that it will need a further grace in addition to that which brought it into existence. St. Augustine finds fault with him for ignoring the fact of sin, but not for saying that nature is a grace. It is very certainly a grace—but Pelagius goes wrong when he forgets that there is another. Besides the universal grace whereby all things are what they are, there is the grace that is proper to Christians, the grace of Jesus Christ, which is of such vital importance to us that we reserve the name for it alone ; this is not the grace which is nature, but the grace which saves nature. If Pelagius had only admitted this he might quite properly have hymned the grace, that brought nature into existence, for it is indeed a grace though not the greatest : *excepta ergo illa gratia, qua condita est humana natura (haec enim Christianis Paganisque communis est) haec est major gratia, non quod per Verbum homines creati sumus, sed quod per Verbum carnem factum fideles facti sumus.*[19]

Here, too, the theologians required a long time to adjust to each other two such closely allied conceptions as that of a nature gratuitously created and that of a nature gratuitously restored. The question whether the human race ever existed in a state of pure nature, the definition of that state, and the adjustment of its relation to the other graces that adorn it—all these are purely theological problems and their history need not concern us now ; but the concept of that nature which is presupposed to grace directly concerns philosophy and the history of philosophy. We have made its acquaintance already in the *anima capax Dei* of St. Bernard, St. Anselm, and St. Thomas Aquinas. All that remains to be done is to provide it with a philosophical name and put the finishing touch to its description by reducing this capacity to the obediential power as St. Thomas has drawn it out.[20]

Nature has a two-fold capacity, as we have seen already from St. Bonaventure's analysis. Just as the miraculous order was added to nature, so now we have to conceive the whole supernatural order added in an analogous manner. Besides those things of which nature is capable in itself there is all that it is capable of becoming at the *fiat* of God's will. It is currently objected against mediæval theology that the addition of the Christian grace to the Aristotelian nature rather resembles an attempt to square the circle. If we take up the ordinary position and assume that the theologians did no more than provide Aristotle's nature with a passport into Christian philosophy the objection, doubtless, would hold. It is perfectly true that Aristotle's φύσις is a Necessity, shut down on itself, a closed system, and that any attempt to open it up to the divine influence would be altogether unjustifiable. Moreover, God did not create it, it is no work of His; how then should He meddle with its disposal? Nature as conceived by the Christians has also an essence and a necessity of its own; neither more nor less necessity than that of the Greeks; indeed, if there is any choice in the matter I should say that it has more than the Greek nature, because it leans on the necessity of Being, in which it participates. So true is this that even God could do no violence to things without violating the Ideas, which are Himself. Nor need we wait for St. Thomas to find theologians who grasp this. St. Augustine will hardly be suspected of undue partiality for nature, but he saw clearly enough that the Christian conception of the Ideas as the art of the Word gave a rigorous stability to their finite participations, that is to say to natures. God does not elevate stones or animals to the beatific vision ; but then it has to be added at once that this necessity of created essences, circumscribed and closed by what they can do or suffer in the natural order, remains open to all that God can effect in them, or confer upon them, in the supernatural order. Since they are created they can still obey the will of their Creator if it pleases Him to enlarge their destiny ; and it is precisely the nature of intellect to be able to be thus enlarged without any alteration in its essence, nay, rather, with fulfilment of its essence. The capacity that lies in human nature for the beatific vision is thus something more than a word ; the capacity is part of human nature itself made to the image of God, from Whom it derives its power of knowing. The capacity for grace is also something more than a word ; if human souls were not susceptible of grace God Himself could not bestow it. But when all is said we must observe the limits of nature. It will obey *ad nutum* if God so commands, but it can do no more than obey.[21] There is nothing in it at all that already belongs to the supernatural, nothing to attract it, still less anything that demands it ; the obediential power, no matter how real it be, remains absolutely passive ; above all, it expresses the distinctive character of a Christian nature, open that is to say towards its Creator.

Such, then, was the nature that the mediævals knew. It contained no less than did the nature of the Greeks or that of modern science, but it had larger hopes. At a time when religious life penetrated everything with its influence, imagination might well take pleasure in overstepping the limits of these hierarchical orders, at the risk occasionally of confusing them ; but philosophical reason was none the less there to mark them clearly. If we require a formula in which to condense the results of more than a millennium of meditation on this problem, St. Thomas Aquinas, perhaps,

will suggest one. Speaking of non-rational nature, he frequently observes that it is as an instrument in the hands of
God.[22] The good workman uses his tools according to their
nature ; for him, nevertheless, they are but tools, to be used
for his own ends. When man is in question this " instru-
mental " view of nature no longer suffices. A reasonable
being is endowed with will ; God Himself does not use a
person as an instrument ; he is free, and God respects the
freedom He has Himself created.[23] But He can move it
also from within, call to it, solicit it ; if brute natures
are treated as instruments, rational natures can become
His " collaborators." The metaphysical study of the Christian conception of Providence has led us to conclude to the
fact ; that of the conception of nature shows its possibility,
and it will become still more apparent when we replace the
human individual in society and in history, where the
harmony between his nature and his end is progressively
revealed.

CHAPTER XIX

THE MIDDLE AGES AND HISTORY

In orienting nature, and man who is but a part of nature, towards a supernatural end, Christianity necessarily modified the received historical outlook, and even indeed the very conception of history. It is commonly asserted, however, that the Middle Ages remained a complete stranger to every kind of historical preoccupation, and indeed, to use a well-worn expression, was altogether lacking in historical sense. Illustrious scholars have vouched for the fact. The age we call " middle," which we regard, that is to say, as essentially transitional, is supposed, by a queer paradox, to have had no feeling at all for the transitory character of human affairs. Quite the contrary : " Its deepest characteristic is a faith in the immutability of things. Antiquity, especially later antiquity in its last centuries, was dominated by the idea of a perpetual decadence ; modern times since their dawn are no less animated by a faith in indefinite progress ; the Middle Ages knew neither this discouragement nor this hope. For the men of that time the world had always been such as they saw it—and that, by the way, is why their paintings of scenes from antiquity seem so naïve—and it would doubtless be the same at the last judgment."[1] We might perhaps feel a little surprised at these massive affirmations did we not know how thoroughly indifferent some philologists are when it comes to dealing with ideas. They seem to regard them as insufficiently real to be regarded as facts calling for genuine historical treatment, and manage sometimes to combine with a most exacting scientific rigour about what the mediævals wrote, a perfectly unscrupulous arbitrariness about what they thought.

The truth here, as elsewhere, is that if we seek our modern conception of history in the Middle Ages we may make up our minds at once that we shall not find it there ; and if the absence of this modern conception amounts to the absence of any conception, then we may as well admit that the Middle Ages had none. We might just as easily prove, by the same process, that the Middle Ages had no poetry, or again, as it used to be thought, even in the face of the cathedrals, that it had no art, and, as is still maintained in the presence of all its thinkers, that it had no philosophy. The true question is, on the contrary, whether there was not a specifically mediæval conception of history, other than that of the Greeks, and other than that of the moderns, but nevertheless real.

We might well have guessed *a priori* that this was the case when we deal with a time when all minds lived on the memory of an historical fact, of an event to which all previous history led up, from which was dated the beginning of a new era ; a unique event, which might almost be said to mark a date for God Himself ; the Incarnation of the Word and the birth of Jesus Christ. The men of the Middle Ages were possibly unaware that the Greeks dressed otherwise than they did themselves ; more probably they knew it well enough and cared very little ; what they did care about was what the Greeks knew and what the Greeks believed, and still more what they were able neither to know nor to believe. In the distant past, after the history of the creation and the fall, there was just a multitude of men without faith or law ; somewhat later the Chosen People, living under the Law, went through their long series of adventures ; still more recently came the birth of Christianity, inaugurating a new era, already marked with many great events, the fall of the Roman Empire, for instance, and the foundation of Charlemagne's. How could a civilization believe in the fixity and permanence of things when its own sacred books—that is to say the Bible and the Gospel—were history books? It would be simply waste of time to enquire of such a society whether it was changing and was aware of the change; nevertheless, we might well enquire how it conceived itself to be changing, that is, whence it came, whither it was going, what exact point it imagined itself to occupy on the road that runs from the past to the future.

Christianity had put the end of man beyond the limits of this earthly life; it had affirmed at the same time that a creative God allows nothing to fall outside the designs of His providence; it therefore had to admit also that everything, both in the life of individuals and in the life of societies of which individuals form a part, is ordered to this supraterrestrial end. Now the first condition of any such ordering is that there should be a regular unfolding of events in time, and first of all, of course, that there should be a time. This time is no abstract framework within which things endure, or, at any rate, it is not only that. Essentially it is a certain mode of existence proper to contingent things, unable to realize themselves all at once in the permanence of a stable present. God is Being, and there is nothing which He can become because there is nothing which He is not; change and duration have therefore no existence for Him. Created things, on the contrary, are finite participations of Being; fragmentary so to speak, always incomplete, they act in order to fulfil their own being, and therefore they change and consequently endure. That is why St. Augustine considers the universe as a kind of unfolding, a *distensio*, which imitates in its flowing forth the eternal present and total simultaneity of the life of God.

Man's state, indeed, is neither that of God nor that of things. He is not simply carried forward on an ordered stream of becoming like the rest of the physical world ; he is aware that he stands in the midst of it and grasps in thought the flux of becoming itself. Successive instants, that would otherwise simply arrive and pass away into the void, are gathered up and held in his memory, which thus constructs a duration, just as the sense of sight gathers up dispersed matter into a framework of space. By the mere fact that he remembers, man partially redeems the world from the stream of becoming that sweeps it along, and redeems himself along with it. In thinking the universe, and in thinking ourselves, we give birth to an order of being which is a kind of intermediary between the mere instantaneity of the being of bodies and the eternal permanence of God. But beneath the frail stability of his memory, which would founder into nothingness in its turn did not God support and stabilize it, man himself passes away. Wherefore, far from ignoring the fact that all things change, Christian thought felt almost to anguish the tragic character of the *instant*. For the instant alone is real ; here it is that thought gathers up the débris saved from the shipwreck of the past, herein live all its anticipations of the future ; nay, it is here that it simultaneously constructs this past and this future, so that this precarious image of a true permanence that memory extends over the flux of matter, is itself borne on by that flux, and with it all that it would save from collapse into pure nought. Thus the past escapes death only in the instant of a thought that endures, but the *in-stans* is something that at once stands in the present and presses on toward the future where likewise it will find no resting-place ; and at last an abrupt interruption will close a history and fix a destiny for ever.

Thus for all mediæval thinkers there existed men that

pass in view of an end that does not pass. But this was not all. In proclaiming the " good news " the Gospel not only promised the just an individual beatitude, but also announced their entry into a Kingdom, that is to say, a society of the righteous, united by the bonds of their common beatitude. The preaching of Christ was early understood as the promise of a perfect social life, and the constitution of this society came to be looked upon as the last end of the Incarnation. Every Christian realized therefore that he was called to enter as a member into a far vaster community than any human one to which he belonged already. A stranger to every nation, but recruiting its members from all nations, the City of God will gradually build itself up while the world lasts, and the world itself has no other reason for lasting than the expectation of its final fulfilment. In this celestial, that is to say invisible and mystical, city, men are the stones, and God is the Architect. Under His direction it grows, towards it tend all the laws of His providence, to assure its advent He made Himself Legislator, expressly promulgating the divine law He had already written on men's hearts, and carrying that law beyond what the due order of a merely human society would require, but which would be an insufficient basis for a society in which man should dwell with God. If there were certain Christian virtues, humility for example, not easily to be found a place in the catalogue of Greek virtues, it was precisely because ancient morals were ruled above all by the exigencies of human social life considered as the last end, while Christian morals, on the other hand, were ruled by a society higher than that which subsists between man and man, the society of intelligent creatures with their Creator. What did not exist at all for the Greeks became the necessary foundation of all Christian life ; humility is the recognition of the divine sovereignty and the absolute dependence of creatures, and there you have the fundamental law of what St. Thomas so forcibly calls "the republic of men under God." [2] How much was owing to this notion in the "republic of minds," the "eternal society," or even the "humanity" and the "realm of ends" in the philosophies of Leibniz, Malebranche, Comte and Kant, we may easily guess : the dream of a universal society of purely spiritual essence is nought but the phantom of the City of God haunting the ruins of metaphysic. For the moment, however, we have only to consider what we can learn from this concerning the Middle Ages, and the place it conceived itself to occupy in the history of civilization.

Its first consequence, if we look at it from this standpoint, is the substitution of a new sense of duration, quite other than that of the cycle and eternal return to which Greek necessitarianism so readily lent itself. Man has an individual history, a true "natural history" unfolding itself in linear and foreseen sequence from stage to stage, until at last death comes and cuts it short. This regular process of growth and growing old is also a constant progress from infancy to age, limited, however, by the span of human life. As he advances in age each man accumulates a certain intellectual capital, perfects the cognitive faculties whereby he acquires it, and augments his being, so to speak, as long as his powers permit. When eventually he vanishes from the scene his efforts are not on that account lost, for what is true of individuals is true also of societies that survive them, and of intellectual and moral disciplines which survive the societies themselves. For this reason, as St. Thomas often notes, there is a progress in the political and social order, just as there is in the intellectual order of science and philosophy, each new generation becoming the beneficiary of all the truths accumulated by its predecessors, profiting even by their very errors, and transmitting a growing heri-

tage to posterity. But then, for Christians, it does not suffice
to consider the results acquired by individuals, societies
or sciences. Since there exists an end promulgated by God
Himself, towards which, as we know, His will directs all
men, how should we not gather them all up under one and
the same idea, and order the whole sum-total of their
progress towards this single end ? How should such progress
be measured save in relation to such an end ? Apart from
an end what meaning would " progress " have ? For this
reason the Christian thinkers would naturally come to
conceive, with St. Augustine and Pascal, that the entire
human race, whose life resembles that of a single man,
passes from Adam till the end of the world through a series
of successive states, grows old in regular sequence, laying up
meanwhile a store of natural and supernatural knowledge
until it shall attain the perfect age, which shall be that of its
future glory.

So, if we are to conceive it as the Middle Ages conceived
it, must we represent the history of mankind. It is no
history of a continuous decadence, since, on the contrary,
it affirms a regular collective progress of humanity as such ;
nor is it the history of an indefinite progress, since, on the
contrary, it affirms that progress tends towards its perfection as towards an end ; rather is it the history of a progress
oriented towards a definite term. There is nothing here in
any event to authorize the view that the mediævals thought
that all things had always been what they then were, and
so would remain till the day of judgment. The idea of
progressive change, as just defined, was formulated in the
most forcible manner by St. Augustine and those Christian
thinkers he inspired. It was new ; neither in Plato nor in
Aristotle, not even in the Stoics, do we find the now so
familiar notion of humanity conceived as an unique collective being, made up more of dead than living, always in

progress towards a perfection, drawing ever nearer and
nearer. Ordered and penetrated through and through by
an internal finality, almost we might say by an unique
intention, the succession of generations in time has not only a
real unity, but, being now offered to thought as something
more than an accidental succession of events, it acquires an
intelligible meaning ; and therefore, even if the Middle Ages
is to be taxed with a lack of historical sense, we must at
least grant it the merit of assisting at the birth of a philosophy of history. Nay, more, let us say that it had one,
and that in so far as it still exists, our own is more penetrated
with mediæval and Christian principles than we usually
imagine.

One might very well conceive a philosophic history in
the manner of Voltaire and Hume, that is to say altogether
free—or almost so—from Christian influence, and nothing
would forbid us to call its conclusions a philosophy of history.
It may perhaps be doubted whether, in this sense, there was
ever any great historian who had no philosophy of history
of his own ; even if no effort were made to make it explicit,
it would be none the less real, and possibly the more
effective as less conscious of itself. The Christians, for their
part, were obliged to make their philosophy explicit, and
to give it a determinate orientation. They differed, in the
first place, from other historians inasmuch as they conceived
themselves to be well informed about the beginning and
end of history—ignorance of these two essential factors would
make it impossible for any infidel to grasp the meaning of
history, or even so much as to suspect that it had one.
Because they put faith in the Bible and the Gospel, in the
story of creation and in the announcement of the Kingdom
of God, Christians were able to venture on a synthesis of
the totality of history. All subsequent attempts of the same
kind merely replaced the transcendent end that assured the

unity of the mediæval synthesis, by various immanent forces that served as substitutes for God : but the enterprise remained substantially the same, and it was the Christians who first of all conceived it : namely, to provide the totality of history with an intelligible explanation, which shall account for the origin of humanity and assign its end.

However ambitious it be the mere design is insufficient in itself ; if we are going to realize it we shall have to accept the necessary conditions. If we are well assured that a God Who cares for the least blade of grass will not leave the rise and fall of empires to chance, if He forewarns us of the designs of His wisdom in ruling them, then we shall feel ourselves competent to discern the hand of Providence in detailed historical facts and explain them accordingly. It will be one and the same thing then to construct history and disentangle its philosophy ; all events will fall of themselves into the place assigned them by the divine plan. Such and such a people will live in a particular territorial setting, with such and such a character, such and such virtues and vices ; it will appear at a given moment in history, and endure for a determinate time as the economy of the providential order may demand. And not merely such a people, but such an individual, or religion, or philosophy.[3] Following St. Augustine's lead, the Middle Ages therefore represented the history of the world as a great poem, which takes on a complete and intelligible meaning as soon as we know the beginning and the end. Doubtless at many points the hidden sense will escape us ; we may suppose that the " ineffable musician " would often keep his secret back ; however, we shall decipher enough to be sure that all has a meaning, and to be able to conjecture how each event stands to the unique law that rules the whole. The task is doubtless arduous and full of pitfalls, but it is neither mistaken in principle nor altogether impossible. And so among Christian philosophers there appeared historical works of an amplitude hitherto undreamt of, embracing the totality of accessible facts and systematizing them all in the light of a single principle. St. Augustine's *City of God*, with its sequel in Paul Orosius' *History*, makes no secret of an ambition, which, indeed, it could hardly hide, since it was the *raison d'être* of the whole. Looking back on it at the time of the *Retractations*, St. Augustine sums up its aim and plan in a few words : " The first four of these twelve books describe the birth of two Cities, that of God and that of the world, the next four set forth their progress, and the last four their ends." The same design is manifest in the *Discours sur l'histoire universelle*, wherein Bossuet takes up Augustine's work for the use of a future king of France. It has not escaped the attention of an excellent historian of Bossuet that the idea of this work is closely akin to Pascal's notion of humanity conceived as an unique man, and thereby, through St. Augustine, to the Christian conception of history : " The idea of the *Histoire universelle* was not only to be found in Pascal, but since the early days of the Church it was everywhere, in St. Augustine, in Paul Orosius, in Salvian ; it is even to be found in Balzac's declamations. The chief difficulty did not lie in the conception but in the execution ; for it demands an incredible mass of knowledge, intellectual power, logic and skill. To have this outlook on human affairs one only needs to be a Christian, but to build up such a work on the idea one would have to be a Bossuet." [4] The thing could not be better put ; let us simply add that the conclusion may be turned the other way about : to build such a work on the idea one only needs to be a Bossuet, but to achieve the idea itself one would have to be a Christian.

So durable has been the influence of Christianity on the conception of history that it is still discernible, after the

seventeenth century, in thinkers who appeal to it no longer
or even oppose it. It was not Scripture, to be sure, that
guided the thought of Condorcet, but none the less he
conceives the idea of drawing a " comprehensive picture
of the progress of the human mind " ; his philosophy of
history has been run through a Christian mould of *tempora
et aetates*, as if the succession of " epochs " was now the work
of " progress " without the Christian God Who assures the
progress. It is a typical case of a philosophic conception
issuing from a revelation, appropriated by a reason that
imagines itself to be the sole and true inventor, and then
turned as a weapon against the revelation whence it came.
Comte and his " three states," leading up to the religion of
humanity, almost makes one think of an Augustine turned
atheist, and a City of God brought down from heaven to
earth. Schelling's " pantheism," setting out to assure a
determinate succession of world-ages—*die Weltalter*—from
within, posits a divine immanence at the metaphysical
heart of things, and all that history does is to explicitate its
development through time. Hegel goes still farther. This
bold genius saw clearly that a philosophy of history involves
a philosophy of geography ; he therefore includes it in his
powerful synthesis, which rules the whole dialectical movement of reason. The Greeks had early felt that even the
physical world was ruled by a mind ; Hegel does them justice
here, but neither does he fail to perceive that the application of the same idea to history only took place later on and
was the work of Christianity. He twits the Christian idea of
providence, in the first place, as essentially theological, as
something put forward as a truth of which the proofs do
not lie in the rational order ; and also as too indeterminate
to be really useful even to those who accept it as such :
mere certitude that events are ruled by a divine plan that
escapes us in no way helps us to link up those events in

intelligible relations. It is none the less true that if the Hegelian philosophy of history refuses to vouch for the truth—*die Wahrheit*—of the dogma of providence, it undertakes nevertheless to demonstrate its correctness—*die Richtigkeit*.[5] It could do nothing else, let us add, for it was simply by this that it lived. What Hegel here offers us is once more a *Discourse on Universal History* in which the dialectic of reason has taken the place of God. His ambition to provide us with an intelligible interpretation of history as a whole bears the evident mark of a time in which reason is so profoundly saturated with Christianity that what, without Christianity, it would never have even dreamed of undertaking, it imagines itself able to effect, and to effect from its own resources.

The study of the mediæval conception of history leads us naturally to ask how the Christian thinkers themselves conceived of their own position with respect to those who had preceded them and those who would come after. When it attained this degree of systematization the philosophy of history would necessarily be wide enough to embrace the history of philosophy. Here, then, the circle of our enquiry begins to close upon itself. Coming back to our starting point, and putting the question with respect to the doctrines that have formed the object of these studies, I would enquire of the Christian thinkers whether the relation in which they stood to Greek philosophy seemed to them to be purely accidental, or whether they conceived it to respond to intelligible necessities, and as occupying a definite place in the unfolding of a divine plan.

The Middle Ages has left us no *Discourse on the Universal History of Philosophy*, but it produced a few fragments, and above all it marked its own position within the totality of this possible history with much more care than might be imagined. These men would not have been in the least astonished to hear that they were living in an age that would be called " middle," that is, an age of transition. The Renaissance, that early invented the term, was also a " middle " age. The same may be said of our own ; and indeed the only one that could be conceived otherwise would belong less to history than to eschatology. Nor would they have been in any way mortified to be told they were a generation of heirs. Neither in religion, nor in metaphysics, nor in ethics, did they suppose they had invented everything ; their own conception of the unity of all human progress would have stood in the way of that. Quite the contrary ; even in their capacity as Christians, and in the supernatural order itself, they recognized the whole Old Testament in the New, and so felt themselves ruled by the providential economy of revelation. The result is that when we speak of Christian philosophy it is impossible to separate the Bible from the Gospel, for the latter incorporates the former, appeals to it everywhere even for the " great commandment," giving it completion at the same time. We could not possibly base a Christian philosophy on the Gospel alone, for even where it does not cite the Old Testament it everywhere presupposes it. And so in the providential plan, as the men of those times conceived it, the preaching of the Gospel inaugurated an " age " of the world which carried on the work of preceding ages, gathered up their fruits and added to them ; and this was the age in which they felt themselves placed. Was it not, moreover, on the religious plane, the final age ? Could any other follow save the eternal Kingdom of God ?

No less did the mediæval philosophers avow themselves to be heirs in the field of natural knowledge ; and, moreover, they knew why they were such. None of them doubted but that from generation to generation there is progress in philosophy. One might apologize for mentioning

that they did not overlook this fairly evident fact had they
not been accused of doing so. There was quite enough of
the history of philosophy in Aristotle to teach them that the
Presocratics, "like children just trying to talk and only
succeeding in babbling," had left their successors merely
formless attempts at the explanation of things. St. Thomas
himself recalls it, and delighted to retrace the history of
philosophical problems and to show how men, occupying
the ground step by step—*pedetentim*—came slowly nearer to
the truth. The mediævals felt it incumbent upon them to
gather up the spoils of this always incomplete success and
push on the advance. They saw themselves providentially
placed at the crucial point where the whole heritage of
ancient thought, absorbed by Christian revelation, was now
to multiply a hundredfold. The age of Charlemagne struck
men's minds as the coming of an era of enlightenment :
hoc tempore fuit claritas doctrinae, wrote St. Bonaventure in the
full thirteenth century. Then was effected that *translatio
studii* which, handing on to France the learning of Rome and
Athens, entrusted to Reims and Chartres and Paris the
task of adapting this heritage to, and integrating it with,
Christian Wisdom. None better than the poet Chrétien de
Troyes has uttered the pride felt by the men of the Middle
Age in being the guardians and transmitters of the civilization of the Ancients (Cligès, 27–39). The glory of his native
land, which a French poet of the thirteenth century thus
delighted to express, was none of his own imagining. The
old tradition of the anonymous chronicler of St. Gall preceded him, Vincent of Beauvais followed him, a cloud of
witnesses surround him. When an Englishman, John of
Salisbury, saw Paris in 1164, that is to say before the extraordinary doctrinal flowering-season of the future University,
he had no hesitation about the providential character of the
work that there went forward : *vere Dominus est in loco isto,*

et ego nesciebam ; the Lord is surely in this place and I knew it not.

Thus by its own philosophy of history the Middle Ages was led to conceive itself placed at a decisive moment of the drama that opened with the creation of the world. It never imagined that learning had always been what it had become since Charlemagne, nor that further progress was impossible. Nor did it believe that the world, having progressed as far as the thirteenth century, would progress indefinitely by the play of purely natural forces and in virtue of a kind of acquired momentum. In accordance with its own proper outlook it considered rather that humanity had never ceased to change from the days of its infancy, that it would still go on changing, but also that it was on the eve of the great final transformation. Joachim of Flora might announce the new Gospel of the Holy Spirit ; those who followed him for a while soon came to recognize that after the Gospel of Jesus Christ there will never be another. Long prepared by the Ancients, the true philosophy had just attained, in all essentials, its definitive form. It was neither Plato's nor Aristotle's, but rather that which had been evolved from both by integration with the body of Christian Wisdom. And many another was there incorporated along with them ! The philosophy of Christians was not alone in invoking the Bible and the Greeks : a Jewish philosopher like Maimonides, a Mussulman like Avicenna, had carried out, on their own side, a work that was parallel to that of the Christians themselves. How, then, should there not be close analogies, a genuine kinship even, between doctrines that worked on the same philosophic materials, and appealed to the same religious sources ? Thus it is not with Greek philosophy alone that Christian philosophy was bound up in the Middle Ages ; like the Ancients, the Jew and the Mussulman also did it service.[6] Nevertheless, the work is about to attain a form which, in all essentials, will be definitive. Roger Bacon himself, all unsatisfied as he was, thought that the " great work " would soon be accomplished. Then the Christian thinker looked forward to nothing but an age of light, in which society, more and more thoroughly Christianized, would become more and more incorporated with the Church, and in which philosophy would become more and more completely itself in the bosom of Christian Wisdom. How long would that time last ? No one pretended to know, but all knew that it would be the penultimate act of the great drama that went forward. Afterwards would come the catastrophic reign of Antichrist. Whether the Emperor Charles was to be the last defender of the Church, or whether another was to come after him, none could guess at all. The only thing certain was that after the advent of the supreme champion, whoever he had been or might be, the great tribulations would begin : *post quem fit obscuritas tribulationum.* But they would endure only for the appointed time. Just as the Passion of Christ was as a darkness between two days, so the last assault of evil on good would meet with its defeat. Soon would open the seventh age of humanity, like the seventh day of creation, a prelude to the eternal repose of a day that should have no end : " Then shall descend from Heaven this city—not yet the city that shall be on high, but the city here below—the militant city, as conformable to the triumphant city as may be in this life. It will be reconstructed and restored as it was in the beginning, and then also shall reign peace. How long that peace will endure is known to God." [7]

Apocalyptic considerations these, and less important in their detail than in their spirit and in the promise that concludes and crowns them. *Pax,* the peace from the shadow of the Cross ; God's own promise—*pacem relinquo vobis, pacem meam do vobis*—a peace that philosophy is wholly incapable of giving, but whose triumph at least it can promote in integrating itself with Christian Wisdom. Thus, in its own way, it works for the fulfilment of the divine plan, and, as much as in it lies, makes smooth the path for the City of God. For it teaches Justice and opens the way for Charity. In this sense mediæval philosophy appears to do something more than merely occupy a place in history ; in establishing itself in the axis of the divine plan, it works to forward it. There, where social justice reigns, it is possible to have an order and a factual accord of wills. Let us even say, if you like, that it is possible to have a sort of concord ; but peace is something much more, for where there is peace there is concord, but concord does not suffice for the reign of peace. What men call peace is never anything but a space between two wars ; a precarious equilibrium that lasts as long as mutual fear prevents dissension from declaring itself. This parody of true peace, this armed fear, which there is no need to denounce to our contemporaries, may very well support a kind of order, but never can it bring mankind tranquillity. Not until the social order becomes the spontaneous expression of an interior peace in men's hearts shall we have tranquillity. Were all men's minds in accord with themselves, all wills interiorly unified by love of the supreme good, then they would know the absence of internal dissension, unity, order from within, a peace, finally, made of the tranquillity born of this order : *pax est tranquillitas ordinis*. But if each will were in accord with itself all wills would be in mutual accord, each would find peace in willing what the others will.[8] Then also we should have a true society, based on union in love of one and the same end. For to love the good is to possess the good ; to love it with undivided will is to possess it in peace, in the tranquillity of a stable joy that nothing can disturb. Mediæval philosophy put forth all its powers to prepare the reign of a peace which of itself it could not give. In labouring for the unification of minds by the constitution of a body of doctrine acceptable by every man's reason it set out to assure the interior unity of souls, and their accord with one another. In teaching that all things desire God, and in asking men to look beneath the infinite multiplicity of their actions for the secret spring that puts them all in motion, Christian philosophy prepares their minds to welcome the order of charity in themselves and to long for the extension of its reign over all the earth. Where is true peace? In the common love of the true good : *vera quidem pax non potest esse nisi circa appetitum veri boni*.[9]

If mediæval philosophy worked for peace it was because it was itself the workmanship of Peace. All history tends towards the supreme tranquillity of the divine republic as towards its term, because God, the Creator of human beings who move towards Him through time, is Himself Peace. No trembling precarious concord like our peace ; not acquired at the price of any internal unification, however perfect it may be supposed to be : He is Peace because He is One, and He is One because He is Being. As, therefore, He has created the finite being and the finite unity, so also He has created the finite peace. In ordering intelligences and wills toward Himself by knowledge and by love, of which, as we have seen, He is the supreme Object, God bestows on consciences the tranquillity that unifies and, in unifying, unites. This tranquillity, as a created effect of the divine peace—*quod divina pax effective in rebus producit*—attests therefore in its own way the creative efficacy of a supreme and subsistent Peace whence its own existence derives : *causalitatem effectivam divinae pacis*.[10] Doubtless the divine peace itself escapes our gaze, as God Himself with Whom it is altogether one, but we behold its finite participations in the unity of essences, in the harmony of the laws that link up physical things with each other, and the harmony which social laws would produce rationally amongst men. For Peace reaches through all things from end to end, uniting them with might, and ordering all things sweetly.

So to regard Christian philosophy is not simply to take it at its own valuation, to see it as historically it saw itself, but to see it at work in history, for indeed it went to work. It did not accomplish all it would, nor always even all it ought, for after all it was but a philosophy, that is to say a human effort labouring at a superhuman task. But at any rate all the greatness of which it can legitimately boast came from whatever fidelity it showed to its own proper essence. The spirit of mediæval philosophy was one with the spirit of Christian philosophy. It was fruitful and creative in so far as it was willing to be incorporated with a Wisdom that lived itself by faith and charity. The Christian thinkers felt almost to anguish the narrow limits of mediæval Christendom : *boni igitur paucissimi respectu malorum Christianorum, et nulli sunt respectu eorum qui sunt extra statum salutis.*[11] However, even in these narrow bounds, Christian philosophy could live. It died, primarily, of its own dissensions, and these dissensions multiplied as soon as it began to take itself for an end, instead of serving the Wisdom which was at once its end and source. Albertists, Thomists, Scotists, Occamists, all contributed to the ruin of mediæval philosophy in the exact measure in which they neglected the search for truth to exhaust themselves in barren controversies about the formulæ in which it was to be expressed. The multiplicity of these formulæ would have constituted no drawback, rather the reverse, if the Christian spirit that kept them in unity had not been too often obscured, sometimes lost. When this happened mediæval philosophy became no more than a corpse encumbering the soil it had dug and on which alone it could build. For it was the great workman of a Christendom that could not live without it, and without which it could not live. Failing to maintain the organic unity of a philosophy at once truly rational and truly Christian, Scholasticism and Christendom crumbled together under their own weight.

Let us at least hope that the lesson will not be thrown away. It was not modern science, that grand uniter of minds that destroyed Christian philosophy. When modern science was born there was no longer any living Christian philosophy there to welcome and assimilate it. The architect of peace had died of war ; the war came of the revolt of national egoisms against Christendom, and this revolt itself, which Christian philosophy should have prevented, came of the internal dissensions that afflicted it because it had forgotten its essence, which was to be Christian. Divided against itself, the house fell. Perhaps it is not too late to attempt its reconstruction ; but if Christian philosophy is to start on a new career, a new Christian spirit will have to be everywhere diffused, and philosophy too will have to learn to absorb and retain it. That is the only atmosphere in which it can breathe.

CHAPTER XX

THE MIDDLE AGES AND PHILOSOPHY

At the end of our long enquiry we should naturally wish to pass it in review and mark its chief results ; not in order to justify them once more, it is too late now for that, but rather to fix them finally, and especially perhaps to point out what they could never be and therefore also are not.

I have put the problem of the spirit of mediæval philosophy in an essentially historical setting, and so far the conclusions that I trust to have reached are historical conclusions. Whether true or false, they are historically true or false, and in their immediate result should more especially affect the teaching of the history of the mediæval philosophies. Perhaps it has not been sufficiently noticed that the legitimacy of such teaching is, in itself, a real problem. If St. Augustine merely re-edited Plato, if St. Thomas and Duns Scotus are merely Aristotle misunderstood, it is quite useless to study them at all—unless, at least, from the point of view of the history of philosophy, and the existence of a yawning gap between Plotinus and Descartes will have to be admitted. The historian can only recount what happened ; if in philosophy nothing happened between the end of the Greek period and the opening of modern times, then there is a gap in history only because there was a gap in things, and history is certainly not to blame for that. But if the preceding conclusions are, in all essentials, true, the case is very different. On the supposition that St. Augustine added something to Plato, and that St. Thomas Aquinas and Duns Scotus added something to

Aristotle, the history of mediæval philosophy will have a proper object. I shall be forgiven for seeming to attach a certain importance to the matter if I merely observe that there are people whose business it is to teach it.

Let us admit, however, that this point is only of secondary importance—the professors of history not being the final cause of history, whatever illusions they may occasionally entertain on that score. The true question is whether, wishing to set forth a spirit of mediæval philosophy,. and having identified it with the spirit of Christian philosophy, I am not bound to impose a double limit here on the scope of my conclusions. On the one hand they are bound up with the validity of the historical considerations on which they are based ; on the other, granting that these are valid, they still leave open a philosophic problem of wider import, which they cannot of themselves resolve, but whose existence subtracts nothing but their own validity. Now, speaking historically, it is obvious that I have made no attempt to draw up a comparative table of mediæval philosophy and Greek philosophy. It seems to me, moreover, that history itself would avail nothing here and that such a table would be useless. How should we successfully compare Christianity with the later Stoicism, seeing how extremely difficult it is really to demonstrate, even in the case of contemporary forms of thought, in which direction influence has been exerted ? The case would be the same with Plotinus and his master Ammonius Saccas. In attempting to collate texts, taken precisely in their materiality, we should arrive at facts whose interpretation would be too largely arbitrary to deserve the name of proof. Even, moreover, if we con-fined ourselves to what the mediæval philosophers really knew of Greek thought prior to the Christian era, a prac-tically indefinite number of particular problems would arise. Suppose, finally, that we made a provisional selection of some of the master problems, it would be almost impossible to show simultaneously, and with equal clearness, what Greek thought gave and what Christian thought added. The very method of such an enquiry would inevitably convey the impression of a grave injustice committed against Greek thought, though that might be far from the intentions of the author. If, on the other hand, on the points in discussion, Plato and Aristotle had already taught, and taught in the same sense, all that is attributed to the influence of the Bible, then indeed a real injustice would have been committed against Antiquity. If Plato and Aristotle were really monotheists, if they identified God and Being and taught the creation of matter, if, as some have been bold enough to maintain, the unmoved movers of Aristotle, although uncreated, eternal and necessary, are nothing but the Thomist angels created in time and contingent ; if, along with his being, the intellect and will of man stand to God as those of a creature to his Creator, of beings to Being —if all this was as true for the Greeks as it is for the Christians, then the central thesis of this book is historically false and is simply to be rejected. Historians have the right to decide, provided they consent to discuss the question on the basis on which it has been put.

But that is not all. Although the question has been opened up from the standpoint of facts, the answer I have given proposes to bring a positive contribution to a much larger problem than that of the spirit of mediæval philosophy, that, namely, of Christian philosophy. Here again it would be well to state the case with all possible precision, since it is far from having been everywhere put as clearly as I might have hoped. The conclusion that arises out of this study, or rather the thread that runs through it from end to end, is that everything happened as though the Judeo-Christian revelation were a religious source of philosophical development, the witness *par excellence* to this development in the past being the Latin Middle Age. The thesis may be charged with being a mere piece of apologetic, but if it is true, the fact that it may serve the turn of apologetic does not prove that it is false ; and if it is false, it is not because it can be used for apologetic purposes that it is false. The only question to be settled is, therefore, is it true ?—and after that anyone is free to make what use of it he will. What on the other hand, would really put it outside the limits of philosophy, or even the history of philosophy, would be to maintain that if it is true, yet everything which either directly or indirectly undergoes the influence of a religious faith ceases, *ipso facto*, to retain any philosophic value. But this is a mere " rationalist " postulate, directly opposed to reason ; for, finally, if there really was a mediæval philosophy it appealed to nothing but its own rational evidence ; it is not to be got rid of by ignoring it *a priori*, or by setting up some distorted substitute as a butt for criticism. A philosophy may invoke a revelation and be false, but if false it is not on account of the revelation, but because it is bad philosophy ; the errors of Malebranche, deeply and genuinely Christian as he was, would be a sufficiently good example. But a philosophy may draw inspiration from a revelation and be true, and if true, it is because it is good philosophy. When reason starts making these arbitrary exclusions it loses its right to judge.

Need I say that I have no illusions as to the efficacy of my remark ? It will in no way change the accepted outlook, but at least it justifies a demand that the situation be frankly faced, and acceptance of principles followed up by acceptance of their consequences. In the name of the postulate I dispute, the title of philosophy may be refused to the systems elaborated by the thinkers of the Middle Ages, but if it be established that their chief positions are really their own, and are by no means a simple heritage from the Greeks, it must be recognized at the same time that all that the classical metaphysics inherited from the Middle Ages from the beginning of the seventeenth century, is enough to exclude them also, *ipso facto*, from the field of philosophy. That a metaphysical thesis should become rational it is not enough that it should have forgotten its religious origin. Therefore there will have to be expelled from philosophy, as also from its history, the God of Descartes and those of Leibniz, Malebranche, Spinoza and Kant, for these would no more have existed without the God of the Bible and the Gospel than would the God of St. Thomas. And then Auguste Comte would be in the right : modern metaphysic is merely a shadow cast by mediæval theology, and both, having been found to be insufficient, may be relegated, without serious inconvenience, to some appropriate branch of mental archæology. What, on the other hand, would not be frank at all would be to pretend that the value of modern metaphysics is due to their being completely cut off from religious influences, when in fact they were born and bred of religion, and then to refuse a hearing to mediæval metaphysics because they had the honesty to avow their parentage. If contemporary disregard for the philosophies of the Middle Ages has no other motives than these, it might very well be extended to cover more modern examples. I would even venture to predict, that by means of a little judicious elimination from Greek thought of all that was due to religious influence, it might even find some further conquests in the past.

Let us therefore take mediæval philosophy as it stands, and let us recognize moreover that it presents a sufficient problem, for truly it offers an astonishing spectacle to the historian of ideas who would grasp it all. It is impossible to open the *Summa Theologicas* or the *Commentaries on the*

Sentences without finding them full of texts of which the Greek origin is not doubtful, and of glosses on these texts. The library of a mediæval theologian contained first the Bible, then Aristotle, then the Commentaries on Aristotle such as those of Albert the Great or St. Thomas ; then, if its resources permitted, the supercommentaries, the commentaries on the commentaries, such as that which John of Jandun made on Averroes, and, to crown all, a mass of *Quaestiones Disputatae* to explain what it all meant. At the same time, when closely examined, these commentaries leave behind such an impression of freedom, that they have often been charged with unfaithfulness. It could hardly be otherwise, seeing that even when they bear upon the same text they are often in contradiction with each other. In the inexhaustible arsenal of Aristotle everybody finds the phrase required to justify his own position. " Authority's nose," as an old author puts it, " is made of wax ; you can turn it in any direction you like." Who, then, is here deceived ? Either the Middle Ages took Greek philosophy seriously, and then it has to be recognized that its interpretation is neither philosophically coherent nor historically faithful ; or else the work that it had in hand bore no relation to that of the Greeks, and then to what purpose do St. Thomas and St. Bonaventure, instead of going forward boldly with the constitution of a specifically Christian philosophy which shall be original and new, prefer to deck themselves out in a collection of old rags torn out of Greek thought, and risk being taken for Christians dressed up as Greeks, that is to say, for men who are neither Greeks nor Christians ? Whether the less Christian or the less Greek, we can hardly be expected to take them for philosophers.

The Middle Ages as a whole was quite sufficiently complex to furnish the crucial instances needed to verify hypotheses of this kind. We may easily guess what the teaching of philosophy would have been in the Universities of the thirteenth century, had it been deliberately withdrawn from the influence of its Christian surroundings. Averroes tried the experiment with Islam, and the Averroists unweariedly renewed it on Christian soil. The result was what we know—total philosophic sterility. We may discuss, and doubtless always shall discuss, the fidelity of their interpretation, for we in our day have almost as many various versions of Aristotle as the Middle Ages itself knew : but that is not the question, for these men, whose whole ideal was a total absence of originality, would have been struck with consternation if you had proved to them that they had added anything. Call to mind Averroes' invectives against Avicenna, caught in the barefaced act of " inventing " something ! Thus Greek philosophy, cut loose from Christian revelation, survived in this medley of Aristotelianism and Neo-Platonism ; it lasted for several centuries, from the thirteenth to the sixteenth, and not a single original idea that we know of ever came out of it. If we regret that the Middle Ages, turning Greek philosophy to account, permitted itself to abandon the letter, or if, on the other hand, for the convenience of a history rather overfond of linear simplifications, we would find a Middle Ages vowed to the most systematic psittacism, the Averroists are there and will give us every satisfaction. " Ipsedixitism " is their speciality ; not, however, of St. Bonaventure nor St. Thomas Aquinas.

The latter gain nothing unfortunately, for it is for this precisely that they are reproached in the name of the critical spirit. Drawing inspiration from Plato and Aristotle, appealing to their principles, the Christian philosophers drew thence conclusions of which neither Plato nor Aristotle had ever dreamt—nay, for which they could have found no place in their systems without ruining them.

Such is particularly the case of the famous distinction of essence and existence, necessary, in whatever sense it be taken, for Christians, but inconceivable in the philosophy of Aristotle. From the standpoint of a consistent Peripateticism the notion of potency is bound up with that of matter ; everything immaterial is therefore a pure act, that is to say, a god. For a Christian philosopher a being that is immaterial is not yet a pure act, because it is in potency with respect to its own existence ; that is why St. Thomas, expanding the notion of potentiality, disassociates it from that of materiality ; instead of identifying it with a certain mode of being, that namely of matter, he extends it to existence itself. Duns Scotus goes further. Anxious to assure a positive reality to all the elements of composite beings, he effects the same disassociation on the notion of act as St. Thomas had done on that of potency. Matter, as he conceives it, has an actuality of its own, and, as its distinction from form demands that it be immediately and radically different, it must be credited with an actuality that is none other than that of potency itself ; an unavoidable consequence, thinks Duns Scotus, of the Christian concept of creation ; for if matter is created, it is of being, and if it is of being it is act, or in act. There, then, we have an Aristotelianism in which, in the name of a metaphysic of being to which it is altogether a stranger, potency is no more bound up with materiality than form is with actuality. But is not that to base metaphysics on a pure equivocation, and along with metaphysics, all the physics, psychology and ethics it inspired ?

The fact is undeniable, but remains to be interpreted. If anyone would say that in themselves, and from a dogmatic standpoint, the conclusions of St. Thomas or Duns Scotus are open to discussion, he is perfectly at liberty to do so. As philosophers they are amenable to the common law, and since their doctrines are given out as rational they must submit to be judged by reason. But whoever would assume the right to judge must first make some effort to understand ; nothing is less flattering, for anyone who plunges into it, than a commentary on a misreading. Supposing, however, that the doctrine discussed is really understood, how should we find fault with it for not being entirely shut up within the limits of a ready-made system, which it quite openly took as a starting point ? We might as well say that Malebranche and Spinoza were not philosophers because, basing themselves on Descartes, they drew out of his method consequences he had not foreseen himself. Is Fichte's doctrine to be treated as a simple blunder as to the authentic sense of Kantism, and Schopenhauer's attempted objectification of willing as a misapprehension of the *Critique of Practical Reason* ? To set out to make a critique of systems which shall be historical and philosophical at the same time and in the same respect, is to attempt a contradiction in terms. Every philosophy derives from another, and differs from the one it derives from ; historical criticism may occupy itself with the mode of derivation and the nature of the difference, but it would destroy both itself and its object if it refused it the right to be different.

If the doctrinal evaluation of a system in history is one thing and its history itself is another, nothing is proved against Christian philosophy by showing that it set the Greek philosophies from which it derived in a new light. To condemn it as philosophically inconsistent with these you would have to prove that the problem of being and becoming was not fundamental in the doctrines of Plato and Aristotle. If it could be shown that neither of these thinkers attempted the distinction of the necessary and the contingent, associated the notions of the necessary and the real, of the contingent and the possible, subordinated the order of contingence to an order of necessity which at once explains its reality and forms the basis of its intelligibility, then much would have been done to put the coherence of Christian philosophy in doubt. But if it be true, on the contrary, that the Greeks had already raised the problem of the source of being, how can it be denied that the Christian philosophers were precisely in line with them when they pushed the problem of reality to the level of existence and thereby, for the first time, gave its full significance to the concept of actuality? Such, it seems to me, was the distinctive work of the Christian philosophy; thence came all its audacities, and thereby it is a genuine philosophy, the influence of which was exerted beyond the limits of the Middle Ages, and will continue to be exerted as long as there are men to believe in the possibility of metaphysics. Only, to be convinced of the existence of a mediæval philosophy, we must study its representatives. Whoever is content to go on repeating on the faith of authority that the men of that time were slaves to authority, or to find fault with them in the name of intellectual liberty for refusing to be slaves to the letter of Aristotle, will always be in the right; it is always easy to be right when one has decided to maintain a thesis at the price of any incoherence. The critical spirit, thus conceived, assumes every kind of licence except against itself. We may possibly prefer another ideal; to possess a mind that is free enough to yield to the evidence of facts and not to seek to be right against all reason.

But we shall escape the one half of the objection only at the risk of succumbing to the other. If mediæval thought rejoiced in this independence and lived a genuine philosophic life, what real connection could it have with religion? In one word, if it did not betray philosophy, how could it avoid betraying Christianity? It would be vain to conceal the gravity of the charge, although perhaps the vindication of mediæval thought will not prove impossible. But we might in the first place demand of those who deny its existence that they cease to do so in the name of two contradictory principles. That the essence of philosophy was dissolved and allowed to be lost in religion is at least a possibility ; that the essence of Christianity, on the other hand, was let fall and reduced to a mere philosophy, that is also a perfectly intelligible objection ; but it is not possible to make both objections at once. If the mediæval thinkers were so philosophic that for the sake of philosophy they compromised the essence of Christianity, why should their systems not be taken seriously, and by what right are they eliminated from the histories of philosophy ? But if, on the contrary, they sacrificed philosophy to the religious exigencies of Christianity, why are they still rebuked for not having been Christians ?

Between these two contradictory objections lies the historical truth—true because in conformity with the facts. The charge of having sacrificed too much to philosophy is at once the oldest and the most banal of all the objections that have ever been directed against the Christian philosophers. Protestantism, even to-day, considers it its duty to " react against the invasion of the Church by the pagan spirit," considers moreover that this was one of the chief ends that the Reformers of the sixteenth century proposed to themselves. Very true : there are plenty of texts of Luther to witness to it if they are wanted. But the objection, although it can be taken in a specifically Protestant sense, is not necessarily Protestant in essence. Malebranche was not a Protestant, but poured out bitter enough complaints about the pagan character of scholasticism, this " philosophy of the serpent." Erasmus was no Lutheran, nor ever wished to be one, but that did not prevent him from protesting, with Luther, against the mixture of Aristotle and

Gospel that proceeded from Albert the Great, St. Thomas and Duns Scotus ; for him, too, the " philosophy of Christ " is the Christ without philosophy, that is to say, simply the Gospel. But even in the Middle Ages itself St. Peter Damian, all the anti-dialecticians, even the Popes, had no need to wait for the Reform to warn theologians solemnly of the way in which they imperilled faith when they turned philosophers. With what vigour does not Gregory IX remind the Masters in Theology of the University of Paris that philosophy, this handmaid of theology, is bidding fair to become the mistress ! These theologians, who ought to be " theologues," have they not become mere " theophants " ? With them, nature takes precedence of grace, the text of the philosophers replaces the inspired word of God, the Ark of the Covenant stands next door to Dagon, and, by dint of wishing to confirm faith by natural reason, faith itself is rendered useless, since there is no longer any merit in believing what is demonstrated. Let not, then, these theologues become philosophers—*nec philosophos se ostentent*—such is the solemn warning issued to the University of Paris by a Pope who did all he could to protect it and save it from itself.[1] The testimonies are too numerous, and come from too many sides, not to point to the presence of some real difficulty, rooted in the nature of things. Let us therefore try to see exactly where and what it was.

Taken in its essence, the objection is put above all from the standpoint of religion, and whatever the conception of Christianity that inspires it, it is a Christian objection. The danger against which it would put the Christian philosopher on his guard is the danger of putting man back on the plane of ancient Greek naturalism, by a fatal forgetfulness of the Gospel and of St. Paul, that is to say of grace. In itself, nothing could be more appropriate ; no one with the interests of Christianity at heart would dream of denying the utility, nay, the permanent necessity, of such a warning. *Ne evacuetur crux Christi ;* to eliminate it would be to eliminate Christianity ; but then we must be careful not to deduce exigencies from this truth that would be incompatible with the exigencies of others, and, notably, the truth of philosophy. Erasmus protests, in the name of humanism, against the contamination of the Gospel by the philosophy of the Middle Ages ; he would take us back beyond Duns Scotus, St. Thomas and St. Bonaventure, to the letter of the sacred text. Let us suppose he is right, let us even admit the desirability of reminding the Christian philosophers that philosophy should never be allowed to lead to forgetfulness of the Gospel ; by what right does Erasmus add that the Gospel itself is a philosophy ? If Christians had no right to anything outside the Gospel and the Church, doubtless there would be nothing lacking to their Christianity, but is it quite certain that they would still possess a philosophy ? Either one will see nothing in the Gospel but a mere natural moralism, which would be to suppress its religious character and annihilate Christianity under pretext of saving it ; or else the supernatural and religious character of the Gospel will be maintained, and how, then, can it be taken for a philosophy ? If we would be consistent we shall have to pass over Erasmus and seek another position.

Luther's was much stronger ; he had at least the merit of frankness. In his view, " there is no greater enemy of grace than Aristotle's ethics." Henceforth, no compromise. Not only is Christianity not a philosophy, but there never will be any philosophy (this *stultitia*) fit to call itself compatible with the Gospel. Very few, even of his own disciples, went to this length. However, even at the moment when we reject his conclusions interiorly, or when we reject them with all our forces, we can never forget the old truth that

Gregory IX and so many mediæval theologians took good care not to let drop ; the radically transcendent character of Christianity, as a religion, with respect to all philosophy in general, and Christian philosophy itself in particular.

It is desirable to insist on this point, for its disregard is the source of the most insidious and lasting misconceptions. From the standpoint of Luther and a consistent Lutheranism —*rara avis*—no misconception is possible, since Christianity, as he conceives it, bars him from any philosophy of nature. But of all those who, without denying the legitimacy of philosophy, would incorporate Christianity whole and entire therein, the idea of a Christian philosophy will always be a dream hovering for ever on the horizon of the possible, and fated by its nature never to be realized. Christian philosophy, say they, has never existed, exists not yet, but it might exist, soon doubtless it will come to birth ; soon, soon—provided life does not fail the dreamer before the dream becomes reality. Perhaps the object of their aspirations does not belong to the order of the possible. These seekers do not find what they want in mediæval philosophy, but they will never find it in themselves either— the thing simply cannot be. They would have a philosophy which shall be genuinely philosophy, that is to say, purely and rigorously rational, but in which shall be integrated all those transcendent and supernatural experiences which belong of right to Christianity and constitute its religious essence. The interior and intimate drama of nature and grace, the hidden life of charity, the mysteries of the divine life in God and in the soul, all these are things without which Christianity would be no longer itself ; and these are the things which they try in vain to integrate with this Christian philosophy which is always on the eve of birth, and the absence of which from the Christian philosophy already born, forbids them to take it for Christian.[2]

It is precisely because they are absent that it is a philosophy and nothing but a philosophy, and it is because the integration of these things with a philosophy would be contradictory, that the one they seek for will never exist. They must make the best of it ; a philosophy, even a Christian philosophy, can well find a place in the edifice of Christian wisdom—it contains already so much else !— but it will never be the equivalent of Christian wisdom. To demand of a part that it absorb the whole, and remain faithful to its essence all the while, is to destroy at once both the part and the whole. If, then, mediæval philosophy is criticized for not having been Christian on the ground that its speculations on nature, on prime matter, on the unity or plurality of forms, contain nothing of essential interest for the work of salvation, or, conversely, on the ground that what is of essential interest for the work of salvation seems there to find no place—if the critic says this, then he simply shows that he understands neither the nature of Christianity nor the nature of philosophy. The mediæval thinkers understood both very well, and that is why, resolutely picking up the threads of Greek speculation, they succeeded in making it progress. If anyone has doubts about it, is it not precisely because mediæval philosophy is sought where certainly it is not to be found, that is in religion ?

In the history of philosophy, it is said, there is not to be found a single authentically Christian principle for which the mediæval philosophers, and through them philosophy itself, were indebted to the Bible or the Gospel. If this were true, what would it prove against the existence and originality of Christian philosophy ? Absolutely nothing. Christians who were, or wished to remain, or aspired to become philosophers, gave proof of the most elementary good sense in not addressing themselves to the Gospel in order to find there something on which to found one more philosophic sect to add to all the others. Historians, on their side, who point out that the Gospel brought about no sudden revolution in the whole state of psychological physical, metaphysical and ethical problems, do no more than admit the evident fact that Christianity modified philosophy, not by adding a new system to the ancient systems, but by transcending them all. The " good news " consisted in the announcement of salvation. What the Gospel put into men's hands was not therefore the key to the problem of prime matter ; after, as before, the coming of Christ, philosophers had no other technical data at their disposal than the patiently accumulated results of the labours of the Greeks ; if they made these their starting-point it was because there was no other possible starting-point for them. The philosophy of Christians was indeed, and could be nothing else than, a continuation of Greek philosophy. The only thing we would add is, that if in their hands and by their efforts it continued, it was precisely because they were Christians.

Their religion brought with it no new philosophy, but it made them new men ; and the new birth effected in themselves was bound to bring about a renaissance of philosophy, which, to say the truth, was sadly in need of it. The thought of Plato and Aristotle, that of Democritus and Epicurus, the moral discipline of Stoicism, had produced their fruits. We will not say that they were dead ; Seneca, Epictetus, Marcus Aurelius, would rise and contradict us ; but they could bear only the same kind of fruit, and each time, during the first Christian centuries, the old tree seemed to become green again with new sap, it was because some new religious life was animating it. Often it was Christian life ; Plotinus himself, who refused Christianity, was, along with Origen, the pupil of Ammonius Saccas, and neither the latter nor his disciples were unaware of the existence of

Christianity. When the Christian philosophers put their hands to the work, the Gospel, in all probability, had already had something to do with the new hues in which pagan thought itself was appearing. It was discussed and refuted, but the pagan philosophers knew it as a force that counted and had to be reckoned with. And what then will it be when the Christians in their turn take up the old problems to infuse some new spirit into them ? Assuredly they will never look there for the secret of their religious life ; its source lies elsewhere ; but neither will they refuse to allow their religious life to bring any contribution to the discussion and help to prepare the upshot. The mediæval philosophers did nothing else, they simply continued the work begun in the second century and brought it nearer to perfection.

If philosophy, even Christian philosophy, is contingent with respect to Christianity, why, then, did it come into existence ? Why, in other words, did the Christian Middle Ages possess a philosophy ? To this it would be easy simply to answer that there was bound to be a philosophy as soon as there were philosophic Christians. There was nothing that forced them to philosophize, but neither was there anything to forbid them. But such a reply would be superficial. The truth is that in fact, if not in right, the formation of a Christian philosophy was inevitable, that it is still so to-day, and will so remain as long as there are Christians, and Christians who think. The inevitability does not flow from the essence of Christianity, which is a grace, but arises from the very nature of the recipient of the grace. This recipient is a nature, and nature is the proper object of philosophy. As soon as a Christian begins to reflect on the subject that carries grace, he becomes at once a philosopher.

If from all the varied analyses I have put before you I may venture to extract a conclusion from their respective conclusions, I should say that the essential result of Christian philosophy is a deeply considered affirmation of a reality and goodness intrinsic to nature, such as the Greeks, lacking knowledge of its source and end, only dimly foreshadowed. I know that this thesis sounds like a paradox, but the facts stare us in the face, and even a classical blunder is none the less a blunder. And is it not an evident blunder to take what the Fathers and mediæval doctors say of nature's corruption, as if they had said it of nature? That, however, is the blunder committed ; nor does it consist in any simple misapprehension of the intensity of a sentiment, but in the attribution of a sentiment which the Christian thinkers did not feel, and the denial of one, which, if they had not felt, they would have been excluded from Christianity as they themselves conceived it. Whence arose this error? No doubt it has many and very complex causes, but one, at least, is indubitable, namely, the changed view of the essence of Christianity originating in the Reformation and largely spread abroad and confirmed by Jansenism.

What in fact is untrue of mediæval Christianity is true of the Christianity of the Reformers. Neither the Jews, nor the Greeks, nor the Romans to whom the Gospel was preached, ever took this preaching as a negation of nature, even fallen nature, or as the corresponding negation of free will. In the first centuries of the Church, on the contrary, to be a Christian was essentially to hold a middle position between Mani who denied the goodness of nature, and Pelagius who denied its wounds, and therewith the need of grace to heal the wounds. St. Augustine himself, although the anti-Pelagian controversy made him the Doctor of Grace, might equally well be called the Doctor of Free Will, for, having begun by writing a *De libero arbitrio* before coming into contact with Pelagius, he judged it necessary to write a *De gratia et libero arbitrio* in the height of the Pelagian conflict. If you would have a *De servo arbitrio* you must look to Luther for it. For the first time, with the Reformation, there appeared this conception of a grace that saves a man without changing him, of a justice that redeems corrupted nature without restoring it, of a Christ who pardons the sinner for self-inflicted wounds but does not heal them. Now as time passed and mediæval Catholicism lost its influence, it began to be confused, by a singular illusion, with this very reformed Christianity which was its absolute negation. And from that moment the whole work of mediæval philosophy could not but appear as an empty pretence which, for the rest, could deceive nobody, or else, on the most indulgent hypothesis, as a naïve misunderstanding.

If in fact we suppose that the Middle Ages denied the stable persistence of nature beneath the wounds inflicted by sin, how could we possibly imagine that its philosophers would concern themselves seriously with the achievement of a physics, an ethics, and finally a metaphysics based on this physics and this ethics ? Since, *ex hypothesi*, all that the Greeks took as the proper object of rational speculation has henceforth ceased to exist, what basis is any longer left for philosophy ? It avails nothing to object that there seem in fact to have been philosophers in the Middle Ages ; they may have considered themselves to be such, but could not have been so in reality, their very Christianity barred the way to philosophy. This would certainly have been the case had their religious position been what it has been made out to be. Where there is no free will there can be no struggle against vices, no painful achievement of virtues, and therefore no place left for morals. If the natural world is altogether corrupted, who would waste time over Aristotle's physics ? As Luther said : you might as well choose Dung as a theme for rhetorical display.[3] In short, there never would have been any mediæval philosophy if the first Christians had understood their religion as Luther understood his : far from pushing forward the work of the Greek thinkers, they would, like Luther, have prohibited them to be read, they would have seen nothing in philosophy but a pestilential scourge, an instrument of righteous vengeance in the hands of an angry God.

The sole difficulty in the way of this thesis, but a very real one, is that the position of the mediæval Christians was exactly the reverse of all that. Luther knew it well, and that is why he never forgave them. According to him, all the philosophers and theologians of the Middle Ages were pagans, they believed that nature subsists in despite of original sin, and that once restored by grace it becomes capable of action, progress, and merit. St. Bonaventure and St. Thomas would have been slightly surprised to find themselves taken for pagans, but they would have readily admitted all the rest ; and when this is once understood, the existence of mediæval philosophy appears as natural as its condemnation by the originator of the Reform. The mediævals needed a philosophy for all the reasons that forbade it to Luther. Ardent defenders as they were of grace, they defended not less the nature that God made, doubly precious since God Himself died to save it. It is not they who will let us forget, along with his misery, the greatness of man. The more he is aware of its dignity the higher feeling he will have for its glory, and indeed, if he is to be able to refer it to God at all, he must perforce know something about it. " To possess what one knows nothing at all about, what glory can there be in that ? " [4] It is St. Bernard who asks, and he, we may be sure, was not the man to err by excessive indulgence for nature. Now if the work of creation is not abolished, nothing could be of greater use to these theologians than to bend over it anxiously and interrogate it concerning its Author ;

like careful physicians, they will seek its original form beneath all its disfiguring ills in order to learn the true remedy. And how should we apply these remedies unless we know something of the anatomy of the soul ? And how shall we know the soul without the body, or the body without the whole universe of which it forms a part ? Assuredly it is not necessary to know all these things in order to preach salvation nor in order to receive it, but if it comes to the constitution of a " saving science," that is to say a theology, how, if it would show us how to save the world, should it repudiate all interest in the world to be saved ?

Now the science of that world to which the Gospel would bring salvation is not to be found in the Gospel ; if the Old Testament made no pretence to supply it, neither did the New. The mediæval thinkers therefore found themselves faced with a double responsibility, that, namely, of maintaining on the one hand a philosophy of nature while at the same time building up a theology of supernature, and of integrating the first with the second in a coherent system. To suppose *a priori* that the work was ruinous to either is once more to forget the very principle that inspired the whole enterprise. How could they have thought that the science of grace could obstruct that of nature, when grace was there only to perfect nature, having first restored it ? But conversely, since nature is there but as the subject and point of application of grace, how could the science of the one, if properly conducted, conflict with the science of the other ? The only thing to do was to attempt the work ; to attempt it, it was at once indispensable to start from the only known philosophy, that is to say, Greek philosophy, and impossible to rest content with it.

The Fathers of the Church and the philosophers of the Middle Ages set out therefore from Greek philosophy, from Plato and Aristotle. Now in setting out from a point you move away from it, but you can also take something along with you. While the man Plato and the man Aristotle stand motionless in the historical past, Platonism and Aristotelianism will live with a new life and collaborate in a work all unforeseen by their authors. It was due to them that the Middle Ages was able to achieve a philosophy. They supplied its idea—*perfectum opus rationis*—they pointed out some of its master problems and, along with them, the rational principles that command their solution and even some of the necessary technique. The debt of the Middle Ages to the Greeks was immense, and is fully recognized, but the debt of Hellenism to the Middle Ages is as great, and nothing is less appreciated ; for even from mediæval religion Greek philosophy had something to learn. Christianity communicated to it some share in its own vitality and enabled it to enter on a new career.

Hence the peculiar character of mediæval philosophy which sometimes causes such surprise. The more one reads the mediæval commentaries on Aristotle the more one is convinced that their authors knew exactly what they were about. St. Thomas can write his pages on the *Metaphysics* without once saying that Aristotle taught the doctrine of creation, nor yet that he denied it. He knew very well that Aristotle does not teach it, but what interests him is to see and make clear that, although Aristotle did not grasp this capital truth, his principles, while remaining precisely what they are, are perfectly capable of bearing its weight. It is true that for this purpose they have to be deepened in a way that Aristotle did not foresee ; but to deepen them thus is merely to bring them into closer conformity with their own essence since it is to make them truer. In this sense it may well be said that it was not at all as historians that the mediævals interested themselves in Greek philosophy. The Aristotle of history carries with him all his

failures as well as all his successes ; he is less made up of the
truths his principles will carry than of the truths he saw in
them himself ; if history takes him in all his greatness, it
takes him also with all his limitations. And Plato is in the
same case. But what, on the contrary, the mediævals seek
in the Greeks is all that makes them true and nothing else ;
and where they are not yet wholly true the means to make
them so. The work of elaboration is delicate, sometimes
subtle, but here St. Bonaventure, St. Thomas, and Duns
Scotus always show the greatest steadiness. There is nothing
artificial about their methods, they never force principles at
the risk of destroying them, but they extend or prolong
them as far as is necessary to make them yield up all the
truth that is in them. The age of commentators, as some are
pleased to call it, was above all an age of commentator-
philosophers. Therefore they are not to be blamed for having
at once the name of Aristotle constantly on their lips, and
for constantly making him say what he did not say. They
would philosophize, not play the historian, and unless it be
demanded, which God forbid, that philosophy should be
exclusively a field for historians of philosophy, not even
history itself can find any fault with them.

What, then, remains in the attitude of the mediæval
masters to offend or embarrass us ? Probably nothing save
their docile modesty in instructing themselves in philosophy before setting out to further its progress. If that is a
crime then they are certainly guilty, and there is no remedy
left. They believed that philosophy could not possibly be
the work of a single man, no matter what his genius might
be, but that it progresses, like science, slowly, as the result
of the patient collaboration of generations, each leaning on
its predecessors in order to surpass their achievement.
" We are like dwarfs," said Bernard of Chartres, " seated
on the shoulders of giants. We see more things than the

Ancients and things more distant, but it is due neither to the sharpness of our sight nor the greatness of our stature, it is simply because they have lent us their own." This proud modesty we have lost. Many of our contemporaries prefer to remain on the ground ; they put their pride in seeing nothing at all unless they can see it by their own efforts, and console themselves for their petty stature by recollecting their advanced age. It is a sad old age that loses all its memories. If it were true, as some have said, that St. Thomas was a child and Descartes a man, we, for our part, must be very near decrepitude. Let us rather hope that truth in its eternal youth shall keep our minds always young and fresh, full of hope for the future and of force to enter there.

NOTES

THE *Gifford Lectures* are a public course, and not addressed solely to an audience of philosophers. There was therefore a double motive for not overloading them with technical details which might be difficult to follow and of little interest to any but specialists. I have printed them substantially as they are delivered. In adding the following notes I do not propose to alter their character ; they remain a course of lectures with all the advantages and inconveniences that go with that form. It seemed useful, however, to indicate to readers, whose attitude might be very different from that of hearers, a part at least of the historical substructure on which these lectures rest, and, in default of a complete justification of the theses advanced, to show at least in what sense, and in what type of text, this justification may be found.

[It has not been thought necessary to reproduce here all the original notes of the Author. Considerations of space have dictated a selection, and this has been made from among those which bear more directly on the main theses of the whole work, and those which may prove to be of fairly wide philosophical interest. It is not possible to hope that everyone will agree with the precise selection made.—TRS.]

NOTES TO CHAPTER I

[1] M. SCHELER, *Krieg und Aufbau*, Leipzig, 1916.

[2] P. MANDONNET, O.P., in *Bulletin Thomiste*, 1924, pp. 132–136, and M.-D. CHENU, O.P., in *Bulletin Thomiste*, January, 1928, p. 244.

[3] E. BRÉHIER, *Histoire de la philosophie*, Vol. I : *L'antiquité et le moyen âge*, Paris, Alcan, 1927.

[4] See C. SIERP in Kleutgen, *La philosophie scholastique*, Vol. I, p. viii.

[5] E. BRÉHIER, *Histoire de la philosophie*, Vol. I, p. 494.

[6] E. GILSON, *Etudes sur la rôle de la pensée médiévale dans la formation du système cartésien*, Paris, J. Vrin, 1930. Cf. *La liberté chez Descartes et la théologie*, Paris, Alcan, 1913.

[7] MALEBRANCHE, *Recherche de la vérité par la raison naturelle*, Preface.

[8] W. P. Montague, *Belief Unbound*, New-Haven, Yale University Press, 1930, pp. 9–10 and p. 97.

[9] Lessing, *Ueber die Erziehung des Menschengeschlechtes*, cited by M. Guéroult, *L'evolution et la structure de la doctrine de la science chez Fichte*, Strasbourg, 1930, Vol. I, p. 15.

NOTES TO CHAPTER II

[1] Justin, *Dialogue with Trypho*, II, 6.

[2] *Op. cit.*, VII.

[3] *Op. cit.*, II, 6. Justin is not alone among the early Christians in claiming the title of philosopher. Cf. Athenagoras, *Legatio pro Christianis*, VII ; Irenaeus, *Adv. Haereses*, I, 3, 6, and II, 27, 2 ; Clement of Alexandria, *Stromates*, I, 1, and I, 18, and VI, 15.

[4] Justin, *2nd Apology*, XIII.

[5] *Op. cit.*, X and XIII. Cf. St. Ambrose' formula : " Omne verum a quocumque dicatur, a Spirito Sancto est."

[6] St. Paul, Rom. vii. 14–25.

[7] Tatian, *Adversus Graecos*, XXV.

[8] Hermias, *Gentilium philosophorum irrisio*, II–X. Note that Hermias assumes the title of *philosopher*.

[9] Lactantius, *Institutiones*, VII, 7, 7. Cf. VII, 7, 4.

[10] E. Bréhier (*Histoire de la philosophie*, Vol. I, p. 493) maintains that early Christianity was not in the least speculative, but only a movement of social mutual aid in the spiritual and material orders. However, there is no need to consider what it had become " after many centuries " to doubt the truth of this. According to the *Dialogue with Trypho* the conversion of Justin proceeded confessedly on philosophical lines. But it is especially interesting to observe that, from the end of the second century, Christians are being rebuked for pretending to philosophical knowledge that had remained hidden from the Greeks. These illiterates seem to suppose themselves wiser than Plato and Aristotle. Not only then did these ancient Christian communities claim to be in possession of a new interpretation of the world, but the fact was cast in their teeth. Notice the appeal to the rights of reason in Minucius Felix, *Octavius* (end of second century) : " Now seeing that he cannot endure (*sc.*, Cecilius the pagan) to see illiterate and poor ignorant people as he calls us, discuss divine things, he must be reminded that all men are born rational, without distinction of age, quality or sex, and owe wisdom, not to fortune, but to nature ; that even philosophers and other celebrated inventors of arts and sciences have been regarded as belonging to the dregs of the people before their wit appeared in their works ; so true is it that the rich, idolaters of worldly treasure, have more respect to gold than to heaven,

and that it was precisely poor people like ourselves who discovered wisdom and manifested it to others." That is a sort of claim to democracy in philosophy and appeal to the universality of reason, and it shows that in those ancient Christian communities the speculative spirit was intensely alive. The pretension, insupportable to the philosophers, that a humble *vetula* knew more about the universe than Plato and Aristotle, was to be expressed anew by St. Francis of Assisi and his disciples ; it is an integral part of the Christian tradition. The title of the lost treatise of Hyppolytus (died about 236–237) : *Against the Greeks and Plato : or on the Universe*, seems also to show that this Roman bishop took an interest in speculation.

11 Maine de Biran, *Sa vie et ses pensées*, E. Naville, Paris, 1857, p. 405.

12 St. Anselm, *De fide Trinitatis*, Praef.

13 It serves no purpose to object that a reason which allows itself to be taken in tow by faith is voluntarily blinded, and that it is only too easy to imagine that one has proved what one believes. If the believer's demonstrations make no impression on the unbeliever, he will not hold himself authorized to appeal to a faith that his opponent does not accept. All that the believer can do, as far as regards himself, is to make sure that he is not the victim of illusion and to criticize himself severely. As regards his opponent, he will not fail to wish him the grace of faith, and all the enlightenment of the intelligence that accompanies it. This point cannot honestly be overlooked. The problem of Christian philosophy is not confined to the question of its constitution, but embraces also that of its understanding. The contemporary paradox of a Christian philosophy evidently true for its defenders, and of no value in the eyes of their opponents, does not necessarily imply that its defenders are blinded by their faith ; it may perhaps be explained by the fact that absence of the light of faith in the opponent leaves truth opaque where it might be transparent. This in no way authorizes the Christian philosopher to argue in the name of faith, but rather to redouble his purely rational efforts until the light thus gained leads other minds also to turn to its source and draw the same enlightenment.

14 M.-D. Chenu, O.P., in *Bulletin Thomiste*, January, 1928, p. 244. Cf. the well-balanced remarks of J. Maritain [*De la sagesse augustinienne*, in *Revue de philosophie* (XXX), 1930, pp. 739–741] to which we wholeheartedly subscribe.

15 See Père Synave's remarkable study, *La révélation des vérités divines naturelles d'après saint Thomas d'Aquin*, in *Mélanges Mandonnet*, Paris, J. Vrin, 1930, Vol. I, pp. 327–365.

16 St. Thomas Aquinas, *De Veritate*, XIV, 10, Resp.

NOTES TO CHAPTER III

[1] CONDORCET, *Tableau historique des progrès de l'esprit humain*, Paris, G. Steinheil, 1900, p. 87.

[2] P. DECHARME, *La critique des traditions réligieuses chez les Grecs*, Paris, 1904, p. 217.

[3] P. DECHARME, *op. cit.*, pp. 233–234.

[4] M.-D. ROLAND-GOSSELIN, *Aristote*, Paris, 1928, p. 97.

[5] On the " tendencies towards polytheism among the ancient Hebrews " see A. LOD's *Israel*, Paris, Renaissance du Livre, 1930, p. 292—As to the supposed monotheism of the Greeks (G. MURRAY, *Five stages of Greek religion*, New York, 1925, p. 92), we may say shortly that it never existed. On this point the Christians were often too generous to the Greeks. True, they had a certain interest in the matter. They were accused of impiety for refusing sacrifice to the gods of the Roman pantheon ; the Apologists defended themselves by trying to prove that Plato was on their side, that he too admitted no more than one divine principle. However, even here, with life and death at stake, the Apologists did not omit to note the difference between their own position and that of the Greeks. One of them points out that Moses speaks of Being, Plato of " that which is " : ʽΟ μὲν γὰρ Μωϋσης, ὁ ὤν ἔφη ὁ δέ Πλάτων, τὸ ὄν. *Cohort ad Graecos*, Cap. XXII. This work, which has been falsely attributed to Justin, is dated by A. PUECH about 260–300 : *Litt. grecque chrétienne*, Vol. II, p. 216. Similarly Athenagoras declares : What the Greeks call the divine principle : ἔ τὸ Θεῖον, we call God : τον Θεόν ; where they speak of the divine : περὶ τοῦ Θείου, we say that there is a God : ἔνα Θεὸν. ATHENAGORAS, *Legatio pro Christianis*, Cap. VIII.

We ought to add, however, that this interpretation has against it the authority of A. E. TAYLOR, *Platonism* (G. Harrap, London, p. 103). He has no doubt of Plato's monotheism because, in moments of religious feeling, he never speaks of " the gods " but of God. That, no doubt, is a fact, but probably not a decisive one. A polytheist may speak of " *the* god," but a monotheist can never speak of " the gods." But we are thus opportunely reminded that a strong tendency to monotheism is discernible in Plato even if he never quite attains it, and we may say as much of Aristotle.

[6] The famous text of the *Republic* (509 b) which places the good beyond essence, is enough to prove that Plato, even had he identified God with the good, would, on that very account, have refused to identify him with being, and still more so with infinite being ; for to put the good above being is to subject being to a limiting determination. An excellent interpreter of Plotinus makes the following very just remarks : " But the unity of measure is necessarily transcendent to the things measured, the things it

serves to evaluate and fix. It is probably in this sense that we must understand the famous text of the *Republic* (509 b) so often cited by Plotinus : ' The good is beyond essence, surpasses it in dignity and power ' ; at any rate that is the sense in which Plotinus understands it. No essence can be what it is save thanks to the measure which exactly fixes its limits ; and here this is called the Good." E. Bréhier, *La philosophie de Plotin*, Paris, Boivin, 1928, p. 138.

⁷ Of course there might be interminable discussions on this point. One of the best apologies for Plato known to us—and we cite it against our own thesis—is that of Père M.-J. Lagrange, O.P. (*Platon théologien*, in *Revue thomiste*, 1926, pp. 189–218). According to this excellent exegete the Idea of the Good referred to in the Republic (VI, 509 b), and of which Plato says that it gives things not only their intelligibility, but their very being, is identical with the Demiurge of the *Timaeus*. Père Lagrange recognizes that " Plato does not expressly say so," but adds that, " he makes it clearly understood " (*art. cit.*, p. 196). It is somewhat surprising, in the first place, that the philosopher did not take the trouble to say expressly in one word that the Demiurge is the Idea of the Good, seeing that if this had been in his mind the whole meaning of his philosophy would have been transfigured. If " nowhere, in any text, did Plato co-ordinate the Idea of the Good and the artificer or Demiurge " (*art. cit.*, p. 197) the reason probably is that he did not co-ordinate them in his thought. And how could he ? For if the Demiurge is the Idea of the Good, why should he have to work with his eyes fixed upon the Ideas—he on whom they would all depend ? Even if we grant that he is the Idea of the Good and bestows being on things, we still have to ask what Plato means by *being :* is it existence, as in Christian thought, or a mere intelligibility which saves the being in becoming from pure non-being ? The latter seems to be the true Platonic view. The sensible world of the *Timaeus* certainly receives its intelligibility, but that does not mean that it receives its existence. Thus, before we can accept the identification proposed by P. Lagrange, we have three fundamental difficulties to overcome : (1) We do not know whether Plato made the identification and we do know that he did not say that he made it ; (2) we know that the Good, even if it be the supreme god, is but the highest among the gods (*art. cit.*, p. 204) ; (3) we see, in consequence, that since he is not Being, he cannot give being, so that in any event we are in a system of ideas that differs widely from the Christian.—Against this identification of the Platonic god with the Ideas see P. E. More, *The Religion of Plato*, Princeton University Press, pp. 119–120, and P. Shorey, *The Unity of Plato's Thought*, Decennial Publications VI, University of Chicago Press, 1903, p. 65.

But the most complete plea in favour of the identification of God and being in Plato is that of A. Dies, *Autour de Platon*, Paris, Beauchesne, 1927, Vol. II, p. 566 *et seq*. If this excellent Hellenist is right in what he says on page 556, if Fénelon and Malebranche are to be taken as legitimate commentaries on Plato, then we may as well admit, without further ado, that the central thesis of these lectures is false. However, until further proof is forthcoming, it seems that Dies has looked with Christian eyes at formulæ which are very far from being Christian, and that even if we leave his historical analysis and conclusions exactly as they stand they nevertheless mean less than he imagines. But his analysis of the Platonic texts is masterly and quite indispensable for anyone who would see the objection put with all its force. Dies' conclusions will be found *op. cit.*, pp. 556 and 561.

On the question consult also Eust. Ugarte De Ercilla, S.J., *Anepifania del Platonismo*, Barcelona, 1929 (discusses Dies' thesis, pp. 278–286)—A. E. Taylor, *A Commentary on Plato's Timaeus*, Oxford, Clarendon Press, 1928, pp. 80–82—R. Mugnier, *Le sens du mot Θειος chez Platon*, Paris, J. Vrin, 1930—J. Baudry, *Le problème de l'origine et de l'eternité du monde dans la philosophie grec de Platon à l'ère chrétienne*, Paris, Les Belles-Lettres, 1931.

[8] Aristotle's god, conceived as an individual "sovereignly real," has been made the subject of thorough studies by O. Hamelin, *Le système d'Aristote*, Paris, Alcan, 1920, pp. 404–405—L. Robin, *La pensée grecque et les origines de l'esprit scientifique*, Paris, 1923, pp. 368–369—J. Chevalier, *La notion du nécessaire chez Aristote et ses prédécesseurs*, Paris, Alcan, 1915, p. 144.

L. Robin applies the term " supernatural " to the Aristotelian order of being as being, but this cannot be unreservedly allowed. The meaning of the word " supernatural " depends on the meaning given to the word " nature." This latter, in Aristotle, means all that is composed of matter and form ; so that anything immaterial is supernatural, for instance all the " separate substances." For a Christian, an immaterial substance may very well belong to the order of natures ; the angels, for instance, are natural intellectual substances : *substantiae perfectae intellectuales in natura intellectuali* (St. Thomas Aquinas, *Sum. theol.*, I, 51, 1, Resp.). A substance does not cease to belong to the natural order merely because it is not a composite of matter and form like the angels, it must be something not composed of essence and existence, and that amounts to saying that it must be God. Thus we return to the Christian idea of being as being : *Ego sum qui sum*, the only being in whom essence and existence are identical. It remains to be seen whether Aristotle achieved this conception.

The strongest text in favour of the affirmative is that referred to by M. L. ROBIN, *Metaph.*, E, 1, 1026 a, 27–32. Aristotle is defining the object of theology : " If there is no substance other than those consisting of a nature Physics will be the first science and, as first, universal ; but if there is an immovable substance, the science of this must be prior and must be first philosophy, and thus universal because it is first. And it will belong to this to consider being *qua* being, both what it is and the attributes that appertain to it *qua* being." At first sight, nothing could be clearer ; but what does it mean from Aristotle's standpoint? No doubt we are here concerned with the First Mover, but not with him alone. The problem is this : is there any place above Physics for another science which would be theology? We ask, according to the usual method, whether there is any specific object to be assigned to such a science. Let it be admitted that Physics deals with " natures " composed of matter and form : then there will be room for a theology if there exist substances superior to these natures; superior, that is, as being immaterial and as being their cause. Now there are several, and Aristotle indeed mentions them in the same chapter : " All the causes must be eternal, but especially those that are separable and immovable, for they are the causes of what is divine in visible things " (*loc. cit.*, 1026 a, 16–18). Thus, then, the substance which is the object of metaphysics is not the substance of *a* being, but that of a plurality of unmoved movers. These are the οὐσία ἀκίνητος ; and if the phrase " being as being " is more eminently applicable to the first mover inasmuch as he is first, it is not applicable to him exclusively.

If anyone feels a difficulty about taking the term οὐσία in this sense, it should be enough to remember that in the same chapter Aristotle uses it to designate the still more numerous class of physical beings : ἡ φυσικὴ ἐπιστήμη τυγχάνει οὖσα περὶ γένος τι τοῦ ὄντος, περὶ γὰρ τὴν τοιαύτην ἐστὶν οὐσίαν ἐν ᾗ ἡ ἀρχὴ τῆς κινήσεως καὶ στάσεως ἐν αὐτῇ (*Metaph.*, E, 1, 1025 a 18–21). Class then is opposed to class, not a class to a being. Many other expressions show the same thing. The τὸ δ'ὡς ἀληθὲς ὄν (*Metaph.*, E, 4, 1072 b, 18. Cf. K, 8, 1065 a 21) is opposed to accidental being, which forms a class, and it is itself a class, that of beings *par excellence* : τῶν κυρίως (*Metaph.*, E, 4, 1027 b 31). Thus, for Aristotle, the proper object of natural theology is not the Christian God, but the divine order : τὸ Θεῖον (1026 a 20), the genus of metaphysical beings : ἐν τῇ τοιαύτῃ φύσει ὑπάρχει, χαὶ τὴν τιμιωτάτην δεῖ περὶ τὸ τιμιώτατον γένος εἶναι (1026 a 20–22).

⁹ Of course we do not maintain that the text of Exodus is a revealed metaphysical definition of God ; but if there is no metaphysic *in* Exodus there is nevertheless a metaphysic *of* Exodus ; and we shall see it developed in due course by the

Fathers of the Church, whose indications on this point the mediæval philosophers merely follow up and exploit. See the texts of St. EPHREM OF NISIBIS in Rouet de Journel, *Enchiridion patristicum*, 4th Edn., Herder, 1922, 729, p. 254—St. GREGORY OF NAZIANZEN, *op. cit.*, 993, p. 370, and 1015, p. 379—St. GREGORY OF NYSSA, *op. cit.*, 1046, p. 393—St. CYRIL OF ALEXANDRIA, *op. cit.*, 2098, pp. 657-658.—It was this text of Exodus that was decisive for Hilary of Poitiers in the midst of his doubts ; see the account at the beginning of his *De Trinitate* (about 356) : " Haec igitur, multaque alia ejusmodi cum animo reputans, incidi in eos libros, quos a Moyse atque a prophetis scriptos esse Hebreorum religio tradebat : in quibus ipso creatore Deo testante de se, haec ita continebantur : *Ego sum qui sum* (Exod. iii. 14) ; et rursum : *Haec dices filiis Israel : misit me ad vos is qui est (ibid.).* Admiratus sum plane tam absolutam de Deo significationem, quae naturae divinae incomprehensibilem cognitionem aptissimo ad intelligentiam humanam sermone loqueretur. Non enim aliud proprium magis Deo, quam esse, intelligitur ; quia id ipsum quod est, neque desinentis est aliquando, neque coepti." *De Trinitate*, I, 5. Hence he deduces the eternity, infinity, perfection and incomprehensibility of God.

As to the primitive meaning of the Mosaic text, M. A. Lods considers it to mean simply that Jähve " is what he is, the Being undefinable by man." This is already an explanation which " does not lack grandeur, but is apparently too theological, too lacking in spontaneity, to express the original idea of the Madianite god " (A. LODS, *Israel*, Paris, 1930, pp. 373-374). This is a question for Hebraists. But M. Lods' own analysis and translation of the text make it very difficult to deny that Jahve really wishes to reveal His name. If we say that " *Ego sum qui sum* " amounts to a refusal to reveal the name, what are we to make of the rest of the text, which M. Lods thus translates : " Then he added ; thus shalt thou answer the children of Israel : He who sends me is I AM." Here, certainly, I AM is Jahve's name, as we see once more in the next verse where Jahve is simply substituted for I AM : " And God said to Moses again, ' answer the children of Israel ; it is Jahve, the Lord God of your fathers . . . who has sent me.' " Exod. iii. 15. It seems, then, that the patristic-mediæval philosophy is a correct development of this text.

[10] St. THOMAS AQUINAS, *Sum. theol.*, I, 13, 11. On the Platonic primacy of good over being see *De Malo*, I, 2 Resp.

[11] St. BONAVENTURE, *Itinerarium mentis in Deum*, V, 4.

[12] In this sense we may say that the term *essentia* is proper to God alone, and that all else belongs to the category of *substantiae* : " Nefas est autem dicere ut subsistat et subsit Deus

bonitati suae, atque illa bonitas non substantia sit vel potius essentia, neque ipse Deus sit bonitas sua, sed in illo sit tanquam in subjecto ; unde manifestum est Deus abusive substantiam vocari, ut nomine usitatiore intellegatur essentia, quod vere ac proprie dicitur ; ita ut fortasse solum Deum dici oporteat essentiam. Est enim vere solus, quia incommutabilis est, idque nomen suum famulo suo Moysi enuntiavit, cum ait : *Ego sum qui sum*, et : *Dices ad eos :* qui est, *misit me ad vos.*" St. Augus-tine, *De Trinitate*, VII, 5, 10.

[13] Duns Scotus, *Opus Oxoniense*, lib. I, dist. 2, qu. 1 and 2, sect. 2, art. 2, n. 2.

[14] Aristotle, *Physics*, III, 6, 206 b, 23.

[15] St. Thomas Aquinas, *Compendium theologiae*, Cap. XX.

[16] *Op. cit.*, I, Cap. XX.

[17] St. Anselm, *Proslogion*, Cap. IV.

[18] St. Bonaventure, *Itinerarium mentis in Deum*, Cap. V, n. 3.

[19] Duns Scotus, *Opus Oxoniense*, lib. I, dist. 2, qu. 1 and 2, sect. 2, art. 2, n. 2. Duns Scotus denies, in a certain sense, that the existence of God is an immediate or self-evident truth *per se notum;* but he adds that St. Anselm himself never thought so, since he demonstrates it. If Duns Scotus found a place in his system for St. Anselm's argument it was only by profoundly modifying its meaning, and bringing it into line with his own metaphysic ; he did not accept it in the form in which Anselm left it. Duns Scotus' proofs of God's existence should occupy a prominent place in any history of Christian philosophy, for they are immediately based on the idea of being and its essential properties, causality and eminence.

[20] Gaunilo, *Liber pro insipiente*, 7.

NOTES TO CHAPTER IV

[1] The Thomist distinction between essence and existence expresses the radical contingence of existence in all that is not God. Substantially contemporaneous with the very beginnings of Christian thought, it was inevitable that this fundamental intuition should find at length its appropriate technical formula. It appears for the first time clearly in William of Auvergne : " Quoniam autem ens potentiale est non ens per essentiam, tunc ipsum est ejus esse quod non est ei per essentiam duo sunt revera, et alterum accidit alteri, nec cadit in rationem nec quidditatem ipsius. Ens igitur secundum nunc modum com-positum est et resolubile in suam possibilitatem et suum esse." (Cited by M.-D. Roland-Gosselin, *Le De Ente et Essentia de saint Thomas d'Aquin*, Paris, J. Vrin, 1926, p. 161 : a fundamental work for the study of the question and its history.) Since the

idea expressed by the distinction is closely bound up with Christianity, which itself deepens the Jewish tradition, there is nothing surprising in the fact that St. Thomas, in spite of all his efforts, failed to find the distinction in Aristotle : see some excellent pages by A. FOREST, *La structure métaphysique du concret selon saint Thomas d'Aquin*, Paris, J. Vrin, 1931, Chap. V, pp. 133–147. In an eternal and uncreated world, such as that of Greek philosophy, an essence is eternally realized and is inconceivable save as realized. It is of some moment, then, to understand that the real distinction of essence and existence, although clearly formulated only at the beginning of the thirteenth century, was always virtually present after the first verse of Genesis. In any created being, however simple, were it a separate and subsistent form like an angel, the essence does not contain in itself the sufficient reason of its existence ; it must have received it ; therefore essence and existence are really distinct. This radical composition, inherent in the status of a creature, suffices to distinguish every contingent being from Being Itself ; cf. ST. THOMAS, *Quodlibet*, II, art. 4, ad. 1 : " Sed quia non est suum esse, accidit ei aliquid praeter rationem speciei, scilicet ipsum esse . . ." The expression *accidit*, which might lead us to confuse St. Thomas' thought with Avicenna's, should be taken in the sense that St. Thomas himself gives it. It does not mean that the essence is something which, without the existence, would not exist ; for what, then, would be this *thing which would not exist ?* It means that the actual existence of the realized possible belongs to this possible only in virtue of the creative action conferring existence. A. Forest puts his finger on the knot of the question and shows thereby the true significance of the Thomist solution : " With St. Thomas the essence does not designate, in the manner of Avicenna, a nature that could be seized as such, independently of its relation to existence ; what here divides the two philosophers is the Greek doctrine of necessity on the one side, and the Christian liberty on the other " (*op. cit.*, p. 154, cf. p. 161). In other words, the real composition of essence and existence does not imply that God could make essences subsist that would not exist, or withdraw existence from created beings and leave their essences—such hypotheses would be absurd—but that He is able not to create them and could annihilate them. Thus " in a general way the distinction of essence and existence harmonizes with the doctrine of creation " (*op. cit.*, p. 182). This is the exact truth of the matter, and I see hardly anything to add to Forest's exposition, except perhaps to refer the whole to Exodus.

[2] On the meaning of the plural *Elohim*, see A. LODS, *Israel*, pp. 290–293. Hebraists are not wholly in agreement as to the primitive character of Jewish monotheism, nor on the exact

significance of the concept of creation in the Biblical narrative (see M.-J. LAGRANGE, *Etudes sur les religions sémitiques*, 2nd Edn., Paris, Gabalda, 1905 ; P. W. SCHMIDT, *Der Ursprung der Gottesidee*, Münster i. West., Aschendorff, 2nd Edn., 1926). It is certain in any case that the idea of creation profoundly influenced the thought of Israel from the sixth century onwards, and we shall give the proofs later on in this study, especially in connection with the idea of providence. The Christians found it not merely affirmed but defined in the second *Book of Maccabees*, 7, 28. Thus they developed a religious datum, the interpretation of which had long been fixed when they received it. See the use made of the text of Maccabees in ORIGEN, *In Joan Comm.*, I, 17, 103. The passage from the order of revelation to the order of knowledge may be observed *in vivo*, so to speak, in the following moving passage from St. Augustine : " Audiam et intelligam, quomodo *in principio* fecisti *coelum et terram*. Scripsit hoc Moyses, scripsit et abiit, transiit hinc ad te. Neque nunc ante me est : nam si esset, tenerem et rogarem eum, et per te obsecrarem, ut mihi ista panderet ; . . . Sed unde scirem, an verum diceret ? Quod si et hoc scirem, num ab illo scirem ? Intus utique mihi, intus in domicilio cogitationis ; nec Hebraea, nec Graeca, nec Latina, nec Barbara vox, sed veritas, sine oris et linguae organis, sine strepitu syllabarum diceret. Verum dicit : et ego statim certus confidenter illi homini tuo dicerem : " verum dicis." *Confess.*, XI, 3, 5. To the truth promulgated from without by revelation, responded the light of reason from within. Faith, *ex auditu*, at once awoke an answering chord.

[3] On this point St. Bonaventure does not hesitate : " Nisi tu sentias, quod totalitas rerum ab ipsa (essentia divina) procedit, non sentis de Deo piissime. Plato commendavit animam suam *factori*, sed Petrus commendavit animam suam *Creatori* " : *In Hexaem*, IX, 24.

Against St. Bonaventure is the opinion of A. E. TAYLOR (*Plato*, pp. 442–444), who maintains that " the Demiurge is a creator in the full sense of the term." His meaning, in the pages referred to, seems to be that Plato's world is not eternal like Aristotle's, but began with time like the Christian. Taylor does not tell us whether he would go so far as to credit Plato with the Christian conception of creation, *i.e.*, as a gift of being by Being. In all probability he would, because the Demiurge, in his view, does not fashion pre-existing matter ; what we call matter amounts to mere non-being in Plato's eyes (A. E. TAYLOR, *A Commentary on Plato's Timaeus*, Oxford, 1928, pp. 79 and 493). Jowett constantly uses the word " creation " in another sense in his translation of the *Timaeus*, but when he says, for instance, that " the elements move in a disorderly fashion before the work of creation begins," he shows very clearly that this is a pseudo-creation ; the elements precede it.—According to P. E. MORE on the contrary : " Creation could not be for a Greek philosopher, as it was to be for the Christians, an evocation of something out of nothing by the mere word *fiat :* ' creation,' indeed, in the sense here taken, is rather a misnomer for what is more properly an act of fashioning or shaping. To Plato, the thought of a creator and a thing created implied necessarily the presence of a substance out of which the object is created " (P. E. MORE, *The Religion of Plato,* p. 203).—It must be admitted that it is rather an excessive simplification of Plato's thought to attribute to him, without further qualification, an uncreated matter, seeing that it is not so much as mentioned in his works. However, it is difficult to explain the ordering activity of the Demiurge without admitting that what he fashions on the model of the Ideas, is at any rate something—whatever this something may be. Whence comes this element other than the Idea ? Plato nowhere says that the Demiurge creates it nor even that he concreates it with the form. Whether it be given prior to the formative activity—as Plato says, indeed, but which need not be taken literally, especially in a myth—or whether it is given contemporaneously with this activity, it is at least something given. It seems difficult, then, to avoid the conclusion that the Platonic universe contains an element not depending on the action of the Demiurge. Even if we put out of account the relation in which the Demiurge stands to the Ideas, his activity is formative rather than creative. See the very firm conclusions of A. RIVAUD, *Timée* (in *Platon, Oeuvres complètes,* Vol. X).—So great was Plato's influence that Philo the Jew, who should have been the first to develop a philosophy of creation *ex nihilo,* never even conceived the idea. On this point see the penetrating observations of E. BRÉHIER, *Les idées philosophiques et religieuses de Philon d'Alexandrie,* 2nd Edn., Paris, J. Vrin, 1925, pp. 78–82. It seems, therefore, that the Jewish religious tradition brought forth no philosophic fruit until it was grafted on to the Christian stock. The first Christian thinkers quite understood what kept them apart from Plato on this point. One of them, starting from Exodus, defines God as Being ; 'Εγω ειμι ὁ ὢν, and goes on to observe that the artificer of the *Timaeus* is not the Creator of the Bible, because he needs something on which to exercise his activity : see *Cohortatio ad Graecos,* XXI–XXII. The same reserve as concerns Plato is found in THEOPHILUS OF ANTIOCH, *Ad Autolycum,* II, 4. Irenaeus again contests the Platonic thesis, but as he found it, deformed, in the Gnostics : *Adversus Haereses,* II, 1–3.

⁴ ST. THOMAS AQUINAS, *In Phys.,* lib. VIII, lect. 2, n. 5— The interpretation of this and similar texts in St. Thomas

cannot be dealt with in a lecture (see R. JOLIVET, *Essai sur les rapports entre la pensée grecque et la pensée chrétienne*, Paris, J. Vrin, 1931, p. 54 *et seq.*). Surprise is sometimes expressed that St. Thomas should attribute the idea of creation to Plato and Aristotle. But in fact he did nothing of the sort.

As far as regards Plato, the case admits of no doubt. St. Thomas very clearly contrasts the Platonic plurality of universal principles (God, matter, ideas) with the Aristotelian and Christian unique principle ; and very well perceives the difference between the *informative* action of the Platonic Ideas and the creative action of the Christian God : *Super lib. de causis*, XVIII, at the end. He notes, furthermore, that according to certain interpreters, Plato considered matter as uncreated, wherefore he admitted no idea of matter, nor of the individuals that depend on it : *Sum. theol.*, I, 15, 3 ad 4. And of course it goes without saying that he understands that Aristotle criticized Plato for considering the Ideas as subsisting separately : *Sum. theol.*, I, 15, 1 ad 1. In such circumstances it is impossible to regard St. Thomas as overlooking the fact that Plato's world contains being that does not derive from God.

Nor is he deceived in the case of Aristotle. Misunderstanding has arisen here for want of observing that St. Thomas' source is St. Augustine, *De civ. Dei*, VIII, 4 : " Fortassis enim qui Platonem, caeteris philosophis gentium longe recteque praelatum, acutius intellexisse atque secuti esse fama celebriore laudantur, aliquid tale de Deo sentiunt, ut in illo inveniatur et causa subsistendi, et ratio intelligendi et ordo vivendi," *Patr. lat.*, Vol. 41, col. 228–229. Thus St. Augustine says that possibly certain interpreters of Plato got as far as that. Probably he is thinking of Plotinus, Proclus or even Porphyry. Armed with this text, to which he refers, St. Thomas reconstructs the history of the question and divides it into three stages : (1) The early pre-Socratics, who looked only for the cause of the accidental transmutations of bodies, and supposed their substances to be without cause ; (2) Plato and Aristotle, who put the question of the cause of corporeal substance. They admitted a matter which has no cause (" distinxerunt, per intellectum, inter formam substantialem et materiam, *quam ponebant incausatam* ") and a universal formal cause (Aristotle) or several (Plato's ideas). Both (*utrique*) attained the idea of a universal principle which makes each particular being, not merely *such* a substance, but *this* substance, they reached a " *principium totius esse*," and we cannot say that Aristotle's God is not " *causa substantiae coeli*." But note that St. Thomas never uses the word *creatio* in connection either with Plato or Aristotle ; for their universal cause of the substance of beings is not a creative cause ; (3) after Plato and Aristotle *others* rose to the consideration of the cause of the very *existence* of these substances : " Utrique igitur (*sc.* Plato and Aristotle) consideraverunt ens particulari quadam consideratione (*sc.* a consideration which goes only to an aspect of being), vel inquantum est *hoc ens*, vel inquantum est *tale ens*. Et sic rebus causas agentes particulares assignaverunt. Sed *ulterius aliqui erexerunt se ad considerandum ens inquantum est ens ;* et consideraverunt causam rerum non solum secundum quod sunt *haec* vel *talia*, sed secundum quod sunt *entia*. Hoc igitur quod est causa rerum, inquantum sunt entia, oportet esse causam rerum, non solum secundum quod sunt talia, per formas accidentales ; nec secundum quod sunt haec, per formas substantiales ; sed etiam secundum omne illud quod pertinet ad esse illorum quocumque modo. Et sic oportet ponere etiam materiam primam causatam ab universali causa entium," *Sum. theol.*, I, 44, 2, Resp. To sum up this decisive passage, it appears that St. Thomas grants that Plato and Aristotle knew the universal cause of the substantiality of beings, but refuses to credit them with knowledge of the cause of the existence of these substances. This very clear text of the *Summa* helps us to interpret that of the *De Potentia*, III, 5, Resp., which refers us precisely to St. Augustine, *De civit. Dei*, VIII, 4. St. Thomas once more recites the history, and notes three stages of philosophic reflection : (1) The explanation of accidental mutation, (2) beginnings of the explanation of substantial forms : " Posteriores vero philosophi . . ." ; (3) consideration of being in general : " Posteriores vero Philosophi ut Plato, Aristoteles et *eorum sequaces*, pervenerunt ad considerationem ipsius esse universalis ; et ideo ipsi soli posuerunt aliquam universalem causam rerum, a qua omnia alia in esse prodirent, ut patet per Augustinum (*De civit. Dei*, VIII, 4, non procul a fine). Cui quidem sententiae etiam catholica fides consentit." Thus St. Thomas considers that Plato, Aristotle *and their successors*, posited a universal cause of things, but does not say that they all attained the idea of a creative cause. Since he refers to the text of St. Augustine we must understand his conclusion in function of that text, and that brings us back to the doctrine of the *Summa* : no creation in Plato or Aristotle, creation, *fortassis* says St. Augustine, in certain neo-Platonists. St. Thomas suppresses the *fortassis* because, writing in the thirteenth century, he has Avicenna in mind, and Avicenna certainly conceived God in Biblical terms : " Est autem ponere aliquod ens quod est ipsum suum esse. . . . Unde oportet quod ab uno illo ente omnia alia sint, quaecumque non sunt suum esse, sed habent esse per modum participationis. Haec est ratio Avicennae (*Metaph.*, VIII, 7 and IX, 4) . . ." *De potentia, ibid.* It is only then that we arrive at the idea of creation properly so called, and it is directly connected with the distinction of essence and existence in all that is not God.—St. Thomas' true position in the *Summa* is very clearly noted by J. MARITAIN, *La philosophie Bergsonienne*, 2nd Edn., Paris, Rivière, 1930, p. 426.

[5] St. AUGUSTINE, *Confessions*, X, 6, 9.

[6] St. THOMAS AQUINAS, *Compendium theologiae*, Cap. LXVIII.

[7] St. THOMAS AQUINAS, *op. cit.*, Cap. LXIX—" Probat enim (Aristoteles) in II *Metaph.* quod id quod est maxime verum et maxime ens, est causa essendi omnibus existentibus : unde hoc ipsum esse in potentia, quod habet materia prima, sequitur derivatum esse a primo essendi principio, quod est maxime ens. Non igitur necesse est praesupponi aliquid ejus actioni, quod non sit ab eo productum." *In Phys.*, lib. VIII, lect. 2, art. 4. In virtue of Aristotelian principles he passes beyond Aristotle's conclusions : there could hardly be a clearer case.

[8] St. THOMAS AQUINAS, *Sup. lib. de Causis*, lect. XVIII.

[9] Sed contra est quod dicitur, Exod. iii. 14, ex persona Dei : Ego sum qui sum. St. THOMAS, *Sum. theol.*, I, 2, 3.

[10] St. THOMAS AQUINAS, *Cont. Gent.*, II, 6.

[11] *Ibid.* The direct proof of creation given by St. Thomas further on, II, 15, falls back expressly on Chapter VI, where he establishes " quod Deo competit esse aliis principium essendi."

[12] " The more universal the effect the higher the cause, because the higher the cause the greater number of effects to which it extends. But to be is more universal than to be moved. . . . Wherefore above that kind of cause which acts only by moving and changing, there must be a cause which is the first principle of being, and this we have shown to be God. God, therefore, does not act solely by moving and changing things." St. THOMAS, *Cont. Gent.*, II, 16. It is hardly necessary to recall that not the Augustinians only, but also their irreconcilable adversaries the Averroists, clearly realized the difference between Greek and Christian thought on this point : see MANDONNET, *Siger de Brabant (Les philosophes Belges*, VII), *De erroribus philosophorum*, p. 4, n. 4 and p. 8, n. 2. The suspicion of the Augustinians with regard to St. Thomas is explained in part by their mistaken attitude to the historical relation between St. Thomas and Aristotle. By dint of emphasizing his borrowings from Aristotle and merely lightly suggesting what he added, having even at times all the air of borrowing what he really added, St. Thomas made it difficult for himself to show that, for him, Aristotle's principles were not bound up with Aristotle's conclusions. Thus, not perceiving the new significance he gave to the principles themselves, the Augustinians and Averroists regarded his doctrine at first as nothing but an Aristotelianism that dared not follow its own path to the end, and in fact as a shamefaced Averroism. But it is something very different.

Let us add finally that it was altogether natural that in this field, Christian thought ·should have been preceded by Jewish thought, since they both possessed the Bible. On Maimonides doctrine see the important Chap. II in A. FOREST, *La structure métaphysique du concret selon St. Thomas d'Aquin*, Paris, J. Vrin, 1931, pp. 50–51.

[13] See *Avicenne et le point de départ de Duns Scotus*, in *Archives d'hist. doctr. et littéraire du moyen âge*, II (1927), pp. 98–99. We have tried to demonstrate the thesis with reference to St. Thomas, because it is in his case that it is most misunderstood. In the case of Duns Scotus the task would be easy. We know his suspicion of the *physical* proofs of God's existence ; his distaste for the proof by the First Mover is due precisely to the fact that it wears too much the air of a physical proof. If there is question of the first of the natural movers, this will not be God ; if this mover is first, not merely in the order of movement, but in the order of being, it will certainly be God ; but then the proof appertains not to the physicist but to the metaphysician. As Duns Scotus says in a striking formula : " How should a physicist prove that a mover is first without being more of a metaphysician to prove it first than of physicist to prove it a mover ? " (*In Metaph.*, lib. VI, qu. 4). At the moment he wrote these lines Duns Scotus sounded the last depths of Christian philosophy ; and he did that often enough. We must add that at bottom his doctrine is in no contradiction here with that of St. Thomas ; rather should we say that one Christian philosophy throws light on another. Aristotle's proof of the existence of the first mover is quite in its place in his *Physics*, lib. VII, it is reached directly as the cause of movement, the proper object of Physics. In St. Thomas, on the other hand, the proof is developed on the plane of being, and consequently is altogether metaphysical, the contingence of movement being here nothing but a particular and very evident case of the radical contingence of the created being. To be convinced of this we have only to remember that the proof of the first unmoved and therefore immutable mover implies for St. Thomas that this being is eternal, necessary, *habens esse per seipsum ;* whence it follows : *quod essentia divina, quae est actus purus et ultimus, sit ipsum esse,* and finally : *quod Deus est primum et perfectissimum ens, unde oportet quod sit causa essendi omnibus quae esse habent* (see *Compend. theologiae*, Cap. LXVIII). Such a first mover is evidently more metaphysical as first than physical as a mover.

[14] Pére Laberthonnière had a very keen sense of the radically new turn taken by philosophy under the influence of the Christian revelation (*Le réalisme Chrétien et l'idéalisme grec*, Paris, 1904). It amounted, in his view, to a " radical opposition " between Hellenism and Christianity, an opposition which leaps to the

eye in the Fathers, but is less obvious in the Middle Ages owing to the efforts made to cover it up. Naturally enough, there was an opposition in the order of religion ; there Christianity made an altogether fresh start ; but it is by no means certain that this religious revolution carried with it anything more than a philosophical *progress*. The Christians never considered themselves as merely completing Greek religion, but always thought that they were merely completing Greek philosophy. There may well have been novelty in religion without opposition in philosophy, because the conflict of conclusions, wherever it existed, could be resolved by a deeper understanding of principles. H. RITTER seems to strike a juster note (*Histoire de la philosophie chrétienne*, Vol. I, p. 47) : " Scholastic works were influenced by Aristotle only as regards their exterior form, in their intimate depths they responded much more to the thought of the Fathers." And further on : " Whence it follows that scholasticism may be considered as a mere continuation of Patristic philosophy." But that is to say too little, for Aristotle furnished the mediævals with a whole technique and with principles which, although not yet realized in their full import, were already true. Christian thought brought new wine ; but the old bottles were still sound.

[15] GILBERT MURRAY, *Five Stages of Greek Religion*, 2nd Edn., 1925, p. 7.

[16] Understood in this sense, the real distinction of essence and existence is essential, not only to Thomism, but to all Christian metaphysics. As far as the meaning is concerned, apart from the formula, it is everywhere present in St. Augustine. The formula itself has been criticized, notably by Suarez ; but Suarez does not deny what is substantially affirmed in it, namely that God alone exists in virtue of Himself and that no other things hold their existence from themselves. For further acquaintance with this controversy, which leaves the fundamental question intact, consult P. DESCOQS, *Thomisme et suarezisme*, in *Archives de philosophie*, Vol. IV, Paris, G. Beauchesne, 1926, pp. 131–161, especially p. 141 *et seq.* If the " real distinction " is taken as a *physical* distinction between combinable and separable elements, Scotists and Suarezians would quite rightly deny, not only the distinction itself, but also that St. Thomas admitted it. If we take it on the contrary in a *metaphysical* sense, as it is understood here, no Christian philosopher denies what the formula affirms, even when rejecting the formula itself. That is rightly pointed out by DESCOQS and established by FR. DEL PRADO, *De veritate fundamentali philosophiae christianae*, Fribourg (Switzerland), 1911, especially Chap. V, pp. 33–37.

NOTES TO CHAPTER V

[1] I. KANT, *Kritik der reinen Vernunft*, Transcendentale Elementarlehre, 4th Antinomy.

[2] ST. THOMAS AQUINAS, *In Metaphys.*, lib. V, lect. I.

[3] This has not been overlooked by the moderns. Hume's critical analysis of the idea of causality consisted precisely in showing that it results from an extension of our own psychological experience to reality ; we believe that one phenomenon produces another because we have internal experience of the fact that our idea of one phenomenon conjures up another ; subjective psychological habits are unduly erected into objective causal relations. This very anthropomorphism, which in Hume is made to justify his critique of causality, is used by Maine de Biran to justify his positive doctrine of causality. According to him we may be certain that there is real efficient causality in nature, because of our direct experience of the hyperorganic efficacy of the will in ourselves. It is therefore well recognized that the classical conception of causality rests on an inference from man to nature, and even in the nineteenth century there was at least one philosopher who recognized the legitimacy of this kind of inference. It would be easy to cite more recent cases, and the Bergsonian philosophy would be a good example, for creative evolution supposes the extension to the universe of the human experience of freedom.

[4] " Actus autem est duplex : primus et secundus. Actus quidem primus est forma et integritas rei : Actus autem secundus est operatio," ST. THOMAS AQUINAS, *Sum. theol.*, I, 48, 5, Resp.

This mode of rooting causality in the very actuality of the being is common to all the great mediæval doctors. They differ, on the other hand, in the way in which they regard the relation of the faculty which operates to the substance to which it belongs. The question arises especially *à propos* of the soul.

(*a*) Some refuse to distinguish the soul and its operations. The soul is a simple spiritual substance, participating in God's simplicity as it participates in His spirituality. The soul therefore operates directly by its essence, just as God does. Thus it may be said that it is the actuality of the essence which directly produces the operation, so that the connection in question (between being and causing) is here at its maximum. Cf. WILLIAM OF AUVERGNE, *De anima*, Cap. III, pars. 6 ; reproduced in *Archives d'hist. doctrinale et litt. du moyen âge*, Vol. I, p. 55, note 2.

(*b*) Others consider that the simplicity of a created essence cannot possibly be of this type. In order to distinguish such essences from God they admit a distinction between the soul and its faculties, and hold that the former operates through its faculties, to which it communicates its own actuality ; but this distinction is then reduced to a minimum.—ALCHER DE CLAIRVAUX in the *De spiritu et anima*, calls them functions : *officia*. For Hugh of St. Victor they are accidents (*De sacramentis*, I, 3, 25). For St. Bonaventure they are instruments of the substance, but consubstantial with the substance itself (I. *Sent.*, 3, 2, 1, 3, Concl.). The continuity between the act of the substance and that of its faculties, the radical community of the actuality of the being and that of the operation, here remains evident (cf. *La philosophie de saint Bonaventure*, pp. 331–332).—According to Duns Scotus the distinction between the soul and its faculties is not real but only formal ; texts in H. DE MONTEFORTINO, *J. D. Scoti, Summa theologica*, I, 77, 1 ; Vol. III, p. 533.

(*c*) According to St. Thomas there is a *real* distinction between the substance and its faculties, and he makes them accidents not, as Hugh of St. Victor does, to mark that they are hardly distinguishable, but to emphasize the reality of the distinction. However, as we have seen, the operation is only a second act, which, like the first act of being, is due to the actuality of the form. The real continuity is thus maintained in spite of the metaphysical distinction. St. Thomas himself affirms it elsewhere when he says that : *ipsa anima secundum quod subest suae potentiae, dicitur actus primus ordinatus ad actum secundum* (*Sum. theol.*, I, 77, 1, Resp.). To emphasize the intimate character of this relation he calls the faculties " propers," that is, natural properties of the soul (*loc. cit.*, ad 1) ; that they are all in the soul as in their principle : " *sicut in principio* " (*loc. cit.*, art. 5, ad 2) ; that they flow from the essence of the soul as from their cause : " *fluunt ab essentia animae sicut a principio*," " *sicut a causa* " (*loc. cit.*, art. 6) ; that they result from it as colour results from light : " *per aliquam naturalem resultationem sicut ex uno naturaliter aliud resultat, ut ex luce color* " (*loc. cit.*, art. 6, ad 3, and art. 7, ad 1). Thus the actuality of the soul is the principle of the faculties and of their operations, in St. Thomas as in his predecessors.

[5] The whole synthesis of creationist metaphysics, including its principle, which is being and its actuality, is contained in the following few lines : " Whatever belongs to a subject *per se* inheres necessarily in that subject, as rationality in man and upward movement in fire. Now the actual production *per se* of any effect whatever belongs to the being in act, for every agent acts inasmuch as it is in act. Therefore every being in act can make something actually to exist. But God is being in act as has been shown (lib. I, Cap. XVI). Therefore it belongs to Him to produce actual being and to be the cause of its existence." ST. THOMAS AQUINAS, *Cont. Gent.*, II, 6. Briefly : God is cause because He is Being, and as He is the Being that presupposes no other being, He is also a cause that presupposes no other cause :

now the first cause produces the first effect, and the first effect, presupposed by all the others, is existence ; therefore it belongs to Being to cause existence, *i.e.*, to create. That is why creation is an act proper to God : *Cont. Gent.*, II, 21, and *Sum. theol.*, I, 45, 5, Resp.—On the central place held by the idea of being in St. Thomas, see Fr. Olgiati, *L'anima di san Tommaso*, Milan, undated ; and, concerning creation, *op. cit.*, pp. 80–82.

[6] This explains in part why, as the mediæval Augustinians clearly saw, the Greek philosophers did not attain to the idea of creation. Here, then, we have one of those cases where an idea, in itself rational, escapes the grasp of reason because it lacks the aid of revelation : " Haec autem veritas, etsi nunc cuilibet fideli sit aperta et lucida, latuit tamen prudentiam philosophicam, quae in hujus quaestionis inquisitione largo tempore ambulavit per devia. . . . Utrum autem posuerit (Aristoteles) materiam et formam factam de nihilo, hoc nescio ; credo tamen quod non pervenit ad hoc, sicut melius videbitur in problemate secundo : ideo et ipse etiam deficit, licet minus quam alii. Ubi autem deficit Philosophorum peritia, subvenit nobis sacrosancta Scriptura, quae dicit, omnia esse creata et secundum omne quod sunt in esse producta. Et ratio etiam a fide non discordat, sicut supra in opponendo ostensum est." St. Bonaventure, *In II Sent. dist.*, 1, p. 1, a.1, qu. 1, Resp.

[7] E. Gilson, *Introduction a l'étude de saint Augustin*, Paris, J. Vrin, 1929, p. 243.

[8] St. Augustine, *De doctrina christiana*, I, 32, cited by St. Thomas, *Sum. theol.*, I, 19, 4, ad 3. Cf. *Cont. Gent.*, I, 86.

[9] J. Durantel, *Saint Thomas et le pseudo-Denis*, Paris, 1919, p. 154.

[10] Plato, *Timaeus*, 29 E.

[11] A. E. Taylor, *Plato*, London, 1926, pp. 441–442.

[12] J. Durantel, *Saint Thomas et le pseudo-Denis*, p. 138.

[13] J. Chevalier insists that the idea of creation keeps Greek and Christian thought quite distinct (*La notion du nécessaire chez Aristote et chez ses prédecesseurs*, Paris, Alcan, 1915). We entirely agree, and cannot do better than send the reader to his book. At the source of the Christian universe lies the radical contingence of a being sovereignly free. But it ought perhaps to be added that if Christian thinkers succeeded in conceiving God as both *necessary* and *free*, they did so precisely because they had first succeeded in conceiving Him as identical with Being. This identification once achieved, God remains the necessary being, but in a sense quite other than Aristotle's ; from the fact that He is the integration of the whole nature of being, and that He alone is so, *it is to Him alone that necessity properly belongs*. No effect of God can be necessary in Christian philosophy because no effect of God is God, no being is Being. It is just because God

is absolute being and absolute goodness that He has this inclination to communicate Himself by creating analogues of His being : but no matter how powerful this inclination may be, infinite if you will, it remains on a plane that wholly excludes necessity. A relation of necessity between beings and Being is a contradiction in terms. Outside God there is nought but the contingent : " Dicendum quod sicut divinum esse in se est necessarium, ita et divinum velle, et divinum scire ; sed divinum scire habet necessariam habitudinem ad scita, non autem divinum velle ad volita. Quod ideo est, quia scientia habetur de rebus secundum quod sunt in sciente, voluntas autem comparata ad res secundum quod sunt in seipsis. Quia igitur omnia alia habent necessarium esse secundum quod sunt in Deo, non autem secundum quod sunt in seipsis habent necessitatem absolutam, ita quod sint per seipsa necessaria ; proper hoc, Deus quaecumque scit, ex necessitate scit ; non autem quaecumque vult, ex necessitate vult." St. Thomas, *Sum. theol.*, I, 19, 3, ad 6. Thus the necessary being wills only Himself necessarily ; with respect to the contingent, *precisely because it is contingent in the order of being*, He is free.—This sequence of ideas is strongly marked in Duns Scotus and his school : " Deus, cum sit a se, est infinitae necessitatis ; ergo quocumque alio non existente, non sequitur non esse. Si autem necessario aliquid a se distinctum causaret, posset non esse ex defectu minus necessarii : ergo cum id sit plane impossibile, oportet Primum esse causam non necessario, sed libere et contingenter causantem." Hier. De Montefortino, *J. D. Scoti, Summa theologica*, I, 19, 3, Resp. Cf. *De primo principio*, V, n. 71b. Thus the freedom of the creative act rests directly on the metaphysics of *Exodus*, and we have to go beyond the Judeo-Christian idea of omnipotence to the Judeo-Christian idea of Being. In a word, the perfection of *a* being, even were it Aristotle's pure thought itself, always implies necessity, but the perfection of Being implies necessity, infinity and freedom at once.

14 The concept of analogy is one of those that offers the greatest difficulty to the modern reader of a mediæval treatise ; the mediæval thinkers themselves were far from agreement as to its definition, and even St. Thomas does not seem to have attempted a full elucidation. We may regard it as having two functions, the one unitive, the other separative. An analogue is always drawn towards its principle in virtue of being an analogue, and at the same time departs from its principle in virtue of being no more than an analogue. When, in particular, we speak of the analogy of being, what does it really mean ?

(1) It means to start with, that being, first and foremost and as of full right, is, and can be nothing other than an analogous concept. Every being that exists, in so far as it is really existent,

is distinct from every other being. As soon as we leave the case of the particular existence, and rise to those elements which are common to several beings, such as essences, species and genera, we pass from that which exists to that which does not exist. Not that species and genera are nothing at all, but they have no existence which is proper to themselves. In other words, a thing's existence is proper to itself as by definition, since, if it were not proper to it, it would not be its, and consequently the thing would not exist. Hence when we say of a thing that it is, the word " being " can designate only the act of existence appertaining precisely to this thing. That amounts to saying that since the word " being " never means the same being twice over when applied to two different beings, it is not *univocal*.

(2) Nor, on the other hand is it *equivocal*, for though existences are irreducible to each other, they all agree nevertheless in this, that they are acts of existing. That is why we say that between being and being there is *analogy*. In what does this relation of analogy consist ? It is a relation between proportions, or as it is called, a proportionality. What is common to the concept of being as applied to God and the concept of being applied to man, is this : that just as God's being is that by which He is, so a man's being is that by which he is. That does not mean that the relation between God and His being is the same as the relation between man and his being ; on the contrary, they are infinitely different ; but in both cases the relation exists, and the fact that it exists within the bosom of each being sets up an analogy between them all. That precisely, is the *analogy of being*, and we see why it is only an analogy of proportionality, since it can be set up between beings which are out of all proportion, provided only that each of these is to itself what the others are to themselves. Causal analogy, based on the relation of an effect to its cause, is a different relation, which we shall study next in order in the text of the lecture. It should always be remembered that the likeness of the creature to God is merely the likeness of an effect the being of which is only an analogue of the being of its cause, and although it is analogous it is nevertheless *infinitely* different. See ST. THOMAS AQUINAS, *Sum. theol.*, I, 13, 5. *Quaest. disp. de Veritate*, Q. II, art. 11, and Q. XXIII, art. 7, ad 9.—It is well known that Duns Scotus maintained the univocity of being, but here we have to do with another systematization and another terminology, which leaves intact, however, the purely analogical character of the real relation between creatures and God. He calls every concept *univocal* which has sufficient unity to afford a basis for a contradiction. In this sense, to say that the concept of being is univocal means that it is really being that is attributed to God just as it is really being that is attributed to creatures ; thus we have a *univocity of concept*, but radical diversity of nature.

The texts will be found conveniently grouped in HIER. DE MONTEFORTINO, *J. D. Scoti, Summa theologica*, I, 13, 5 ; Vol. I, pp. 318–322.

(3) Without the doctrine of analogy the identification of God and being leads to pantheism. This happened more than once in the Middle Ages. See, for instance, David de Dinant, in G. THÉRY, *David de Dinant* (Bibl. thomiste, VI), Paris, Vrin, 1925, particularly the text cited on p. 132 and text 5, p. 135, where God is identified with being in potency, that is to say with matter, and precisely because He is being. The difficulties in which Eckhart entangled himself later on have the same source : " Esse est Deus. . . . Deus igitur et esse idem." See G. DELLA VOLPE, *Il misticismo speculativo di maestro Eckhart nei suoi rapporti storici*, Bologna, 1930, pp. 151–152. He too falls back upon the *Ego sum qui sum*, but his deductions are incautious, not to say contradictory with the principle since, if God is Being, nothing else can be so in the same sense. In his view, on the contrary, the essence of things is to the being of God what potency is to act, so that they are one, just as the union of potency and act is one (*op. cit.*, p. 179). Cf. " Deus enim est esse. Constat autem quod esse est indistinctum ab omni eo quod est, et quod nihil est nec esse potest distinctum et separatum ab esse." G. THÉRY, *Le commentaire de Maître Eckhart sur le livre de la sagesse*, in *Archives d'hist. doctrinale et litt. du moyen âge*, Vol. IV (1929–1930), Paris, J. Vrin, 1930, p. 257, note : " Rursus vero et hoc notandum quod nihil tam unum et indistinctum quam Deus et omne creatum. . . . Primo quia nihil tam indistinctum quam ens et esse (essence and existence) potentia et actus ejusdem, forma et materia. Sic autem se habet Deus et omne creatum." *Ibid.*, p. 255. On this point consult the very important note 3, p. 256, where Fr. Théry carefully analyses the various senses of the Eckhartian indistinction. See also G. THÉRY, *Edition critique des pièces relatives au procès d'Eckhart*, in *Archives . . .*, Vol. I, p. 172 and text 5, p. 193, where Eckhart expressly appeals to Exodus.—On the other hand it might be shown that if the concept of *divine image* lies at the heart of so much mediæval mysticism, St. Bernard's for instance, it is precisely because it allows of a mystical deification without confusion of substance ; man is here no more than a subject informed by a likeness to God under the assimilative action of grace.

[15] ST. THOMAS AQUINAS, *Sum. theol.*, I, 45, 7, Resp.

[16] FRANCIS BACON, *De augmentis scientiarum*, III, 5.

[17] If we would grasp the exact meaning of mediæval finalism certain important points must be noted :

(1) It is not for philosophy to dictate scientific method : science must judge for itself what procedure in research and mode of explanation is adapted to its own purposes. If, then, scientists

renounce finalism—and it may well be that they do so less than they profess—they have every right to do so without remark from philosophy. If, on the other hand, there are good philosophic reasons for admitting the presence of finality in the universe, the scientist has no right to forbid the philosopher to take it into account on the pretext that finality cannot be scientifically analysed or reconstructed.

(2) From a philosophical standpoint finality is always to be admitted whenever, in a given whole, neither each element separately, nor, in consequence, their sum, suffices to account for the existence of the whole. The analysis and reconstitution of the whole by mechanistic methods might suffice perhaps for the purposes of scientific explanation, but an exhaustive analysis of this kind, even were it possible, would still leave two questions open : that of the existence of the elements adapted to enter into the synthesis and that concerning the order of these elements in the synthesis. That is why St. Thomas considered it possible to prove the existence of God as cause of the existence of order ; and we may even add that it is demanded also as cause of the aptitude of the elements to enter into the order.

(3) If anthropomorphism is inevitable in our conception of finality because the finality we know best, and know from within, is the finality which synthetizes the elements of the human act, we are not on that account committed to any uncritical anthropomorphism. Neither scientifically nor philosophically is there any ground for admitting a conscious finality in the inorganic world, or in non-cognitive organisms ; finality appears only with knowledge, that is, in germ in the animals and in full clearness only in man. When therefore, the subject of conscious finality seeks the cause of unconscious finality, he cannot avoid conceiving it as analogous to his own. Now here we have to recollect the nature of an analogy of proportionality : God is to His finality as we are to ours, but what God is to His finality we do not know, save, perhaps, that He is identical with it. Oblivious of our inevitable ignorance on this point, we believe ourselves able to make out His finality, and credit God with various " designs," sometimes reasonable in themselves, but quite uncertain, sometimes frankly unreasonable or even absurd from a human standpoint. The mediæval conception of finality did not fall into deviations of this kind ; which, on the contrary, presuppose a complete misconception of its true significance.— On the idea of finality, see R. DALBIEZ, *Le transformisme et la philosophie*, in *Le transformisme*, Paris, J. Vrin, 1927, pp. 174–179 ; and R. COLLIN, *Réflexions sur le psychisme*, Paris, J. Vrin, 1929, pp. 192–198.

(4) It would be of interest to consider whether the mechanistic explanation is not just as anthropomorphic as the finalist. If

homo faber conceives ends, he applies mechanism to their realization, and his conception of nature is just as much saturated with his mechanism as with his finalism. As M. LANGEVIN recently observed : " The doctrine of mechanism is faced with a new crisis ; I would specially note the anthropomorphic and ancestral elements contained in its generalized form, and in some of the fundamental ideas it introduces, such as that of the material point conceived as a limit, of the individualized object, that of force, &c." (*Bulletin de la Société française de philosophie*, 30th year, n. 2, April–June, 1930, p. 58). We have no desire to turn M. Langevin's idea to ends which are not his ; for he would rather purge science of all lingering traces of anthropomorphism than do anything to add to it (*ibid.*, pp. 61–62) ; the suggestion, however, may be usefully meditated. What rather darkens the problem is the difficulty of imagining what science would be like were all anthropomorphism successfully eliminated from mechanism—unless perhaps some more subtle anthropomorphism were substituted in its place. When M. Langevin adds : " There is no question of abandoning, but only of modifying the image " (p. 70), we may well ask whether the necessary presence of the image will not always imply some residue of anthropomorphism, and even whether the idea of a non-anthropomorphic science evolved by a human intellect has any conceivable meaning at all.

[18] B. PASCAL, *Pensées et opuscules*. Ed. L. Brunschvieg, 4th Edn., p. 215.

[19] MAINE DE BIRAN, in *Maine de Biran, sa vie et ses pensées*, E. Naville, Paris, 1857 ; Journal intime du 15 Mars, 1821, p. 349.

NOTES TO CHAPTER VI

[1] IRENAEUS, *Adv. Haereses*, III, 25, 5 ; V, 18, 2.

[2] *Op. cit.*, I, 22, 1 ; II, 10, 4 ; II, 30, 9 ; II, 26, 3.

[3] PLOTINUS, *Enneads*, I, 8. On the doctrine of evil in Plotinus, as contrasted with that of St. Augustine, see R. JOLIVET, *Essai sur les rapports entre la pensée grecque et la pensée chrétienne*, pp. 102–111.

The only doctrines here considered are those with which Christian thought came into contact during its period of formation. As regards earlier doctrines which may have had an indirect influence, we may say that Plato left us no systematic solution of the problem of the origin of evil. Aristotle credits him with a frank dualism, that is to say with a recognition of two principles of things, one good and the other bad, and on this point ranges him with Empedocles and Anaxagoras (ARISTOTLE, *Metaphysics*, A, 6, 988 a 7–17). In this view, matter would be the principle of evil in Platonism. It is not easy to find a clear text in Plato to justify the assertion. The *Timaeus* contains nothing precise on matter, which it does not even name. One might, however, be tempted to consider the Chaos that the Demiurge orders as its equivalent, and possibly that is what Aristotle is thinking of. This view is suggested by a text in the *Politics* (273 b), where, without mention of the word, the idea of matter is suggested (τὸ σωματοειδὲς τῆς συγχράσεως αἴτιον), with application to something very like the Chaos of the *Timaeus*, and invoked to account for the presence of disorder in nature. We might also cite in support the text of the *Republic* (II, 379 c), where Plato says that the god is the cause of the few goods that happen to us, but that we must seek out another cause for the great number of ills ; but he does not say what this cause is, and, since it may be man's own fault, the text is not decisive. Perhaps the *Letter to Dionysius* (II, 313 a), whether authentic or not, strikes the true note when it says that Plato had reflected much on the question of evil in general, but had never resolved it.

Aristotle himself often adopts expressions that resemble those of the Christian thinkers. Evil, he says (*Metaph.*, 9, 1051 a, 17–18), does not exist in nature. Act is being, and being is good. Since, however, he introduces the idea of matter and defines it as potentiality as opposed to act, he finds himself led to connect evil with matter. He does not say that it is bad, but it is the principle of mutability and contingence opposed to form and its necessity. It is on account of the potentiality of matter, that we have corruptions, alterations, monsters, etc., etc. We may say that, without being evil in itself, matter makes disorder possible, and, in a sense, inevitable ; it contains, then, a mischievous element (κακοποιόν—in *Phys.*, A, 9, 192 a, 15). But what separates Aristotle and Christianity on this point is, for St. Augustine, the fact that matter is not even the cause of the possibility of evil, nor the reason for its existence ; in itself it implies no tendency to disorder. As God created it the material world was excellent : *valde bonum*, and so it would have remained had not a sin that came to birth in the realm of the spirit, and not in that of matter, brought disorder into matter. Thus the eternal and uncreated universe of Aristotle, in which matter opposes an eternal resistance to the perfection of form, is profoundly different from the created universe of the Christian philosophers, in which matter directly participates in the perfection of the divine being, and is, in however humble a way, a participated likeness to God.

⁴ ST. AUGUSTINE, *Confessions*, VII, 11, 17, to 16, 22. This development is usually interpreted as though Augustine attributed his discovery that all that is is good, to his reading of the Platonic books : " Ergo si omni bono privabuntur, omnino nulla erunt. Ergo quamdiu sunt, bona sunt. Ergo quaecumque

sunt, bona sunt. Malumque illud, quod quaerebam unde esset, non est substantia, quia, si substantia esset, bonum esset " (*op. cit.*, VII, 12, 18). However, Augustine does not say that he had read in the Platonic books that matter is good, or even that it is not bad ; he does not even say that he read there the truths he then discovered touching the problem of evil. The résumé of what he owes to his reading stops at *Confess.*, VII, 9, 15 : " Inveni haec ibi et non manducavi." Then he deals with the conclusions obtained after this reading, but by personal reflection and the help of God : " Et inde ad monitus redire ad memetipsum, intravi in intima mea duce te et potui, quoniam *factus es adjutor meus.*" It is God, therefore, and no longer Plotinus who is his guide here. That is why the problem, otherwise so ingeniously dealt with by M. R. JOLIVET, *Essai sur les rapports entre la pensée grecque et la pensée Chrétienne*, p. 113, does not even really arise.

⁵ ST. AUGUSTINE, *De natura boni*, Cap. XIX. In the same treatise, Cap. XXXIV, Augustine appeals to this other Scriptural text : " Omnis creatura Dei bona est," I Tim. iv. 4. It need not be said that this doctrine became the common heritage of all Christian philosophers. See, for example, ST. THOMAS AQUINAS, *De malo*, I, 1, Sed contra : " Praeterea, Joan I, 3, dicitur : *omnia per ipsum facta sunt.* Sed malum non est factum per Verbum, ut Augustinus dicit (Tract. I in Joan., a med.). Ergo malum non est aliquid."

⁶ *Op. cit.*, X ; cf. *Enchiridion*, Cap. XI–XII.

⁷ ST. AUGUSTINE, *Enchiridion*, Cap. XII, 4.

⁸ *Op. cit.*, XIV.

⁹ ST. AUGUSTINE, *De natura boni*, Cap. VIII. See the parallel arguments of ST. THOMAS, *Cont. Gent.*, III, 71 : Quod divina providentia non excludit totaliter malum a rebus, and *Sum. theol.*, I, 48, 2 : Utrum malum inveniatur in rebus.

¹⁰ ST. BERNARD, *In Cant. Canticorum*, sermo 80, art. 5.

¹¹ ST. AUGUSTINE, *De natura boni*, Cap. XX, cf. XXXIV.

¹² ST. AUGUSTINE, *De genesi ad litt.*, I, 3. See ST. THOMAS, *Sum. theol.*, I, 48, 5, Resp. But elsewhere St. Thomas adds the important remark that this formula is applicable only to moral evil and its moral or physical consequences. If the mere inequality of physical beings, resulting in no happiness or misery for these beings themselves, is to be called evil, then it is an evil that exists independently of sin. Thus : " Haec divisio non est mali nisi secundum quod in rationali natura invenitur, ut patet ex auctoritate Augustini inducta." *De malo*, I, 4, Resp.

¹³ ST. AUGUSTINE, *De civitate Dei*, XXII, 24, 3–5.

¹⁴ It subsists even in the damned, and it is for this reason that they can feel remorse : ST. THOMAS, *Sum. theol.*, I a, II ae, 85, 2, ad 3.

[15] St. Thomas, *Sum. theol.*, I a, II ae, 85, 1, Resp.

[16] St. Thomas, *Sum. theol.*, I, 98, 2, Resp.

[17] See St. Augustine's critique of Porphyry's pessimism : " Sed corpus est omne fugiendum. . . . Omne dixit (Porphyrius), quasi omne corpus vinculum aerumnosum sit animae. Et prorsus si corpus qualecumque est fugiendum, non est ut laudes ei corpus, et dicas quomodo Deus docente fides nostra laudat corpus : quia et corpus quod modo habemus, quamvis habeamus hinc poenam de peccato, et corpus, quod corrumpitur, aggravat animam (*Sap.*, IX, 15) ; tamen habet corpus istud speciem suam, dispositionem membrorum, distinctionem sensuum, erectam staturam, et caetera quae bene considerantes stupent. Verumtamen illud omnino incorruptibile, omnino immortale, omnino ad movendum agile et facile erit. Sed ait Porphyrius : sine causa mihi laudas corpus ; qualecumque sit corpus, si vult esse beata anima, corpus est omne fugiendum. Hoc dicunt philosophi ; sed errant, sed delirant." St. Augustine, *Sermo.*, 242, VII, 7. There is the authentic Christian spirit : altogether different from the gloomy asceticism to which certain mediæval authors succumbed.

NOTES TO CHAPTER VII

[1] St. Thomas Aquinas, *Cont. Gent.*, III, 66, 67. Cf. *De Potentia*, III, 7. The principal Scriptural basis is Isaiah, xxvi. 12 : *Domine, dabis pacem, omnia enim opera nostra operatus es nobis*, and John v. 17.

[2] St. Thomas Aquinas, *Cont. Gent.*, III, 69 : *Sum. theol.*, I, 115, 1. *De Pot.*, III, 7.

[3] St. Paul, in *Act. Apost.*, XVII, 28.

[4] St. Thomas Aquinas, *Cont. Gent.*, III, 70.

[5] *Introduction à l'etude de Saint Augustin*, p. 299.

[6] St. Augustine, *Confess.*, XI, 1.

[7] St. Augustine, *Confess.*, XI, 4, 6.

[8] St. Thomas, *De Veritate*, XI, 1, Resp.

[9] St. Augustine, *De Trinitate*, III, 9, 16 ; cf. Leibniz, *Principes de la nature et de la grace*, art. 15—Malebranche, *Entretiens métaphysiques*, XI, 1-2.

In the Middle Age the doctrine of seminal virtues is met with in St. Bonaventure who had a lively sense of the intimate accord of the doctrine with the general spirit of Augustinianism. The following text clearly shows that the question for him lay in explaining the causality of second causes without attributing to them a creative action which belongs only to God : " Supponamus nunc quod natura aliquid agat, *et illud non agit de nihilo*, et cum agat in materiam, oportet quod producat formam. Et cum materia non sit pars formae, nec forma fiat pars materiae, necesse est aliquo modo formas esse in materia antequam producantur ; et substantia materiae est praegnans omnibus : ergo rationes seminales omnium formarum sunt in ipsa." St. BONAVENTURE, *In IV., Sent.* 43, I, 4, Concl. ; edn. QUARACCHI, Vol. IV, p. 888. The same preoccupation appears in the following : " Solus igitur ille potest seminales illas rationes facere, qui potest creare ; quoniam ipsae non sunt ex aliis, sed ex nihilo, et ex ipsis fiunt omnia quae naturaliter producuntur. Igitur nec pater est creator filii, nec agricola segetum ; quia licet pater operetur interius, sicut natura, tamen operatur exterius, et circa aliquid et ex aliquo, non ex nihilo, licet non operetur adeo exterius, sicut agricola." *In II Sent.*, 7, dub. 3 ; Vol. II, p. 207. See : *La philosophie de saint Bonaventure*, p. 280 *et seq.*, particularly p. 290, note 2, for the texts of Augustine.—The doctrine will be vigorously maintained by John Peckham against St. Thomas : *Chartular. Univers. Parisiensis*, Vol. I, p. 186. However, this Augustinian master thesis is eventually abandoned by the Franciscan school itself, without doubt as a result of the Thomist criticism. In P. J. OLIVI (*In II Sent.*, qu. 31, Resp. edn. QUARACCHI, 1922, Vol. I, pp. 515–551) it appears in full decomposition. From the end of the thirteenth century it is abandoned by Richard of Middleton, as it had been already by St. Thomas ; see : E. HOCEDEZ, *Richard de Middleton*, Louvain, 1925, pp. 197–199, and the texts cited by D. E. SHARP, *Franciscan Philosophy at Oxford in the Thirteenth Century*, Oxford University Press, 1930, p. 233. Duns Scotus merely falls in with the movement when he abandons it in his turn ; although he conceives matter as a positive potency he will not allow the seminal virtues to be inserted into it ; see *Opus Oxoniense*, II, 18 qu. unica.

[10] St. AUGUSTINE, *De gen. ad litt.*, IX, 15, 26–27 ; *De Trinitate*, III, 8, 14.

[11] St. AUGUSTINE, *Confess.*, I, 13, 21.

[12] St. AUGUSTINE, *De civit. Dei*, X, 3, 2.

[13] St. AUGUSTINE, *De libero arbitrio*, II, 10, 29.

[14] " Utraque autem istarum opinionum est absque ratione. Prima enim opinio excludit causas propinquas, dum effectus omnes in inferioribus provenientes, solis causis primis attribuit ; in quo derogatur ordini universi, qui ordine et connexione causarum contexitur ; dum prima causa ex eminentia bonitatis suae rebus aliis confert non solum quod sint, sed etiam quod causae sint. Secunda etiam opinio in idem quasi inconveniens redit . . . ; si inferiora agentia nihil aliud faciunt quam producere de occulto in manifestum, removendo impedimenta, quibus formae et habitus virtutum et scientiarum occultabantur, sequitur quod omnia inferiora agentia non agant nisi per accidens." St. THOMAS AQUINAS, *De Veritate*, XI, 1, Resp.

The theses that St. Thomas criticizes in this article may be thus set out :

I. Radical Extrinsicism (Avicenna)	Natural forms	: Dator formarum.
	Natural sciences	: Dator formarum.
	Natural virtues	: Dator formarum.
II. Radical Intrinsicism (Anaxagoras)	Natural forms	: seminal virtues.
	Natural sciences	: innate ideas.
	Natural virtues	: innate virtues.

Of the three Augustinian theses referred to the first, that of seminal virtues, belongs to intrinsicism, the other two to extrinsicism. It should be added, however, that even the seminal virtues may be brought under the head of extrinsicism in St. Augustine, since, if all is ready-made in nature, it is only because all has been given at once by God. That is a further reason for holding that these theses " in idem quasi inconveniens redit."— On the co-ordination of the Thomist critiques of St. Augustine and of Avicenna see : *Pourquoi saint Thomas à critiqué saint Augustin* in *Arch. d'hist. doctr. et litteraire du moyen âge*, Vol. I (1926), pp. 5–127.

[15] It is in fact to this participation of the divine light in virtue of the possession of an active intellect that St. Thomas reduces the doctrine of illumination : " Alio modo dicitur aliquid cognosci in aliquo sicut in cognitionis principio : sicut si dicamus quod *in* sole videntur ea quae videntur *per* solem. Et sic necesse est dicere quod anima humana omnia cognoscat in rationibus aeternis, per quarum participationem omnia cognoscimus. Ipsum enim lumen intellectuale, quod est in nobis, nihil est aliud quam quaedam participata similitudo luminis increati, in quo continentur rationes aeternae," St. Thomas Aquinas, *Sum. theol.*, I, 84, 5, Resp.—This interpretation of St. Augustine which is, in fact, an altogether new doctrine, is rejected by the thirteenth-century Augustinians. It is directly aimed at by Matthew of Aquasparta, *Quaest. disput. de cognitione*, qu. 2, Resp., " Hanc positionem. . . ." It is attacked by R. Marston, *De humana cognitionis ratione anecdota quaedam*. Assuming an air of agreement with St. Augustine and appealing to his authority, St. Thomas could not but exasperate the Augustinians—hence Marston's invectives : " Patet igitur, quod dicentes omnia videri in lumine aeterno, quia videntur a lumine ab ipso derivato, doctrinam Augustini pervertunt, truncatas ejus auctoritates ad proprium sensum non sine sancti injuria convertentes, antecedentibus et consequentibus praetermissis, in quibus Sancti intentio plenius in hac materia elucescit." John Peckham is against St. Thomas, see *Quaest. disputata*, 1a obj. and ad 1. But when we come to P.-J. Olivi the doctrine of special illumination is as fully in decomposition as that of the seminal virtues.

Olivi says that he holds to it, but avows that he does not really know why : *In II Sent.*, QUARACCHI, 1926, Vol. III, pp. 500–517. After this, the Augustinian illumination, though more or less feebly surviving, does not recover vitality until the seventeenth century with Malebranche. The doctrine is expressly condemned by Duns Scotus.

[16] ST. THOMAS AQUINAS, *De virtutibus in communi*, VIII, Resp.

[17] ST. THOMAS AQUINAS, *Sum. theol.*, I, 84, 5, Resp.

[18] ST. THOMAS AQUINAS, *De Potentia*, III, 7, Resp.

[19] ST. THOMAS AQUINAS, *Cont. Gent.*, III, 19, *Praeterea*.

[20] ST. THOMAS AQUINAS, *ibid.*, III, 20, *Inter partes*.

[21] *Ibid*, III, 20, end.

[22] *Ibid.*, III, 21, *Praeterea*.

[23] *Ibid.*, III, 69.

[24] *Ibid.*

[25] *Ibid.*

[26] " Anima magis laetatur in gloria et plus gaudebit de Dei gloria et honore quam de sua glorificatione, et plus jucundabitur in laudando quam in considerando proprium bonum. Et ideo patet quod ille finis est ulterior," ST. BONAVENTURE *In II Sent.*, I, 2, 2, 1 ad 4.

[27] ST. THOMAS, *Sum. theol.*, I, 44, 4, Resp. Cf. DUNS SCOTUS, texts brought together in HIER. DE MONTEFORTINO, *J. D. Scoti, Summa theologica*, I, 44, 4.

[28] ST. BONAVENTURE, *In II Sent.*, I, 2, 2, 1, 3m fund.

NOTES TO CHAPTER VIII

[1] There is still no agreement as to Aristotle's doctrine on this point. Most commentators, since Zeller, rely on *Met.*, XII, 9 (1074 b, 15 *et seq.*), and refuse all knowledge of the world to Aristotle's god. This is clearly put by W. D. ROSS, *Aristotle*, p. 183 : " God as conceived by Aristotle, has a knowledge which is not knowledge of the universe, and an influence on the universe which does not flow from His knowledge." Other interpreters, following Brentano, credit Aristotle's God with knowledge of things, see for example, E. ROLFES, *Aristoteles Metaphysik*, Leipzig, 1904, I, p. 186, note 61, and J. MARITAIN, *La philosophie bergsonienne*, 2nd edn., Paris, Rivière, 1930, pp. 420–421. It seems difficult to maintain that Aristotle either *denied* or *affirmed* that God knows the universe. What he does affirm is that the object of the divine knowledge is God ; and he affirms also, with respect to Empedocles (*De anima*, I, 5, 410 b, 4–7 ; *Metaph.*, III, 4, 1000 b, 2–6, in J. MARITAIN, *op. cit.*, p. 421), that God knows all that mortals know ; but it is difficult to go further. We cannot cite a single text of Aristotle's attributing knowledge of the world to God, or show that Aristotle made the attribution because St.

Thomas does so on the bases of certain Aristotelian texts ; for the question is precisely whether St. Thomas did not go further than Aristotle on his own ground. From the affirmation that Aristotle's God knows all that mortals know, it cannot be deduced that he knows mortals, and the objects which these mortals know. These are indeed the things that the Christian God knows in knowing Himself, and He knows them in knowing Himself, as to their very existence. Aristotle's God knows all that is or can be in knowing himself, but does he know the existence of beings corresponding to his knowledge ? The texts directed against Empedocles do not say so. The most that we can do is to compare Aristotle's God with a Christian God who is creator neither in potency nor in act, and credit him with the knowledge which appertains to the pure act of a thought that would embrace all—save existences, real or possible, outside himself. If, in this case, we admit with Aristotle that there is something real outside God which he has not created, there is nothing real outside God of which the essence is not implied in the divine knowledge, and notwithstanding this God's knowledge of himself does not imply that he knows the existence of any other thing but himself. If we push the pure act of thought to the plane of the pure act of being, as St. Thomas does, creation becomes possible and the divine knowledge goes further than the order of essences and reaches to that of existences. But Aristotle did not attain to the doctrine of creation. (J. CHEVALIER, *La notion du nécessaire*, pp. 186–187), and that is why he never asserted that God knows the universe.—The problem was discussed in the Middle Ages ; see *De erroribus philosophorum*, Cap. II, n. 15 (in MANDONNET, *Siger de Brabant*, Vol. II, p. 7, and *De quindecim problematibus*, Cap. XI (*op. cit.*, pp. 48–49)).

[2] ATHENAGORAS, *De resurrectione mortuorum*, XVI.

[3] IRENAEUS, *Adv. Haereses*, II, 27, 2.

[4] ST. PAUL, *ad. Hebr.*, I, 3.

[5] ST. AUGUSTINE, *De diversis quaestionibus*, 83, qu. 46, 1–2 : *De genesi ad litteram*, II, 6, 12 ; cf. *Introduction à l'étude de St. Augustin*, p. 109–110, and p. 259.

[6] ST. AUGUSTINE, *De lib. arbit.*, II, 17, 45.

[7] ST. AUGUSTINE, *De vera religione*, XVIII, 35.

[8] ST. AUGUSTINE, *De div. quaest.*, 82, XXIV.

[9] ST. THOMAS, *Sum. theol.*, I, 15, 1, ad 1.

[10] ST. THOMAS, *Sum. theol.*, I, 15, 1 ad 2.

[11] ST. THOMAS, *Sum. theol.*, I, 15, 2.

[12] ST. THOMAS, *De Veritate*, III, 5 ; *Sum. theol.*, I, 15, 3, ad 3.

[13] St. Thomas shows himself more severe on Plato here than usual : " Individua vero, secundum Platonem, non habant aliam ideam quam ideam speciei ; tum quia singularia individuantur per materiam, quam ponebant increatam, ut quidam

dicunt, et concausam ideae ; tum quia intentio naturae con-
sistit in speciebus, nec particularia producit, nisi ut in eis species
salventur. Sed providentia divina non solum se extendit ad
species, sed ad singularia, ut infra (qu. 22, a. 2) dicetur." *Sum.
theol.*, I, 15, 3 ad 4. Elsewhere he declares that Plato admitted
in God no ideas of accidents, and refutes this by the observation
that accidents are also created by God and must have their
ideas : *De Veritate*, III, 7, Resp. Thus in his own thought the
notion of creation appears as the line of demarcation between
Plato and Christian philosophy. It is because Plato did not admit
the creation of matter that he and the Christians conceive the
ideas differently : " Et eadem ratione Plato non ponebat ideas
generum, quia intentio naturae non terminatur ad productionem
formae generis, sed solum formae speciei. Nos autem ponimus
Deum esse causam singularis et quantum ad formam et quantum
ad materiam. Ponimus etiam, quod per divinam providentiam
definiuntur omnia singularia ; et ideo oportet nos singularium
ponere ideas." *De Veritate*, III, 8, Resp.

[14] One of the clearest Thomist definitions of " idea " is as
follows : " Dico ergo, quod Deus per intellectum omnia operans,
omnia ad similitudinem essentiae suae producit : unde essentia
sua est idea rerum, non quidem ut essentia, sed ut intellecta.
Res autem creatae non perfecte imitantur divinam essentiam ;
unde essentia non accipitur absolute ab intellectu divino ut
idea rerum, sed cum proportione creaturae fiendae ad ipsam
divinam essentiam, secundum quod deficit ab ea, vel imitatur
eam," *De Veritate*, III, 2, Resp. The idea, then, is certainly the
divine essence itself conceived under a certain aspect, in relation,
that is, to its possible participations. Cf. " ipsa divina essentia,
cointellectis diversis proportionibus rerum ad eam, est idea
uniuscujusque rei," *Ibid.*, cf. ad 8. This divine knowledge of
the ideas, from the very fact that it bears on the relation of possible creatures to the divine essence, is a practical knowledge.
If there is question of things which will be, or are, effectively
realized, the knowledge of them by the ideas is actually practical ; if there is question of the ideas of things which might be
realized, but will not be, there is a knowledge that is virtually
practical, for God knows them as objects of a *possible* action
(*De Veritate*, III, 3 ; *Sum. theol.*, I, 15, 3). That, moreover, is
why the only ideas which are absolutely determined in God
correspond to beings which His will has decided to create ; as
for the others, He " wills to be able to produce them and to have
the knowledge necessary to produce them " ; He conceives
them as things capable of being made, but not as things made or
to be made. In Him, therefore, they are " *quodammodo inde-
terminatae.*" *De Veritate*, III, 6, Resp. and ad 3. In stating St.
Thomas' position on this point it must not be forgotten that, if

it is true that the ideas are merely the divine essence known as capable of participation, it remains true also that this kind of knowledge is oriented towards action, and bears altogether upon the *creatable*. Cf. " Unde cum idea, proprie loquendo, sit forma rei operabilis hujusmodi . . ." *De Veritate*, III, 7, Resp. " Ideae ordinantur ad esse rerum." *De Veritate*, III, 8.

[15] The generation of the divine ideas in St. Bonaventure may be put schematically thus : God knows Himself, and this adequate knowledge of Himself is a perfect expression of His being, consubstantial with His being ; it is the Word. By the Word, Who is subsisting Truth, He knows Himself, not only in His being, but in all the possible participations of His being. What distinguishes the two doctrines is the special emphasis that Bonaventure puts on the part played by the Word, conceived as Expression and Truth, in the generation of the Ideas. According to St. Thomas God possesses the ideas in virtue of the fact that He knows His essence as capable of being participated ; according to St. Bonaventure He possesses them in virtue of the fact that His Truth *expresses* them as integrally and totally as His power can produce them. Cf. " Quia enim ipse intellectus divinus est summa lux et veritas plena et actus purus : sicut divina virtus in causando res sufficiens est se ipsa omnia producere, sic divina lux et veritas omnia exprimere ; et quia exprimere est actus intrinsicus, ideo aeternus ; et quia expressio est quaedam assimilatio, ideo divinus intellectus, sua summa veritate omnia aeternaliter exprimens, habet aeternaliter omnium rerum similitudines exemplares, quae non sunt aliud ab ipso, sed sunt quod est essentialiter." St. Bonaventure, *De scientia Christi*, qu. 2, Resp. This *rôle* of the " Veritas exprimens " is important enough to characterize the whole Bonaventurian definition of the ideas. These in fact may be defined as " expressions of the divine truth as far as it concerns things," *ipsas expressiones divinae veritatis respectu rerum*—cf. *In I Sent.*, 35, un., I.

[16] Thus in Duns Scotus there is an eternal generation of the intelligible being of things which will one day be created, prior to their creation itself. The divine production of the idea is a kind of eternal prelude to temporal creation. This doctrine almost recalls the eternal generation of Wisdom, the principle of the world's creation, in the Sapiential books : " Verum in mente divina nihil esse potest nisi incommutabile ; ergo quicquid fieri formarique potest, Deus efficere valet, ac reipsa producit juxta propriam cujusque rationem aeternam atque incommutabilem, atque hanc Ideam appellamus ; necessario sunt ideae in mente divina admittenda. Nec aliud sane videntur, *quam ipsa objecta ab aeterno a Deo intellecta*, quaeve per actum intelligendi primum esse intelligibile acceperunt, ad quorum

similitudinem alia effingi atque efformari potuerunt, uti rerum universitatem esse in effectu accepisse constat." Hier. De Montefortino, *J. D. Scoti Summa theologica*, I, 15, 1, Resp. That is why Duns Scotus says that the idea of stone is the stone itself as apprehended by the intellect : *lapis intellectus potest dici idea*. His doctrine, therefore, supposes a realism of the idea much more marked than that of St. Thomas. Duns Scotus is well aware of the fact and openly invokes Plato. With St. Thomas the *intelligible world* is reduced to very little ; there is God's essence, God's knowledge of His essence, and that is all. With St. Bonaventure one might say, speaking rigorously, that the intelligible world consists in the " expressions " generated by which God knows possibles. With Duns Scotus, the expression regains weight, for if the ideas are the things themselves as known in the divine intellect, there really does exist in God a world of intelligible beings. As Duns Scotus says : " Istud videtur consonare cum dicto Platonis . . ." in fact, if *ipsum objectum cognitum est idea*, we may say that there exists in God a universe of essences and that He is peopled with *quidditates habentes esse cognitum in intellectu divino*. Cf. Duns Scotus, *Opus Oxoniense*, I, 35, un.

¹⁷ " Cum Deus sit causa entis, inquantum est ens . . . , oportet quod ipse sit provisor entis inquantum est ens. Providet enim rebus inquantum est causa earum. Quicquid ergo quocumque modo est, sub ejus providentia cadit. Singularia autem sunt entia, et magis quam universalia, quia universalia non subsistunt per se, sed sunt solum in singularibus. Est igitur divina providentia etiam singularium." St. Thomas, *Cont. Gent.*, III, 75. Here we see why Aristotlelianism with its sense of the singular and concrete, is more favourable than Platonism for the development of Christian philosophy, provided only it be transfigured by the metaphysic of Exodus.—One would expect rather a *praesertim* than an *etiam* in the last phrase of the cited text. The equivalent is present in this still firmer declaration : " Ideae ordinantur ad esse rerum. Sed singularia verius habent esse quam universalia, cum universalia non subsistunt nisi in singularibus. Ergo singularia *magis* debent habere ideam quam universalia." *De Veritate*, III, 8, *Sed contra*, and ad 2.

¹⁸ St. Thomas, *Cont. Gent.*, III, 1. Cf. the text cited in the preceding note, and *Cont. Gent.*, III, 94. *Primo namque.*

¹⁹ St. Thomas, *Cont. Gent.*, III, 64, *Amplius ostensum est.*

²⁰ St. Thomas, *Cont. Gent.*, III, 75, *Adhuc, si Deus.*

²¹ St. Thomas, *Cont. Gent.*, III, 97, *Ex his autem.*

²² St. Thomas, *Cont. Gent.*, III, 111.

²³ St. Thomas, *Cont. Gent.*, III, 112.

²⁴ St. Thomas, *Cont. Gent.*, III, 113.

²⁵ St. Thomas, *Cont. Gent.*, III, 113.

NOTES TO CHAPTER IX

[1] St. Paul, I Corinth., xv. 12–19.

[2] *Op. cit.*, xv. 52–53.

[3] *De resurrectione*, VIII.

[4] On this subject we may note two interesting points : I. Even those Fathers who admit the immortality of the soul, will not concede to Plato its *natural* immortality. For Plato the soul *is* life ; for Christian thinkers if it *is* life it is God ; wherefore the soul can be immortal only because it has *received* life and in virtue of a divine decree. In this sense the human soul more nearly resembles the indestructible gods of the *Timaeus*, indissoluble by decree (41 A), than the naturally immortal soul of the *Phaedo*. See Justin, *Dialogue with Trypho*, VI, 1 ; there is some echo of this view in St. Augustine : E. Gilson, *Introduction à l'étude de saint Augustin*, p. 69 ; 170, note 1 ; and 186.—II. It is difficult to be sure at times whether the older Fathers are speaking of the immortality of the resuscitated soul, or that of the soul between bodily death and the resurrection. Justin's expressions are obscure : "Just as man does not exist in perpetuity and the body is not always united to the soul, but when this harmony is to be broken up the soul leaves the body and the man exists no more, so also *when the soul must cease to be* the spirit of life leaves it ; *the soul exists no more* and returns whence it was taken." Justin *op. cit.*, VI, 2. Does it cease to exist altogether or can it return, and does Justin mean only that it exists no more *as soul?* It is difficult to say. Tatian is similarly obscure, *Discourse to the Greeks*, Chap. XIII : "The soul, O Greeks, is not immortal, but mortal. Yet it is possible for it not to die. If, indeed, it knows not the truth it dies and is dissolved with the body, but rises again at last at the end of the world with the body receiving death by punishment in immortality. But, again, if it acquires this knowledge of God it dies not, although for a time it is dissolved." It seems difficult not to admit that Tatian believed in a kind of death of the soul followed by a resurrection for eternal reward or damnation. As to Irenaeus, he represents souls as surviving their bodies, but imagines them as recognizable phantoms, taking the form of their bodies like frozen water in a vase (*Adv. haereses*, II, 19, 16 and II, 34, 1). This brings us to Tertullian, whose thoroughgoing materialism marks an extreme limit of possible variation in Christian philosophy ; but on so manifest an aberration there is no need to dwell : cf. Tertullian, *De anima*, VI.

[5] St. Augustine, *De moribus ecclesiae*, I, 27, 52 : "Homo igitur, ut homini apparet, anima rationalis est mortali atque terreno utens corpore."

[6] St. Augustine, *In Joan Evang.*, XIX, 5, 15.

[7] St. Augustine, *De quantitate animae*, XIII, 22.

8 ST. AUGUSTINE, *De moribus ecclesiae*, I, 4, 6.

9 ARISTOTLE, *De anima*, I, 1.

10 *Op. cit.*, I, 2.

11 *De gener. anim*, 736 b, 28.

12 ARISTOTLE, *De anima*, III, 5.

13 ALBERT THE GREAT, *Summa theologica*, II, tr. 12, qu. 69, membr. 2, art. 2.

14 ST. THOMAS, *Sum theol.*, I, 75, 2, *Sed. contra*.

15 This at the same time settles the problem of the souls of animals. Animals have a soul, but it is not an intellect and therefore not a substance ; and so the question of its immortality does not arise.

16 St. Thomas himself notes this fundamentally important connection of ideas : see *Sum. theol.*, I, 75, 4, Resp.

17 ST. THOMAS, *Sum. theol.*, I, 76, 1.

18 ST. THOMAS, *Sum. theol.*, I, 75, 2, Resp.

19 The doctrine may be summed up in somewhat more technical and, at bottom, clearer formulæ. *Being* is the act of existence. Posited by this act, the being is posited in and for itself. Since it is, it is by definition itself and no other : *indivisum in se et divisum ab aliis ;* the being conceived in its undivided unity is called precisely *substance,* and its property of existing as substance, that is to say, for itself and without substantial dependence on any other, is called *subsistence*. Thus the act of being causes the substance and its subsistence. If we consider it further as making the being to be this rather than that, we call it *formal act,* and, considering its formality separately, we say that the act is *form*. Thereby we add to its property of causing subsistence that of determining as form the kind of substance. Among forms, some have an actuality sufficient to enable them to subsist alone, and these are *pure forms, separate forms*. Others cannot exist save in a matter to which they communicate their actuality ; these are *substantial forms*. Among substantial forms some are subsisting principles of operations proper to themselves as forms, and these are *rational souls ;* others are bound up with matter both as to their being and as to their operation, and these are *material forms*. Man, consequently, is a concrete substance, that is to say a substance in which there are parts which may legitimately be considered separately ; but his being is one, first because the substantial elements of his being, that is to say body and soul, cannot subsist apart, and secondly because it is due to the subsistence of one only of these, that is to say the soul, that the substance, man, subsists. The different *rôle* played by the two parts is quite clear, inasmuch as the soul, once it has obtained the necessary aid from the body, can subsist without the body, as in fact it does after the death of the man, while on the other hand the body can in no case subsist without

the soul, to which latter it owes all its actuality ; as the very dissolution of the corpse is enough to prove.

NOTES TO CHAPTER X

[1] That, moreover, is why there can be no science of the individual as such. For Aristotle, as we know, there is no science save of the universal. This celebrated formula should be taken absolutely ; *in itself* the universal alone is the object of science. When the Christian philosophers repeat the formula they give it another meaning. When St. Thomas, following Aristotle, says that there is no science of the particular, he does not mean, as Aristotle does, that *in itself* the particular is not a possible object of science, but that it is not so *for us*. In itself, and absolutely, a science of the particular is perfectly possible, since God in fact possesses it. See J. CHEVALIER, *Trois conférences d'Oxford*, Paris, éditions Spes, 1928, pp. 22–27, where the point is elaborated.

[2] ATHENAGORAS, *De resurrectione mortuorum*, Cap. XV.

[3] ARISTOTLE, *Metaph.*, Λ 1071 a, 27–29.

[4] ST. THOMAS AQUINAS, *Sum. theol.*, I, 47, 2.

[5] " Individuum compositum ex materia et forma habet quod substet accidenti ex proprietate materiae. Unde et Boetius dicit in lib. *De Trinitate*, Cap. II : *forma simplex subjectum esse non potest*. Sed quod per se subsistat (*scil*. individuum) habet ex proprietate suae formae, quod non advenit rei subsistenti, sed dat esse actuale materiae, ut sic individuum subsistere possit." ST. THOMAS AQUINAS, *Sum. theol.*, I, 29, 2 ad 5.—" Anima illud esse in quo subsistit communicat materiae corporali, ex qua et anima intellectiva fit unum, ita quod illud esse quod est totius compositi, est etiam ipsius animae ; quod non accidit in aliis formis, quae non sunt subsistentes." *Sum. theol.*, I, 76, 1 ad 5.— This declaration, of capital importance, was not forgotten by subsequent representatives of the Thomist school ; they maintained, against the Augustinians, *quod esse animae communicatur corpori :* " Quaestione I quaerit (Thomas) utrum anima possit esse forma et hoc aliquid ; et in responsione principali dicit quod idem esse animae communicatur corpori ut sit unum esse totius speciae. Et in I super Sententias, distinctione 8, quaerens utrum anima sit simplex, dicit hoc idem planius in solutione 3 argumenti, scilicet quod unum esse quod est animae per se fit conjuncti et non est ibi esse nisi ipsius formae." P. GLORIEUX. The " *correctorii coruptorii quare* " (*Bibliothèque thomiste*, IX), Paris, J. Vrin, 1927, p. 361. St. Bonaventure has the same doctrine with another terminology : " Individuum enim habet esse, habet etiam existere. Existere dat materia formae, sed *essendi actum* dat forma materiae." *In II Sent.*, dist. III, p. 1, art. 2, qu. 3, Resp.

[6] " Unde sicut diversitatem in genere vel specie facit diversitas materiae vel formae absolute, ita diversitatem in numero facit haec forma et haec materia ; nulla autem forma, in quantum hujusmodi est *haec* ex seipsa (note the rejection by anticipation of the Scotist *hecceitas*). Dico autem in quantum hujusmodi propter animam rationalem, quae quodammodo ex seipsa est hoc aliquid, sed non in quantum forma." ST. THOMAS *In Boet. de Trinitate*, qu. 4, art. 2, Resp. This remarkable text reminds us that the case of the rational soul is unique. As St. Thomas says in the second text cited in the previous note, only intellectual souls subsist, and that is why it is that of these souls alone we can say that their being is the being of the whole composite. Such is not the case with the souls of brutes. Thus it is as a consequence of their own subsistence that human souls possess an individuality. Nevertheless, it remains true to say that they are not individual as forms, but only as subsisting and forms of *this* substance which, apart from matter, would not exist.

[7] This will be more clearly grasped if we compare the problem of individuation with that of the diversification of beings. It is a principle common to Aristotle and St. Thomas that matter is always for the sake of the form, not the form for the sake of the matter. It is therefore impossible to imagine the diversity of forms as due to the necessity of adapting them to the diversities of the matters ; the truth is the other way about ; diverse matters are required to allow diverse forms to enter into union with them so as to make concrete subjects. Systematically applied, as St. Thomas applies it, this principle becomes an essential part of the metaphysical armature of the universe. Differing herein from Aristotle, St. Thomas links it up with the concept of creation. Thence it results that the Christian God creates forms for themselves, and creates diverse matters only in the measure demanded by the diversity of the forms. " Causa autem diversitatis rerum non est ex materia nisi secundum quod materia ad rerum productionem praeexigitur, ut scilicet secundum diversitatem materiae diversae inducantur formae. Non igitur causa diversitatis in rebus a Deo productis est materia.—Adhuc, secundum quod res habent esse, ita habent pluralitatem et unitatem, nam unumquodque secundum quod est ens, est etiam unum ; sed non habent esse formae propter materiam, sed magis materiae propter formas, nam actus melior est potentia ; id autem propter quod aliquid est, oportet melius esse. *Neque igitur formae ideo sunt diversae, ut competant materiis diversis, sed materiae ideo sunt diversae, ut competant diversis formis.*" ST. THOMAS AQUINAS, *Compendium theologiae*, Pars. I, Cap. LXXI. This principle enables us to understand how, in Thomist doctrine, matter can be the principle of individuation, without individuality being thereby subjected to matter ; for the individual

supposes a matter, but since the matter is there only in view of the diversity of the forms, it is finally due to the form that the concrete substance is endowed with individuality. On this point A. Forest observes, very rightly, that the whole misunderstanding arises from a failure to put the problem of individuation on its proper metaphysical basis. The Thomist doctrine means that matter is precisely a *principle* and nothing else. Whence, then, came the original and individual differences of each concrete being? They are *made possible* by its matter, they *proceed from* its form, which alone gives actuality. *La structure métaphysique du concret*, pp. 255–256.

[8] St. Thomas Aquinas, *Sum. theol.*, I, 29, 1, Resp.

[9] St. Thomas Aquinas, *Sum. theol.*, I, 98, 1, Resp.

[10] St. Bonaventure, *In I Sent.*, dist. 25, art. 1, ad 2.

[11] As to this continuing influence of Christian philosophy upon the history of modern philosophy, see the very true remarks of H. Ritter, *Histoire de la philosophie chrétienne*, Vol. I, pp. 20–22.

[12] St. Bonaventure, *Com. in Joan.*, VIII, 38.

[13] St. Thomas Aquinas, *Sum. theol.*, I, 29, 3, Resp.

NOTES TO CHAPTER XI

[1] Xenophon, *Memorabilia*, IV, 2, 24–25.

[2] Gregory of Nyssa, *De hominis opificio*, Cap. XVI.—St. Bonaventure, *In II Sent.*, 16, 1, 2, Resp.—St. Thomas Aquinas, *Sum. theol.*, I, 93, 4.

[3] St. Ephrem of Nisibis, *Interpretationes in sacram scripturam*.

[4] St. Bernard, *De gratia et libero arbitrio*, Cap. IX., n. 28.

[5] St. Augustine, *Enarr. in Ps.*, 42, n. 6.

[6] St. Thomas Aquinas, *Sum. theol.*, I, 93, 8, *Sed contra* and *Resp.*

[7] Duns Scotus, *Op. Oxon.*, I, 3, 11, 7.

[8] St. Bernard, *De consideratione*, lib. II, Cap. III, n. 6.

[9] St. Augustine, *De civ. Dei.*, XIX, 13.

[10] St. Bernard, *Brevis expositio in C.C.*, Cap. XXII.—Cf. *De gradibus humilitatis*, Cap. II (relation between humility and self-knowledge).

[11] Hugh of St. Victor, *De Sacramentis*, I, 6, 15.

[12] St. Bernard, *De diligendo Deo*, Cap. II.

[13] Pseudo-Bernard, *Meditationes . . . de cognitione humanae conditionis*, Cap. V, art. 14–15.

[14] St. Thomas Aquinas, *Sum. theol.*, I, 91, 1, Resp. ; I, 96, 2, Resp. *Qu de Anima, qu. un.*, art. 1, ad Resp.

[15] St. Augustine, *De symbolo*, I, 2.

[16] St. Augustine, *Confessions*, lib. X, Cap. VIII–XXVII.

[17] Richard of St. Victor, *Benjamin minor*, Cap. LXXVIII.

[18] *Ibid.*, Cap. LXXV.

[19] *Ibid.*, *Benjamin major*, lib. III, Cap. XIV.

[20] The Augustinians teach that the soul knows itself directly, but through " species." See, for instance : " Sic ergo dico, quod anima semetipsam et habitus, qui sunt in ipsa, cognoscit non tantum arguendo, sed intuendo et cernendo per essentias suas objective, sed formaliter per species ex ipsis expressas, unde formatur acies cogitantis sive intelligentis." The reason is : " Anima enim rationalis est imago Dei." MATT. OF AQUASPARTA, *Quaest. disp. de cognitione*, qu. V, Resp ; P. J. OLIVI teaches a similar doctrine, *In lib. II. Sent.*, qu. 76 ; edit. B. JANSEN, Quaracchi, 1926, Vol. III, pp. 148–149, where the memory plays the part of *species*. This text is important because it sends us to ST. ANSELM, *Monologion*, Cap. XXXIII and to ST. AUGUSTINE, *De Trinitate*, IX, 11–12, for the sources of the doctrine. But where the relation between the Augustinian statement of the question and that of St. Thomas may be studied most conveniently is the very remarkable edition of ROGER MARSTON, *Quaest. disp. de emanatione aeterna, de statu naturae lapsae et de anima*, Quaracchi, 1932. See especially *De anima*, qu. 1, pp. 206–221, the conclusion being : " Concedo igitur quod res mere spirituales non possunt cognosci ab anima nisi per speciem aliquam a se differentem."

[21] Consult particularly ST. THOMAS AQUINAS, *De Veritate*, qu. XV, art. I, where will be found all the necessary references to St. Augustine and Dionysius. See especially the *Sed contra* and *Resp.*—This point is also studied by A. GARDEIL, *La structure de l'âme*, Vol. I, p. 24.

[22] R. DESCARTES, *Secondes réponses*.

[23] B. PASCAL, *Pensées*, L. Brunschvieg, edit. minor, p. 345, note 2.

[24] *Op. cit.*, p. 531.

[25] *Op. cit.*, p. 367.

NOTES TO CHAPTER XII

[1] ST. AUGUSTINE, *De div. quaest. 83*, qu. 9.

[2] ARISTOTLE, *Metaph.*, I, 9, 991 a 12.

[3] MATTHEW OF AQUASPARTA, *Quaest. disputatae*, qu. I, De cognitione, Resp ; ad 8.

[4] MATTHEW OF AQUASPARTA, *op. cit.*, Contra 5.

[5] MATTHEW OF AQUASPARTA, *op. cit.*, ad 12.

[6] P.-J. OLIVI, *In II Sent.*, Appendix, qu. II. Note Olivi's excellent *résumé* of the Augustinian position (art. 6) : " Praeterea, nullum fallibile et mutabile potest infallibiliter certificare aut actum infallibilem et immutabilem generare ; sed omnis species seu ratio creata est fallibilis et mutabilis ; ergo impossibile est quod intellectus per speciem creatam aut lumen creatum

infallibiliter et immutabiliter certificetur ; et ita oportet quod certificetur per rationem aeternam et lumen aeternum."

[7] St. Thomas Aquinas, *Quaest. disp. de Veritate*, qu. 1, art. 1 et 2.

[8] The fundamental text is in St. Anselm, *De Veritate*, Cap. XIII. The Augustinians make use of it to establish that, without the divine truth, which is unique, there would be no particular truth of things ; see, for example, R. Grosseteste, *De unica forma omnium*, edit. L. Baur, pp. 106–111, and *De Veritate*, pp. 130–143. St. Thomas, on the contrary, discussing Anselm's conclusions, sets out to show that, granted this thesis, there nevertheless remains a proper truth in things : *Qu. disp. de Veritate*, I, 4, Resp.

[9] St. Thomas Aquinas, *De Veritate*, I, 4, ad 3 and 4.

[10] St. Thomas Aquinas, *op. cit.*, I, 9, Resp.

[11] Duns Scotus, *Op. Oxon.*, I, 3, 4, 2.

[12] Duns Scotus, *Op. Oxon.*, I, 3, 4, 2, 9.

[13] St. Bonaventure, *Itinerarium mentis in Deum*, I, 14.

[14] It may in a sense be said that Kantism consists in attributing to human thought that function of creating intelligibility reserved by the Middle Ages for God : " Scientia Dei aliter comparatur ad res quam scientia nostra ; comparatur enim ad eas sicut et causa et mensura. Tales enim res sunt secundum veritatem, quales Deus sua scientia eas ordinavit. *Ipsae autem res sunt causa et mensura scientiae nostrae.* Unde sicut et scientia nostra refertur ad res realiter, et non e contrario, ita res referuntur realiter ad scientiam Dei et non e contrario." St. Thomas Aquinas, *Qu. disp. de Potentia*, VII, 10, ad 5. This suggests another text which shows that, with Kant, our intellect plays that part with respect to natural things which St. Thomas reserves for the divine intellect, and attributes to us only with respect to artificial things—*our creations* : " Intellectus enim practicus causat res, unde est mensuratio rerum quae per ipsum fiunt, sed intellectus speculativus, quia accipit a rebus, est quodammodo motus ab ipsis rebus, et ita res mensurant ipsum. Ex quo patet quod res naturales, ex quibus intellectus noster scientiam accipit, mensurant intellectum nostrum, ut dicitur *X Metaph.* (com. 9) : sed sunt mensuratae ab intellectu divino, in quo sunt omnia creata, sicut omnia artificiata in intellectu artificis. Sic ergo intellectus divinus est mensurans non mensuratus ; res autem naturalis mensurans et mensurata ; sed intellectus noster est mensuratus, *non mensurans quidem res naturales*, sed artificiales tantum." *Qu. disp. de Veritate*, I, 2, Resp.

[15] Kant, *Kritik der reinen Vernunft*, Preface to 2nd edition.

[16] Fr. Bacon, *Instauratio magna*, Praef ; used by Kant as an epigraph for the *Critique of Pure Reason*.

NOTES TO CHAPTER XIII

[1] St. Thomas Aquinas, *Sum. theol.*, I, 88, 2, Resp. and ad 4m—I, 88, 3, Resp.—*Cont. Gent.*, III, 42–43.

[2] St. Thomas Aquinas, *Sum. theol.*, I, 88, 1, Resp. and I, 88, 3.

[3] St. Thomas Aquinas, *Sum. theol.*, I, 89, 1, Resp.

[4] Duns Scotus, *Op. Oxon.*, I, 3, 3, 24.

[5] Duns Scotus, *Quaest. Quodlib.*, XIV, 14–15.

[6] Duns Scotus, *op. cit.*, 16–17.

[7] St. Thomas Aquinas, *Cont. Gent.*, III, 52.

[8] St. Thomas Aquinas, *Cont. Gent.*, III, 49, ad *Cognoscit tamen*.

[9] St. Thomas Aquinas, *Cont. Gent.*, III, 51, init.

[10] St. Thomas Aquinas, *Cont. Gent.*, III, 54, ad *Rationes*. The basis of the doctrine may be found in St. Augustine : " Neque enim omnes homines naturali instinctu immortales et beati esse vellemus, nisi esse possemus. Sed hoc summum bonum praestari hominibus non potest, nisi per Christum et hunc crucifixum, cujus vulneribus natura nostra sanatur. *Ideo justus ex fide vivit.*" *Cont. Julian pelag.*, II, 3, 19. The text cited above, p. 260, is taken from the *Contra Gentes.*, III, 48. I give it in its entirety, because it helps us to divine St. Thomas' heart behind his doctrine : " In quo satis apparet, quantam angustiam patiebantur hinc inde eorum praeclara ingenia, a quibus angustiis liberabimur, si ponamus secundum probationes praemissas hominem ad veram felicitatem post hanc vitam pervenire posse, anima hominis immortali existente." I would very gladly give up all the " Triumphs of St. Thomas " in which great artists have represented Averroes vanquished at his feet for these few deep-toned lines in which, along with the joy of deliverance, he expresses the fraternal pity of a truly Christian soul. Let him then remain in our eyes " the incomparable St. Thomas Aquinas, as great of mind as of heart." I will not be so unkind to those who reproach me for writing an apology as to appropriate the words without warning them of their source ; they come from Auguste Comte, *Système de politique positive*, Vol. III, pp. 488–489.

[11] St. Thomas Aquinas, *Sum. theol.*, I, 5, 2, Resp.

[12] Duns Scotus, *Op. Oxon.*, I, 3, 7, 39.

[13] St. Thomas Aquinas, *Cont. Gent.*, I, 62 end.

[14] St. Albert the Great, *De intellectu et intelligibili*, Tr. I, Cap. II.

[15] St. Thomas Aquinas, *Sum. theol.*, I, 16, 5, ad 3. Cf. I, 88, 3, ad 2.

[16] St. Thomas Aquinas, *Sum. theol.*, I, 16, 5, ad 2. " Res dicitur verae per comparationem ad intellectum divinum.' *Op. cit.*, I, 16, 6, ad 2 ; cf. *Ibid.*, Resp.

NOTES TO CHAPTER XIV

[1] St. Augustine, *Epist.* 140, II, 3. This text is important for the purpose of determining the exact meaning of the *primus gradus amoris* in St. Bernard, *De diligendo Deo*, Cap. VIII.

[2] Guigues le Chartreux, *Meditationes*, II and V.

[3] St. Bernard, *De diligendo Deo*, Cap. VII.

[4] St. Bernard, *op. cit.*

[5] St. Bernard, *op. cit.* Cf. Pascal, *Pensées* : " Console-toi, tu ne me chercherais pas, si tu ne m'avais trouvé."

[6] St. Thomas Aquinas, *In lib. de Divinis nominibus*, lect. XI.

[7] St. Thomas Aquinas, *op. cit.*

[8] William of St. Thierry, *De natura et dignitate amoris*, I, 2.

[9] St. Thomas Aquinas, *Sum. theol.*, I, 60, 5, Resp. Cf. ad 1.

[10] " Unde homo in statu naturae integrae dilectionem sui ipsius referebat ad amorem Dei sicut ad finem, et similiter dilectionem omnium aliarum rerum. Et ita Deum diligebat plus quam seipsum et super omnia. Sed, in statu naturae corruptae, homo ab hoc deficit secundum appetitum voluntatis rationalis, quae, propter corruptionem naturae, sequitur bonum privatum, nisi sanetur per gratiam Dei. Et ideo dicendum est quod homo, in statu naturae integrae, non indigebat dono gratiae superadditae naturalibus bonis, ad diligendum Deum naturaliter super omnia, licet indegeret auxilio Dei ad hoc eum moventis. Sed in statu naturae corruptae, indiget homo, etiam ad hoc, auxilio gratiae naturam sanantis." St. Thomas Aquinas, *Sum. theol.*, I–IIae, 109, 3, Resp. Thus, originally man was naturally capable of loving God more than all things ; the only divine help then necessary was a divine motion exerted on his nature. After the fall, on the contrary, nature must first be healed by grace, before it can receive the help of the divine motion. Thus it is not simply our nature that henceforth loves God above all things, but our nature restored by grace.—As to what charity adds to the natural love of God above all things, see *loc. cit.*, ad 1m. Our natural love of God above all things goes to God as He is naturally known to us ; now our natural knowledge attains Him as first principle and last end of the universe ; and it is therefore as such that we prefer Him to the rest. Faith reveals Him as our ultimate Beatitude, and so enables us to love Him henceforth as such ; the love of God as object of a possible beatifying knowledge and " secundum quod homo habet quamdam societatem spiritualem cum Deo "—that is what charity adds to nature here.

[11] St. Thomas Aquinas, *Sum. theol.*, I, 60, 5, Resp.

[12] St. Thomas Aquinas, *In II Sent.*, dist. I, qu. 2, art. 2, Resp.

[13] St. Thomas Aquinas, *Cont. Gent.*, III, 24 ad *Sic igitur.*

[14] St. Thomas Aquinas, *Cont. Gent.*, III, 25 ad *Adhuc unum quodque tendit.*

[15] WILLIAM OF ST. THIERRY, *Epist. ad fratres de Monte Dei*, II, 16.

[16] ST. BERNARD, *In Cant. Cant.*, 82, 8.—WILLIAM OF ST. THIERRY, *Epist. ad fratres de Monte Dei, ibid.*

NOTES TO CHAPTER XV

[1] On the Greek sources of the mediæval conception of free will we may usefully consult the works of M. WITTMANN, *Die Ethik des Aristoteles*, Regensburg, 1920. *Aristoteles und die Willensfreiheit*, Fulda, 1921—*Die Lehre von der Willensfreiheit bei Thomas von Aquin historisch untersucht*, in *Philos. Jahrbuch*, Vol. XL (1927), pp. 170–188, and 285–305.—As to the mediæval predecessors of St. Thomas, see O. LOTTIN, *La théorie du libre arbitre depuis saint Anselm jusqu'à saint Thomas d'Aquin*, Louvain, abbaye du Mont-César, 1929.

[2] ORIGEN, *De principiis*, I, Praef., 5.—ST. EPHREM, *Hymni de Epiphania*, 10, 14.—GREGORY OF NYSSA, *Orat. catech.*, 31.—JOHN CHRYSOSTOM, *In Genes. homiliae*, 22, 1. *In Epist. ad Ephesios homiliae*, 4, 2. *In Epist. ad Hebraeos homiliae*, 12, 3.—JOHN DAMA-SCENE, *De fide orthod.*, 2, 30.

[3] ARISTOTLE, *Eth. Nic.*, III, 2, 1111 b, 4–1112 a, 17. It is by the choice we make between good and evil that we are morally qualified. This choice belongs to the sphere of the voluntary (τὸ ἐχούσιον), but is only a part of it, for all choice is voluntary, but all that is voluntary is not a choice. Thus choice is always a voluntary act based on a rational deliberation (*loc. cit.*, 1112 a, 15–16).

[4] As to what remains of *nature* in the will as conceived by Aristotle see the profound remark of ST. THOMAS AQUINAS, *Sum. theol.*, 1a–11ae, 10, 1, ad 1.

[5] ST. AUGUSTINE, *De gratia et libero arbitrio*, I, 1 ; II, 2–4 ; X, 22.

[6] *Op. cit.*, I, 12, 26 ; II, 1, 1 ; III, 3, 7 ; *Retract.*, I, 9, 4. ST. BERNARD, *De gratia et lib. arb.*, I, 2.

[7] ST. ANSELM, *De lib. arbit.*, Cap. II. HUGH OF ST. VICTOR, *De Sacramentis*, I, 5, 21.

[8] DUNS SCOTUS, *Op. Oxon.*, II, 25, 1, 22–24.—*In lib. Metaph.*, IX, 15.

[9] BOETHIUS, *In lib. de Interpret : editio secunda.*

[10] In the excellent article to which I have already referred in Note 1 M. Wittmann has noticed certain of these finely shaded expressions : *quoddam judicium :* " . . . quod cum electio sit quoddam judicium de agendis, vel judicium consequatur . . ." *De Veritate*, 24, 1, ad 20. Cf. " Et hoc modo ipsa electio dicitur quoddam judicium," *Sum. theol.*, I, 83, 3, ad 2. After having shown that things have no *arbitrium*, that the animals have one

that is not free because they do not judge their natural judgment, St. Thomas concludes of man : " Et ideo est liberi arbitrii, *ac si diceretur* liberi judicii de agendo vel non agendo." See M. Wittmann, pp. 295–298. The very penetrating remark of this historian on the meaning given by St. Thomas to the formula of Bœthius should be taken into consideration. The *liberum de voluntate judicium* which means for Boethius a free judgment *on* the will taken as an object, means with St. Thomas : the free judgment which comes *from* the will, inasmuch as its *electio* is a kind of judgment. In another sense, however, St. Thomas also admits that the judgment bears *on* the will : " Homo vero per virtutem rationis judicans de agendis, potest *de suo arbitrio* judicare . . ." (*De ver.*, 24, 1, Resp.).—What must be retained is that even when man judges his will, it is the reason that judges, and that when his free will decides, it is the will that chooses.

[11] ST. BERNARD, *De gratia et libero arbitrio*, VIII, 24 : ST. THOMAS AQUINAS, *Sum. theol.*, I, 83, 2 ad 3.

[12] ST. AUGUSTINE, *De spiritu et littera*, XXXI, 53.

[13] *Ibid.*, XXX, 52.

[14] Definition of *potestas* : " Est igitur potestas aptitudo ad faciendum et omnis aptitudo ad faciendum potestas." ST. ANSELM, *De voluntate*—" Peccavit autem (primus homo) per arbitrium suum, quod erat liberum ; sed non per hoc unde liberum erat, id est per potestatem qua poterat non peccare, et peccato non servire ; sed per potestatem quam habebat peccandi, qua nec ad non peccandi libertatem juvabatur, nec ad peccandi servitutem cogebatur." ST. ANSELM, *De lib. arbit.*, II.

[15] " Ergo quoniam omnis libertas est potestas, illa libertas arbitrii est potestas servandi rectitudinem voluntatis propter ipsam rectitudinem . . . Jam itaque clarum est liberum arbitrium non esse aliud quam arbitrium potens servare rectitudinem voluntatis propter ipsam rectitudinem." ST. ANSELM, *De lib. arbit.*, Cap. III. " Sed nunc quomodo est humanae voluntatis arbitrium hac potestate liberum ; . . ." *Op. cit.*, Cap. V.—" Potestas ergo peccandi, quae addita voluntati minuit ejus libertatem et, si dematur, auget, nec libertas est, nec pars libertatis," *De lib. arbit.*, Cap. I.—" Est enim potestas libertatis genus." *De lib. arbit.*, Cap. XIII.

[16] ST. THOMAS AQUINAS, *De veritate*, XXII, 6, Resp.

[17] ST. THOMAS AQUINAS, *Sum. theol.*, ia–iiae, 17, 1 ad 2.

[18] *Ibid.*, I, 19, 10, Resp.

[19] Wherefore the gift of " counsel " remains even in beatitude : ST. THOMAS, *Sum. theol.*, iiae–iiae, 52, 3, Resp., and ad 1.

[20] ST. THOMAS AQUINAS, *Sum. theol.*, I, 59, 3, ad 3, and I, 62, 8, ad 3.

[21] For the history of this influence I may be permitted to refer to my study on *La liberté chez Descartes et la théologie*, Paris,

Alcan, 1913, pp. 286–432. This part of the book may still be of service. Although it calls for many corrections of detail, the thesis maintained still seems to me to be true in all essentials. I did not know at that time, and have only just discovered, that the influence of Gibieuf on Descartes, and the later hesitations of the philosopher, had already been noted by the Scotist J. A. FERRARI, *Philosophia peripatetica*, 2nd Edn., Venice, 1754, Vol. I, pp. 310–312. I would also take this opportunity to say that the lack of coherence for which I amiably forgave Père Petau in this work (p. 404) seems to me now to be wholly attributable to my own ignorance, at that time, of the historical antecedents of the problem. Petau was altogether right in saying that if we call the will's power of choosing *indifference*, it is altogether inadmissible, while maintaining nevertheless that if we call its power of choosing ill indifference, this latter may be eliminated without touching the former. The will retains its indifference inasmuch as it chooses ; to choose always well is to remain always free. The first part of the work, which treated of the divine liberty, would call for many more corrections. Some very useful ones will be found in M. P. GARIN's book, *Thèses cartésiennes et thèses thomistes*, Paris, Desclée de Brouwer, s.d. (1932). I should have many others to make on my own account.

NOTES TO CHAPTER XVI

¹ ARISTOTLE, *Eth. Nic.*, X, 6, 1176 b 8. What is morally good, with Aristotle, is essentially that which merits praise and honour. The social factor is so important that in his eyes the wicked are distinguished from the good, as those who can be kept from evil doing by *fear* are distinguished from those who can be kept from it by *shame*, or *modesty : Eth. Nic.*, X, 10, 1179 b 11.

² CICERO, *De finibus bonorum et malorum*, II, 14, 45.

³ ST. THOMAS AQUINAS, *Sum. theol.*, IIa, IIae, 145, 1 ad 3.

⁴ ST. THOMAS AQUINAS, *Sum. theol.*, IIa, IIae, 145, 1 ad 2.— The assimilation of the Greek idea of beauty to Christian morals was effected by ST. AUGUSTINE, *De div. quaest. 83*, qu. 30. This text, which is of capital importance for the history of the question, was systematized by St. Thomas (with the addition of elements taken from Dionysius the Areopagite) in *Sum. theol.*, IIa–IIae, 145, 2. What is especially to be noted in Augustine's text, is the transformation imposed on the idea of virtue, on which alone the Ancients based their morals. For Cicero, virtue *is* the *honestum* and *vice versâ :* and for him this is sufficient. For Christians, virtue is still the *honestum* and, as such, is distinguished from the *utile* (what is desirable, not for its own sake, but as a means to something else) ; only, the whole order of the *utile* henceforth includes all that divine providence had disposed in

view of the supreme end, and the *honestum* is that which is to be enjoyed (*frui* not *uti* = utile), but not used, that is to say, God. Thus Cicero's virtues are always the *honestum*, but henceforth wisdom, fortitude, justice and temperance are virtues because they pertain to a soul that rejoices in God, and makes use of all the rest in view of God : " Neque enim ad aliquid aliud Deus referendus est " (*loc. cit.*). If then the virtues are to be desired *propter se* (*op. cit.*, qu. 31, 2) it is not because they suffice, or suffice for us, for to the Christian they come from God and lead to God, which is the reason why they are good : " Quid ergo ? Jam constitutis ante oculos nostros tribus, Epicureo, Stoico, Christiano, interrogemus singulos. Dic Epicuree, quae res faciat beatum ? Respondet : voluptas corporis. Dic, Stoice, Virtus animi. Dic Christiane. Donum Dei. . . . Magna res, laudabilis res : lauda Stoice, quantum potes ; sed dic, unde habes ? Non virtus animi tui te facit beatum, sed quod tibi virtutem dedit, qui tibi velle inspiravit, et posse donavit " (*Sermo* 150, 7, 8–8, 9). This text not only affirms the necessity of grace, which does not appertain to the philosophical order, but also marks that virtue, of which the seeds lie in us naturally (*De div. quaest. 83*, 31, 1), depends on God and grace not only as regards its efficacy, but also as regards its existence and worth, since God is both its principle and end : *non virtus animi tui te fecit beatum, sed qui tibi virtutem dedit.* As the supreme moral value Christianity replaces virtue by God, and the whole conception of the *moral end* is thereby transformed.

[5] ARISTOTLE, *Phys.*, lib. VII, Cap. III (lect. V) cited by St. Thomas in the following form : " Virtus est dispositio perfecti ad optimum ; dico autem perfecti, quod est dispositum secundum naturam." *Sum. theol.*, 1a–11ae, 71, 1, Resp.

[6] ST. AUGUSTINE, *De libero arbitrio*, III, 13, 38.

[7] " Nam virtus est animi habitus, naturae, modo, ratione consentaneus." CICERO, *De inventione rhetorica*, II, 53.

[8] ST. THOMAS AQUINAS, *Sum. theol.*, 1a–11ae, 71, 2, Resp.

[9] ST. AUGUSTINE, *Cont. Faustum manich.*, XXII, 27.

[10] ST. AUGUSTINE, *De lib. arbit.*, III, 14, 42.

[11] John iii. 21.

[12] ST. THOMAS AQUINAS, *Sum. theol.*, Ia–IIae, 71, 2, ad 4.

[13] ST. THOMAS AQUINAS, *Sum. theol.*, Ia–IIae, 71, 6, ad 5.

[14] ST. AMBROSE, *De paradiso*, VIII, 39 : PETER LOMBARD, *II Sent.*, dist., 35, Cap. I : ST. BONAVENTURE, *In II Sent.*, 35, dub. 4.

[15] ARISTOTLE, *Eth. Nic.*, III, 1, 1110 b, 28–30.

[16] On the difference between ἀτύχημα, ἁμαρτήμα and ἀδίχημα see ARISTOTLE, *Eth. Nic.*, V, 8, 1135 b, 11–25. It should be noted that the ἁμαρτήμα is not even a moral evil because it does not depend on a vice and does not result from the injustice of the

acting subject (1125 a, 17–19) ; the fault arises either from a purely accidental error, a mere ignorance, or from an error in deliberation (1142 a, 21). Moral malice depends therefore on the presence of a vice (stable bad habit) ; one can commit injustices without being unjust ; but when the judgment is vitiated and the will disordered so that it chooses ill, then we are unjust : ἄδικος ; perverse : μοχθηρός.—Cf. the very just remarks of PÈRE A.-M. FESTUGIÈRE, *La notion du péché présentée par saint Thomas* : "To sin, in Greek, is ἁμάρτανειν, and ἁμάρτανειν, properly speaking, is to miss one's mark. The Greek saw nothing else in it ; he never left the human plane, the reference to man and man's happiness. For a Christian the word instantly calls up the idea of God and of offence against God. Instinctively he refers it to God and His infinite majesty. We see how wide is the difference" ; in *The New Scholasticism*, Vol. V, 4 (October, 1931), p. 337.

[17] PLATO, *Laws*, IV, 713 D.

[18] *Ibid.*, IV, 709 B.

[19] *Ibid.*, IV, 716 CD.

[20] ST. THOMAS AQUINAS, *Sum. theol.*, Ia–IIae, 19, 4, Resp.

[21] " Cum ergo lex aeterna sit ratio gubernationis in supremo gubernante, necesse est quod omnes rationes gubernationis quae sunt in inferioribus gubernantibus a lege aeterna deriventur. Hujusmodi autem rationes inferiorem gubernantium sunt quaecumque aliae leges praeter legem aeternam. Unde omnes leges, inquantum participant de ratione recta, intantum derivantur a lege aeterna. Et propter hoc Augustinus dicit, in I *De lib. arb.*, Cap. VI, quod : in temporali lege nihil est justum ac legitimum, quod non ex lege aeterna homines sibi derivaverunt." ST. THOMAS AQUINAS, *Sum. theol.*, Ia–IIae, 93, 3, Resp. This would be the place to show in what sense, in a Christian political order, authority may be said to be of *divine right*. The classical misconception on the point consists in supposing that all power is legitimate because all power comes from God. The true doctrine is that no power is legitimate save that which comes from God. To have the right to exact obedience, the authority must first itself obey the eternal law ; its whole legitimacy consists in being an expression of that law.

[22] ST. THOMAS AQUINAS, *Sum. theol.*, IIa–IIae, 142, 1, Resp.

[23] ST. BONAVENTURE, *In II Sent.*, 35, 1, 3, fund 2.

[24] ST. BONAVENTURE, *In II Sent.*, 38, 1, Resp.

[25] ST. THOMAS AQUINAS, *Sum. theol.*, Ia–IIae, 19, 4 ad 2.

NOTES TO CHAPTER XVII

[1] JUSTIN, I *Apolog.*, 15.

[2] ATHENAGORAS, *Leg. pro Christ.*, 33.

[3] ST. AUGUSTINE, *Retract.*, I, 9, 4.

[4] St. Augustine, *Enarr in Ps.* 125, 7 ; *Enarr in Ps.* 141.
Cf. St. Thomas, *Sum. theol.*, " Dicendum quod intentio cordis
dicitur clamor ad Deum : non quod Deus sit objectum intentionis semper ; sed quia est intentionis cognitor." It is interesting to compare this Christian conception with the passage from
the *Nichomachean Ethics* (X, 18, 1178 a, 24–34) where Aristotle
establishes that exterior goods are necessary for the exercise of
virtue. The crux of the proof lies in the fact that intentions are
secret : αἱ γὰρ βουλήσεις ἄδηλοί, and that it is impossible to
discern who wills the good and who does not without the means.
This external and social criterion is altogether opposed to the
Christian view ; which maintains that all intentions are manifest
because they are so to God.

[5] Peter Lombard, *II Sent.*, 35, 2.

[6] Abelard, *Scito te ipsum*, Cap. V.

[7] St. Thomas Aquinas, *Sum. theol.*, I, 79, 13, Resp. Cf.
De Veritate, XVII, 1, Resp.

On this point mediæval terminology was not fixed and the
Thomist position was the upshot of a slow elaboration. At the
source of everything St. Thomas puts the intellect, or natural
light, that is, a faculty. This faculty has two natural *habitus ;*
that of the first principles of knowledge : *intellectus principiorum*,
and that of the first principles of action : *synderesis* (*Sum. theol.*,
I, 79, 12). " Natural " here means that they do not need to be
acquired, but belong to every intelligent being as such. Applying
the first principles of the intellect in the theoretical order, we
get *science ;* applying them in the practical order, we get *conscience*. Conscience therefore is a practical judgment bearing on
an act done or to be done.

Ambiguity arises in the first place because a *habitus* may be
designated by the name of the act that it determines. In fact
the *habitus* is the principle of the act flowing from it. Thus the
synderesis itself, the principle of the judgment of conscience, was
and always may be called conscience. For this reduction of
conscience to *synderesis* St. Thomas refers to St. Jerome, *In
Ezech.*, I, 6 and St. John Damascene, *De fide orth.*, IV, 22. St.
Augustine seems to use the word in the same way, *Enarr. in Ps.*
57, 1. Taken in a broad sense, conscience signifies the natural
possession of the principles of the practical reason (*synderesis*),
or, in the strict sense, their application by way of particular
judgments to the detail of moral conduct, which it apprehends,
prescribes or forbids, approves or disapproves.

A second ambiguity, and a much more serious one since it
touches the heart of the question, arises from the opposition
between the voluntarism of the Augustinian school and the
Thomist intellectualism. The main positions may be schematically distinguished.—A. Thoroughgoing voluntarism (Henry

of Ghent). Conscience belongs to the affective, not the cognitive, part of the soul. There are men who know quite well what ought to be done and lack the conscience to do it. Synderesis is then defined as being " quidam universalis motor, stimulans ad opus secundum regulas universales legis naturae," while conscience is " quidam particularis motor, stimulans ad opus secundum dictamen rectae rationis." H. OF GHENT, *Quodlib.*, I, qu. 18. There is therefore, under the practical reason, an affective moving principle which may be either universal (*synderesis*) or particular (*conscientia*). B. Semi-voluntarism (St. Bonaventure). Conscience is an innate *habitus* of the cognitive faculty, in its practical not its speculative function. The word designates either the intellectual faculty itself, or the *habitus* of practical principles, or the principles themselves contained in the *habitus*. (ST. BONAVENTURE, *In II Sent.*, 39, 1, 1, Resp.). In this, then, is included what St. Thomas calls *synderesis*, as St. Thomas himself admits it may be in all vigour ; but St. Bonaventure's *synderesis* itself is quite different. It stands to the affectivity as good sense does to the reason : " affectus habet naturale quoddam *pondus*, dirigens ipsum in appetendis " (*In II Sent.*, 39, 2, 1). We recognize the Augustinian *pondus ;* thus St. Bonaventure puts *synderesis* in the affective part : " Dico enim quod synderesis dicit illud quod stimulat ad bonum et ideo ex parte affectionis se tenet." *In II Sent.*, 39, 2, 1.—C. Transactional Voluntarism (Richard of Middleton) : *synderesis* may mean our natural and necessary inclination to the good in general, and then it is affective ; or the persuasion of reason inclining, but not necessitating us to good, and then it is intellectual. Conscience is the prescription of the practical reason. D. Intellectualism (Thomas Aquinas) ; *synderesis* and conscience both belong to the cognitive order on its practical side : Cf. DUNS SCOTUS, *In II Sent.*, 39, 1 (*synderesis* is in the higher reason where St. Jerome and Peter Lombard would put it) ; *In II Sent.*, 39, 2 : *synderosis* is the " habitus principiorum " belonging of right to the natural practical reason ; conscience is the " habitus proprius conclusionis practicae," deduced from these principles.

[8] ST. THOMAS AQUINAS, *Sum. theol.*, Ia–IIae, 19, 5, Concl.

[9] ABELARD, *Scito te ipsum*, Cap. XIV.

[10] ST. AUGUSTINE, *De mor. eccl.*, I, 14, 24.

[11] CICERO, *De finibus*, III, 6, 20–21.

[12] ST. AUGUSTINE, *Cont. Julian. pelagianum*, IV, 3, 21. Pelagius' doctrine on this point, as on so many others, constitutes a crucial experiment on the relations of Greek and Christian thought ; it marks the saturation-point where the latter is lost in the former. The present text therefore deserves careful study. Augustine objects to Pelagius that the just live by faith ; now the pagans have no faith, therefore they cannot be just. If they have no

justice neither have they any other virtue. Pelagius replies, like a true disciple of the Greeks, that the source of all virtues lies in the rational soul ; prudence, justice, temperance and fortitude have their seat therefore in our mind, that is their natural place, and by them we are good, to whatever end they may be directed. The pagans did not seek the true end, and hence they will not attain it ; that is, they will not be rewarded, but this does not mean that they are not good : they were *steriliter boni*. Here then we have a morality wholly independent of the end and intention. St. Augustine, on the contrary, reasons like a true Christian when he makes the moral worth of virtue, and therefore its essence, depend on its subordination to the true end ; linking up Christian morality with its principle : " Verae quippe virtutes Deo serviunt in hominibus, a quo donantur hominibus : Deo serviunt in Angelis, a quo donantur et Angelis. Quidquid autem boni fit ab homine, et non propter hoc fit, propter quod fieri debere vera scientia percipit, *etsi officio videatur bonum*, ipso non recto fine peccatum est."

[13] St. Thomas Aquinas, Ia–IIae, 19, 10.

[14] " Quod si virtus ad beatam vitam nos ducit, nihil omnino esse virtutem affirmaverim, nisi summum amorem Dei. Namque illud quod quadripartita dicitur virtus, ex ipsius amoris vario quodam affectu, quantum intelligo, dicitur. Itaque illas quatuor virtutis, quarum utinam ita sit in mentibus vis, ut nomina in ore sunt omnium, sic etiam definire non dubitem, ut temperantia sit amor integrum se praebens ei quod amatur ; justitia, amor soli amato serviens, et propterea recte dominans ; prudentia, amor ea quibus adjuvatur ab eis quibus impeditur, sagaciter seligens. Sed hunc amorem non cujuslibet, sed Dei esse diximus, id est summi boni summae sapientiae, summaeque concordiae. Quare definire etiam sic licet, ut temperantiam dicamus esse amorem Deo sese integrum incorruptumque servantem ; fortitudinem, amorem omnia propter Deum facile perferentem ; justitiam, amorem Deo tantum servientem et ob hoc bene imperantem caeteris quae homini subjecta sunt ; prudentiam, amorem bene discernentem ea quibus adjuvetur in Deum, ab iis quibus impediri potest." St. Augustine, *De mor. eccl. cath.*, I, 15, 25.

[15] St. Augustine, *De lib. arb.*, I, 13, 29.

[16] St. Augustine, *Sermo.* 70, 3, 3.

[17] St. Augustine, *Conf.*, III, 6, 11.

[18] V. Brochard, *La morale ancienne et la morale moderne*, in *Revue philosophique*, Jan. 1901, p. 7.—The expression is interpreted by A.-D. Sertillanges, *art. cit.*, March, 1901, p. 280.

NOTES TO CHAPTER XVIII

[1] St. Thomas Aquinas, *In* VIII *Phys.*, 8, 15, 7.

[2] Duns Scotus, *Op. Oxon.*, I, 3, 4, 9.

³ ST. AUGUSTINE, *De civ. Dei.*, V, 1.

⁴ BOETHIUS, *De consolat. philos.*, lib. V, prosa I.

⁵ ST. AUGUSTINE, *De civ. Dei*, V, 1.—This text is the starting-point of the mediæval speculation on the Christian interpretation of the idea of destiny ; and, in this respect, of capital historical importance.

⁶ BOETHIUS, *De consolat. philos.*, IV, prosa 6.

⁷ ST. THOMAS AQUINAS, *Compend. theol.*, I, 138.

⁸ See ST. THOMAS AQUINAS, *In VI Metaph.*, lect. 3, where the problem is clearly discussed in relation to Greek thought.— Cf. *In I Perihermeneias*, lect. 14.

⁹ All the Fathers and mediæval theologians agree in referring the idea of providence to creation. Thus it is the metaphysic of Being that makes it possible to hold a contingence and a liberty, the indetermination of which is, nevertheless, an object of the divine foresight. The principle of the solution is put with incomparable force by ST. AUGUSTINE, *De civ. Dei*, V, 9 and 10. That is a necessary element in all Christian philosophy and thus a definitive acquisition. But what, then, is it in the creative act which sets up contingence and freedom ? The problem is there ; as we may see for instance, in the hesitations of PETER LOMBARD, *Lib. I Sent.*, dist. 38. And so there is a theological and philosophical history of the question which it may be useful to sum up.

1. There is first the solution for which St. Thomas has provided the formula. God is Being, therefore the cause of all being. The divine causality, which nothing escapes, must consequently produce the necessary, but also the accidental, and even the free. Here, then, we see contingency coming from Being in virtue of the creative efficacy taken as such. What St. Thomas seems to wish especially to prove is that God *can* create the contingent : *Sum. theol.*, I, 22, 4, Resp., and *Comp. theol.*, 1, 140. When it is then asked how God can foresee the future contingent as such, it is sufficient to recall that, for God, all is present in His eternity. He does not foresee, but sees all that will happen as happening. This reply is already fully prepared in BOETHIUS, *De cons. phil.*, lib. V, prosa 6.

2. It may very well be that St. Thomas left room for certain misgivings on the point by not clearly enough indicating what he added to Aristotle. His modesty led some of his successors astray. Like Aristotle's, his God is necessary ; but, unlike Aristotle's, His infinite actuality makes Him a free creative will. Thus St. Thomas bases the possibility of a created and foreseeable contingence on the creative efficacy of God ; and bases the actual existence of this contingence on the freedom of God. Duns Scotus insists especially on this second part of the problem. He seems to fear that in not expressly putting God's freedom at the root of contingence there is danger of falling back

into the Greek necessitarianism. Hence his solution : " nulla causatio alicujus causae potest salvare contingentiam, nisi prima causa ponatur immediate contingenter causare, et hoc ponenda in prima causa perfectam causalitatem, sicut Catholici ponunt. Primum autem est causans per intellectum et voluntatem . . ." *Op. Oxon.*, I, 39, 1, 3, 14. For the same reason he sets himself against Boethius' solution as regards the foresight of future contingents. St. Bonaventure had already hesitated (*In I Sent.*, 38, 2, 1 *et seq.*) ; Duns Scotus will not agree that a science which sees futures in the present suffices to define *fore*sight ; there, too, a part must be played by God's knowledge of the eternal determination of His will : *op. cit.*, 8 and 9. Thus Duns Scotus puts the emphasis on the divine freedom : there is contingence, in the sense of rational freedom, at the source of all contingence.

3. The two solutions above mentioned seem to me to be complementary rather than opposed. St. Thomas denies nothing that Duns Scotus affirms, and *vice versâ*. We have a typical case of two Christian philosophies, distinct as philosophies, because each is a definite method of rational exploration of the same truth. In the controversies of the school the feeling for this profound unity was lost. An eighteenth-century Scotist notes : " Extrinsecam nostrae libertatis radicem faciunt plerique divinam omnipotentiam ; aliis ex ipsa Dei libertas." His reply to the difficulty is that : " si mente nostra Deum concipiamus omnipotentem, seclusa libertate, quemadmodum ille necessario tunc ageret, ita et nos." Thus the root of liberty will have to be put, as Duns Scotus wishes, *rather* in God's liberty than in His omnipotence : J. A. FERRARI, O.-M. Conv. *Philosophia peripatetica*, Venice, 2nd Edn., 1754, p. 316. This peripatetic philosophy is therefore opposed to those " e quorum numero habetur Aristoteles, qui cum nos libere operari concederent, Deum necessario agentem constituebant." A strange peripateticism. It is stranger still to see a profound accord transformed into an irreducible opposition. Who does not see here that the very concept of omnipotence, which is essentially Christian, implies that of liberty and suffices of itself to separate any philosophy in which it is recognized from the Aristotelian necessitarianism ? Both sides have fallen into the error of philosophizing on philosophies instead of philosophizing on real problems. As soon as we turn our backs on reality and begin to think about the formulæ that express it we turn our backs on the sole possible centre of unity, and philosophy dissipates itself into an anarchic verbalism. Thomists and Scotists would see more eye to eye if they spoke less of St. Thomas and Scotus and more of those things of which St. Thomas and Scotus spoke.

[10] ST. AUGUSTINE, *De civ. Dei*, V, 9, 3.

[11] E. GEBHART, *La renaissance italienne*, 2nd Edn., Paris, L. Cerf, 1920, pp. 124–125.

[12] ST. AUGUSTINE, *In Joan. evang.*, VIII, 1 and IX, 1 ; *De Trinitate*, III, 5–6, 9, 16–9.

[13] ST. AUGUSTINE, *In Joan. evang.*, XXIV, 1. Cf. the very beautiful text *De civ. Dei*, X, 12.

[14] ST. AUGUSTINE, *De Gen. ad litt.*, VI, 17, 32.

[15] ST. AUGUSTINE, *ibid.*, VI, 13, 24.

[16] ST. AUGUSTINE, *De civ. Dei.*, XXI, 8, 2 ; *op. cit.*, 8, 5.

[17] ST. THOMAS AQUINAS, *Sum. theol.*, I, 105, 6, Resp. I, 105, 7, Resp. I, 115, 2, ad 4.

[18] " Dicendum quod est ordo naturae specialis et generalis. Ordo naturae specialis transmutari potest et destrui, quia potest in alteram differentiam res relabi, sed generalis non. Sic dicendum, quod specialis ordo attenditur secundum potentiam naturae specialis, generalis ordo secundum potentiam obedientiae, quae est generalis ; contra hunc ordinem non facit [*scil.* Deus], sed contra alium." ST. BONAVENTURE, *In I Sent.*, 42, 1, 3, ad 1.—" Ex quo patet, quod potentia passiva simpliciter attenditur secundum causas superiores et inferiores. Et quia secundum quid dicitur per defectum respectu simpliciter, potentia passiva, quae potest reduci ad actum solum secundum causas superiores, deficiente potentia activa disponente vel consonante, est potentia secundum quid et dicitur potentia obedientiae. Et de hac dicit Augustinus, quod in costa erat, non unde fieret mulier, sed unde ' fieri posset,' scilicet potentia obedientiae.—Possibile igitur, quod dicitur a potentia, non dicitur uniformiter, nec dicitur omnino aequivoce, sed analogice, sicut sanum : et ideo ejus acceptio determinatur per adjunctum." *In I Sent.*, 42, 1, 4, Resp. The text of St. Augustine is from *De Gen. ad litt.*, IX, 16, 30–17, 32. To this may be added *Cont. Faustum Manicheum*, XXVI, 3, but it is to be noted that the expression : *potentia obedientiae* occurs in neither.

[19] ST. AUGUSTINE, *Sermo.* XXVI, V, 6.

[20] The true character of the Thomist obediential power and its connection with the concept of nature are themes which have been treated by Fr. M.-D. Chenu, O.P. I had the privilege of hearing the profound lectures he gave in November, 1931, at the Institute of Mediæval Studies of St. Michael's College, University of Toronto. Here I do no more than incorporate his conclusion in my historical synthesis, and if it throws any ray of light on this important question the thanks are entirely due to him.

[21] ST. THOMAS AQUINAS, *Sum. theol.*, III, 1, 3 ad 3.—*Ibid.*, III, 11, 1, Resp .

[22] *Ibid.*, Ia–IIae, 1, 2, Resp.

[23] *Ibid.*, Ia–IIae, 6, 1, ad 3.

NOTES TO CHAPTER XIX

[1] G. Paris, *La littérature française au moyen âge*, 2nd Edn., Paris, Hachette, 1890, p. 30.

[2] St. Thomas Aquinas, *Sum. theol.*, Ia–IIae, 100, 5, Resp.

[3] St. Augustine, *De civ. Dei*, V, 10, 1–2. St. Augustine here shows that Roman morals were determined by God in order to assure the greatness of the Roman Empire, which would become an open field for the spread of Christianity. Everything which a modern historian would explain in terms of the efficient cause is here interpreted from the standpoint of finality. Hence the unified, systematic and philosophical character of St. Augustine's account.

[4] G. Lanson, *Bossuet*, p. 290.

[5] Hegel, *Philosophie der Geschichte*, Einleit., edit. Reclam, p. 46.

[6] I shall not undertake to say here what Christian philosophy owes to Mussulman and Jewish philosophies. My only excuse is that I do not know, and my only consolation is that no one else knows. Probably the debt is considerable, but neither its nature nor extent could be known without at least some such provisional examination as is here attempted in the case of Christian philosophy. I can offer only a few very general considerations.

In making the comparison Christian philosophy is not to be defined with respect to the Gospel alone, for the Old Testament is contained in the New : " Novum enim Testamentum in veteri velabatur, vetus Testamentum in novo revelatur " (St. Augustine, *Sermo* 160). That is why I have always spoken of the Judeo-Christian tradition ; the essence of Christianity would vanish under any other treatment. Now the Arabs and the Jews worked, just as the Christians did, in the two-fold light of the Old Testament and Greek philosophy, and it would therefore be quite natural that their work should have greatly aided the constitution of Christian philosophy. But although mediæval Christian philosophers owe them a great deal in respect of the technical handling of problems they owe them none of their principles, which all come from the Scriptures or the Fathers. There will often therefore be a parallelism without any borrowing. But even where there was borrowing we ought to consider whether there was not also a transformation. The Jews possessed neither the Koran nor the Gospel ; the Mussulmans had the Koran, the Christians had the Gospel, and there we have probable sources of divergence. We must also take into account the wide differences in the conditions under which these various philosophies were elaborated. What I call Christian philosophy was the proper work of Christian theologians who laboured for Christianity and in its name. This was no doubt the case in a measure with the Jewish philosophers ; but it is dubious whether we can say as much for the Mussulmans. The "Arabian" philosophy was not necessarily "Mussulman" : Averroes' philosophy is so only in a very minor degree, and it may be a question to what extent is that of Avicenna. In any event we here stand in presence of a movement of thought very different from Christian thought. Supposing that a "Mussulman philosophy" existed, which I incline to believe, it would be impossible to divine in advance in what respects it differed from, or resembled, "Christian philosophy." That is precisely why we stand so much in need of a history of ideas. It seems to me certain, however, that Christian philosophy should differ less from Jewish or Mussulman philosophy, than the respective religions do from each other. Philosophy, precisely because it is rational, tends to unity ; religious mysteries, although they act on it, are not incorporated in it ; and those of the Gospel, on account of their very profundity, rule it from a higher point. It is for this reason that Christian philosophy may appear at once to be richer than the others, but poorer with respect to the religion that fed it than the others are with respect to theirs. Built on the substance of an unfathomable mystery it can have no illusions as to its own limits.

[7] St. Bonaventure, *Collat. in Hexaemeron*, XVI, 29–30.

[8] St. Thomas Aquinas, *Sum. theol.*, IIa–IIae, 29, 1, ad 1.

[9] St. Thomas Aquinas, *Sum. theol.*, IIa–IIae, 29, 2, ad 3.

[10] St. Thomas Aquinas, *In de Div. Nom.*, Cap. XI, lect. 1.

[11] R. Bacon, *Un fragment inédit de l' " Opus Tertium "* ; Edn. P. Duhem, Quaracchi, 1909, pp. 164–165.

NOTES TO CHAPTER XX

[1] Gregory IX, 7 July, 1228, in *Chart. Univ. Paris*, Vol. I, pp. 114–116 : see E. Gilson, *Etudes de philosophie médiévale*, Strasbourg, 1921, p. 45, note 1.

[2] This has been attempted already more than once, and will be attempted again. A typical example will be found in V. Gioberti's essay *Della filosofia della revelazione*, published by G. Massari, Turin, and in France by Chamerot, 1856.—It is a question, however, to what extent a philosophy can draw inspiration from Christian dogma. Malebranche, so fearlessly rationalist, was not afraid to go far in this direction. (H. Gouhier, *La vocation de Malebranche*, Chap. V : *La notion de philosophie chrétienne*—an extremely important chapter) ; but even he did not believe that the whole body of dogma could become matter for philosophic speculation, and in function of his own philosophy he made a selection. Even in the Middle Ages the list of *credibilia* and *demonstrabilia* were not the same for St. Anselm, St. Thomas, and Duns Scotus, to say nothing of Occam. That was partly because their idea of a demonstration varied.

What I have called " Christian philosophy " is not *one* system, and therefore the question is not susceptible of a simple reply, but perhaps it is possible to base a reply on the collective effort represented by Christian philosophy. A history of the way in which these relations between faith and reason were conceived in the Middle Ages would help us to discern the central axis round which the various oscillations took place. The problem, for the rest, remains open, and it is very possible that such a dogma as the Incarnation, while remaining a dogma and, of course, a mystery, might suggest deeper views of certain little-known aspects of physical, psychological, æsthetic and moral reality. But if these views are to be integrated with philosophy properly so-called, they would have to be rationally self-sufficient, and they would not consist in any rationalization of mystery, but in a rationalization of reality carried out by a reason made acquainted with a mystery by faith.

The part played by the dogma of the Trinity in the psychology and æsthetic of St. Augustine is well enough known. This, in fact, was one of the many matters with which I had hoped to deal, but the necessity of considering the great cardinal theses in the first place forced me to omit it. The dogma of the Incarnation might play a similar part : and hence we see that Christian philosophy is no exhausted vein—rather the fact that it is in touch with mystery makes it as inexhaustible as mystery itself. The sole line of demarcation which, in my view, must be scrupulously maintained, is the line that separates a rationalization of dogma—and an evacuation of mystery—from the fecundation of reason by dogma. The former enterprise I should regard as illegitimate. Theology utilizes rational concepts for the purpose of locating and defining the mystery, it cannot set out to make it comprehensible. The philosophy of Aristotle never served to " explain " transubstantiation, but rather to make clear what is required if transubstantiation is to take place. What seems to me to be reprehensible about Gioberti's attempt, and others of the same kind, is that it suggests that Christianity can become a philosophy. It amounts to an attempt to build up a Christian philosophy on a suppression of Christianity itself. On the contrary, when dogma comes to the aid of reason it loses nothing of its transcendence, no more than grace does when it comes to the aid of man : but, once more, we need to distinguish.

Revelation proposes other things besides mysteries ; it is a doctrine commonly received by theologians that the Decalogue formulates many truths that are naturally knowable by reason or conscience (existence of God, principles of moral justice, etc.). These truths were positively promulgated, but neverthe-less they can be integrally rationalized because they are humanly intelligible. The Decalogue contains also other truths bearing on facts that are essentially mysterious in themselves, but facts in which reason sees the sole possible solution, and therefore also the necessary solution, of its problems. Such, for example, is the fact of creation, something essentially mysterious in itself, but necessarily affirmed by reason as the sole possible solution of the problem of the ultimate origin of being. God keeps His secret ; but the fact that this mysterious act took place is demanded by reason. Here is the field *par excellence* for Christian philosophy, for here it may show itself as at once fully philosophic and fully Christian ; and it is the field within which, without pretending to cover it, I have tried to remain. Speculations on the Trinity, on the Incarnation, or even on certain aspects of creation, would carry us at once beyond it. From a rational standpoint creation *might* be *ab aeterno ;* whether it is so or whether it is not, it remains equally mysterious, but it is rational that it might be so. Going a step farther, there are dogmas of which man is equally incapable of understanding the essence, and of showing that they are rationally demanded as facts. In respect of some, such as that act of supreme freedom, the Incarnation, the very idea of rational necessity is contradictory. We need not suppose that philosophy has nothing to gain by meditating such things ; perhaps precisely in doing so it will attain to a fuller comprehension of its own deepest truths ; but the fact that they open up transcendent perspectives which may go to complete philosophy will never in the least diminish the mysterious character of these revealed truths. Thus, unlike the ideas of creation, conscience, moral law or intention, the Trinity and the Incarnation can never be integrated with philosophy. They belong of right to theology, and therefore I have not touched upon them in this study.

[3] " Deinde quod Physica Aristotelis sit prorsus inutilis materia omni penitus aetati. Contentio quaedam est totius liber super re nihili et, velut assumpto argumento, rhetorica exercitatio nullius usus, nisi velis exemplum rhetoricae declamationis cernere, ut si de stercore vel alia re nihil, ingenium et artem quis exerceat. Ira Dei voluit tot saecula his nugis et eisdem nihil intellectis humanum genus occupari . . . Ejusdem farinae et metaphysica et de anima sunt." LUTHER to Spalatin, 13 March, 1519 ; edn. Weimar, *Briefwechsel*, Vol. I, p. 359. On the 23rd February 1519 (Vol. I, p. 350) he proposes to suppress a course of Thomist logic and to substitute lessons on the *Metamorphoses* of Ovid, Book I.

[4] ST. BERNARD, *De diligendo Deo*, II.

INDEX TO PROPER NAMES

ABELARD, P., 348, 349, 352, 476, 477
Albert the Great, St., 178, 180, 181, 268, 367, 408, 414, 463, 469
Alcher de Clairvaux, 445
Alexander of Aphrodisias, 260
Alexander of Hales, 180
Al-farabi, 179
Ambrose, St., 428, 474
Ammonius Saccas, 404, 418
Anaxagoras, 451
Anselm of Canterbury, St., 5, 6, 33, 34, 35, 40, 41, 52, 59, 60, 61, 62, 85, 233, 238, 245, 309, 317, 328, 379, 429, 435, 467, 468, 471, 472, 483
Aristides, 70
Aristotle, 2, 5, 8, 15, 28, 38, 41, 44, 45, 49, 50, 51, 58, 59, 65, 66, 67, 68, 69, 70, 71, 72, 74, 75, 76, 77, 80, 81, 82, 83, 84, 105, 114, 118, 131, 139, 140, 143, 153, 161, 168, 169, 172, 175, 177, 178, 179, 180, 182, 183, 186, 187, 190, 191, 193, 195, 196, 197, 198, 199, 200, 202, 222, 224, 226, 233, 241, 242, 243, 249, 260, 262, 263, 266, 268, 269, 278, 305, 306, 307, 309, 310, 314, 324, 326, 330, 338, 340, 351, 353, 355, 356, 367, 369, 371, 372, 373, 380, 389, 396, 397, 403, 405, 408, 409, 410, 411, 412, 413, 415, 418, 421, 423, 424, 425, 428, 429, 430, 432, 433, 435, 436, 437, 439, 440, 441, 442, 443, 446, 451, 452, 457, 458, 463, 464, 465, 467, 471, 473, 474, 476, 479, 480, 484, 485
Athenagoras, 40, 52, 153, 192, 193, 428, 430, 458, 464, 475
Augustine, St., 5, 6, 12, 13, 15, 16, 26, 29, 30, 32, 34, 37, 38, 40, 41, 52, 53, 62, 72, 73, 82, 91, 93, 101, 109, 110, 111, 115, 116, 117, 119, 120, 122, 123, 125, 130, 132, 133, 134, 135, 137, 140, 141, 146, 154, 156, 157, 158, 169, 173, 174, 175, 176, 178, 179, 182, 183, 184, 185, 212, 215, 220, 223, 225, 226, 229, 230, 233, 237, 240, 241, 276, 291, 292, 298, 307, 311, 314, 315, 316, 317, 324, 326, 328, 333, 334, 346,

Augustine, St.—*continued.*
357, 358, 368, 369, 370, 372, 374, 375, 378, 379, 380, 385, 389, 391, 392, 393, 403, 420, 435, 437, 439, 440, 441, 443, 446, 451, 452, 453, 454, 455, 456, 458, 462, 463, 466, 467, 469, 470, 471, 472, 473, 474, 475, 476, 477, 478, 479, 480, 481, 482, 484
Averroes, 15, 80, 178, 260, 409, 469
Avicenna, 80, 178, 179, 180, 182, 255, 266, 397, 409, 436, 441, 442, 456

BACON, FRANCIS, 104, 449, 468
Bacon, Roger, 101, 242, 367, 398, 483
Balzac, J.-L. Guez de, 228, 392
Baudry, J., 432
Beethoven, 89
Bergson, H., 66
Berkeley, 15
Bernard, St., 4, 37, 108, 109, 117, 211, 215, 218, 269, 270, 276, 281, 287, 289, 290, 291, 292, 293, 294, 295, 299, 301, 314, 379, 422, 449, 453, 466, 470, 471, 472, 485
Bernard of Chartres, 425
Bernardin de St.-Pierre, 78
Boethius, 201, 204, 310, 311, 312, 313, 369, 371, 471, 472, 479, 480
Bonaventure, St., 5, 37, 51, 57, 60, 62, 71, 99, 101, 109, 156, 158, 159, 160, 161, 169, 178, 180, 204, 211, 225, 226, 230, 231, 248, 340, 378, 380, 396, 408, 409, 415, 422, 425, 434, 435, 437, 445, 446, 454, 455, 457, 460, 461, 464, 466, 468, 474, 475, 477, 480, 481, 483
Bossuet, 38, 228, 392
Bréhier, E., 427, 428, 431, 438
Brentano, 457
Brochard, V., 362, 478
Bruno, St., 108

CALVIN, 122, 125, 321
Charlemagne, 385, 396, 397
Chenu, M.-D., 427, 429, 481
Chevalier, J., 432, 446, 458

487

Chrétien de Troyes, 396
Cicero, 32, 324, 327, 356, 372, 373, 473, 474, 477
Clement, Epistle of, 69
Clement of Alexandria, 428
Collin, R., 450
Comte, A., 388, 407, 469
Condorcet, 42, 109, 393, 430
Cyril of Alexandria, 434

Dalbiez, R., 450
Dante, 75, 133, 289
David de Dinant, 449
Decharme, P., 430
Della Volpe, G., 449
Del Prado, N., 443
Democritus, 153, 218
Descartes, 13, 14, 15, 18, 26, 59, 61, 62, 66, 67, 71, 78, 79, 104, 105, 106, 182, 211, 226, 227, 229, 236, 245, 322, 403, 407, 411, 426, 467, 473
Descoqs, P., 443
Diès, A., 432
Dionysius, 56, 93, 144, 218, 259, 275, 327, 359, 418, 467, 473
Duns Scotus, 51, 56, 57, 58, 60, 71, 81, 109, 123, 146, 158, 159, 160, 169, 195, 196, 197, 211, 213, 226, 229, 235, 240, 241, 242, 248, 249, 251, 252, 253, 254, 255, 256, 257, 258, 263, 264, 309, 312, 313, 314, 321, 366, 376, 403, 410, 414, 415, 425, 435, 442, 445, 447, 455, 457, 460, 461, 466, 468, 469, 471, 477, 478, 479, 480, 483
Durantel, J., 446

Eckhart, 449
Empedocles, 43, 451, 457, 458
Ephrem of Nisibis, St., 51, 211, 434, 466, 471
Epictetus, 190, 418
Epicurus, 270, 418
Erasmus, 27, 413, 415
Ercilla, U. de, 432

Fenelon, 48, 432
Ferrari, J. A., 473, 480
Festugière, A. M., 475
Fichte, 411
Forest, A., 436, 442, 466
Francis of Assisi, St., 126, 169, 429
Francis of Sales, St., 225

Gardeil, A., 467

Garin, P., 473
Gaunilo, 62, 435
Gebhart, E., 481
Gebirol, 180
Gibieuf, 322, 473
Gilson, E., 427, 446, 462, 483
Gioberti, V., 483
Glorieux, P., 464
Goethe, 23
Gouhier, H., 483
Gregory IX., 414, 416, 483
Gregory of Nazianzen, St., 434
Gregory of Nyssa, St., 175, 210, 220, 434, 466, 471
Grosseteste, Robert, 101, 249, 367, 468
Guéroult, M., 428
Guigues le Chartreux, 470

Hamelin, O., 432
Harnack, A., 10
Hegel, 59, 393, 394, 482
Henry of Ghent, 249, 476, 477
Heraclitus, 27, 241
Hermias, Pastor of, 70, 428
Hieronymus of Montefortino, 445, 447, 449, 457, 461
Hilary of Poitiers, 434
Hocedez, E., 455
Homer, 338
Hugh of St. Victor, 99, 309, 445, 466, 471
Hume, 15, 86, 390, 444
Hyppolitus, 429

Irenaeus, St., 110, 154, 304, 428, 438, 451, 458, 462
Isaac Israeli, 238
Isaiah, 33, 454

James, William, 81
Jansen, B., 467
Jansenius, 122
Jeremiah, 345, 349
Jerome, St., 54, 476, 477
Joachim of Flora, 397
John, St. II., 17, 26, 32, 136, 275, 453, 474
John Chrysostom, 471
John Damascene, 337, 471, 476
John of Jandun, 408
John of Salisbury, 2, 396
Jolivet, R., 439, 451
Jowett, 437
Justin, St. II., 23, 24, 25, 26, 27, 30, 31, 34, 346, 428, 430, 462, 475

KANT, 13, 16, 59, 85, 204, 245, 246, 341, 342, 343, 361, 388, 407, 444, 468
Kleutgen, 427

LABERTHONNIERE, L., 442
Lactantius, 31, 32, 34, 428
Lagrange, M. J., 431, 437
Langevin, P., 451
Lanson, G., 482
Leibniz, 16, 17, 18, 59, 67, 71, 77, 388, 407, 454
Lessing, 19, 428
Lods, A., 430, 434, 436
Lombard, Peter, 223, 316, 348, 474, 476, 477, 479
Lottin, O., 471
Lucretius, 153
Luther, 122, 125, 321, 322, 413, 415, 416, 421, 422, 485

MAIMONIDES, 397, 442
Maine de Biran, 34, 429, 444, 451
Malachi, 65
Malebranche, 14, 15, 16, 18, 48, 59, 61, 62, 71, 86, 129, 141, 157, 225, 227, 364, 388, 406, 407, 411, 413, 427, 432, 454, 457, 483
Mandonnet, P., 427, 441, 458
Mani, 32, 110, 120, 420
Marcus Aurelius, 418
Maritain, J., 429, 441, 457
Marston, R., 223, 230, 456, 467
Matthew, St., 346
Matthew of Aquasparta, 223, 230, 231, 232, 233, 234, 235, 239, 456, 467
Maximus the Confessor, 299
Minucius Felix, 154, 428
Molière, 89
Molina, 321, 322
Montague, W. P., 17, 18, 428
Montaigne, 209, 214
More, P. E., 431, 438
Moses, 51, 434, 435, 437
Mugnier, R., 432
Murray, G., 430, 443

NEMESIUS, 175
Newman, 99

OCCAM, WILLIAM OF, 235, 483
Olgiati, F., 446
Olivi, P. J., 235, 455, 456, 457, 467
Origen, 418, 437, 471

Orosius, Paul, 392
Ovid, 269, 282

PARIS, G., 482
Parmenides, 48
Pascal, 16, 106, 109, 171, 209, 214, 216, 227, 228, 275, 276, 389, 392, 451, 467, 470
Paul, St., 17, 20, 21, 22, 23, 26, 28, 29, 30, 72, 108, 119, 144, 151, 166, 170, 240, 258, 315, 316, 354, 414, 428, 454, 458, 462
Peckham, J., 455, 456
Pelagius, 120, 378, 379, 420, 477, 478
Petau, J., 473
Peter Damian, St., 4, 108, 414
Philo the Jew, 25
Philolaus, 43
Plato, 2, 5, 24, 28, 31, 41, 44, 47, 48, 49, 50, 51, 54, 59, 64, 65, 68, 69, 70, 71, 81, 82, 83, 84, 93, 105, 125, 131, 148, 151, 153, 157, 158, 161, 172, 173, 174, 175, 176, 178, 179, 180, 185, 186, 187, 190, 191, 193, 202, 214, 222, 226, 229, 233, 234, 235, 241, 242, 249, 266, 268, 278, 331, 332, 333, 338, 389, 397, 403, 405, 409, 411, 418, 423, 424, 425, 428, 429, 430, 431, 432, 437, 438, 439, 440, 446, 451, 452, 458, 459, 461, 462, 475
Plotinus, 30, 32, 82, 110, 111, 112, 114, 173, 227, 230, 403, 404, 418, 430, 431, 439, 451
Plutarch, 43
Porphyry, 348, 439, 454
Proclus, 439
Puech, A., 430

RAYMOND LULLY, 99
Rembrandt, 98
Richard of Middleton, 455, 477
Richard of St. Victor, 221, 222, 466
Ritter, H., 443, 466
Robin, L., 432, 433
Roland-Gosselin, M. D., 430, 435
Rolfes, E., 457
Ross, W. D., 457
Rousselot, Pierre, 289, 290

SALIMBENE, FRA, 374
Salvian, 392
Scheler, M., 427
Schelling, 393
Schopenhauer, 411
Schmidt, P. W., 437

Scotus Erigenus, John, 220, 224
Seneca, 31, 140, 418
Sertillanges, A. D., 478
Shakespeare, 89
Sharp, D. E., 455
Shorey, P., 431
Sierp, C., 427
Socrates, 27, 31, 190, 195, 209, 213, 214, 227, 330
Spinoza, 59, 407, 411
Suarez, F., 443
Synave, P., 429

TATIAN, 25, 31, 428, 462
Taylor, A. E., 430, 432, 437, 446
Tertullian, 125, 462
Theophilus of Antioch, 70, 438
Théry, G., 449
Thomas Aquinas, St., 4, 5, 7, 8, 12, 13, 15, 16, 37, 38, 39, 40, 41, 49, 51, 57, 58, 63, 66, 68, 69, 71, 73, 74, 75, 76, 77, 80, 82, 85, 86, 91, 93, 94, 95, 99, 101, 109, 115, 116, 123, 124, 125, 130, 131, 132, 134, 138, 139, 140, 141, 146, 156, 157, 158, 159, 160, 161, 169, 178, 182, 183, 184, 186, 197, 198, 202, 204, 211, 212, 213, 219, 223, 224, 225, 229, 235, 237, 238, 239, 240, 241, 248, 249, 251, 252, 257, 258, 260,

Thomas Aquinas, St.—*continued.*
261, 262, 263, 264, 275, 282, 283, 287, 288, 291, 292, 313, 314, 318, 320, 321, 322, 326, 327, 334, 351, 358, 359, 363, 366, 367, 369, 371, 376, 379, 380, 381, 388, 396, 403, 407, 408, 409, 410, 414, 415, 422, 424, 425, 426, 429, 432, 434, 435, 436, 438, 439, 440, 441, 442, 443, 444, 445, 446, 447, 448, 449, 450, 453, 454, 455, 456, 457, 458, 459, 460, 461, 463, 464, 465, 466, 467, 468, 469, 470, 471, 472, 473, 474, 475, 476, 477, 478, 479, 480, 481, 482, 483

VINCENT OF BEAUVAIS, 396
Voltaire, 109, 390

WILLIAM OF AUVERGNE, 435, 444
William of St.-Thierry, 269, 270, 281, 282, 287, 301, 470, 471
Wittmann, M., 471, 472
Wycliffe, 321

XENOPHON, 43, 466

ZELLER, 457